THE BIRD

THE BIRD

A Natural History of Who Birds Are,
Where They Came From, and How They Live

COLIN TUDGE

 THREE RIVERS PRESS • NEW YORK

THREE RIVERS PRESS and the Tugboat design are registered trademarks of
Random House, Inc.

Originally published in Great Britain as *Consider the Birds: Who They Are and
What They Do* by Allen Lane, an imprint of Penguin Books, Ltd., a division of
Penguin Books, London, in 2008, and subsequently published in hardcover in
the United States by Crown Publishers, an imprint of the Crown Publishing
Group, a division of Random House, Inc., New York, in 2009.

Library of Congress Cataloging-in-Publication Data
Tudge, Colin.
[Consider the birds]
The bird / Colin Tudge. —1st pbk. ed.
p. cm.
Originally published: Consider the birds: who they are and what they do.
London : Allen Lane, 2008.
1. Birds. 2. Birds—Behavior. 3. Birds—Evolution. I. Title.
QL673.T77 2010
598—dc22   2010022206

ISBN 978-0-307-34205-8

Design by Lauren Dong

First American Paperback Edition

147468846

To My Grandchildren

CONTENTS

Illustrations

Original illustrations by Jane Milloy.

Preface

As a small boy in South London, just after the Second World War, I recognized only five kinds of birds. There were pigeons and sparrows, which were everywhere; the ducks and swans in the local parks; and a mixed category of "ordinary birds" that flew overhead from time to time and perched on roofs, for no particular reason except that they were birds and that's what birds do. London Zoo soon broadened my horizons, with its Ostriches, Emus, and penguins, a statutory line-up of torpid owls like fluffy Russian dolls with revolving heads, and a huge array of parrots. (London Zoo had two of each species in those days—or sometimes only one of each.) There was also a mad-eyed creature in a tall Gothic cage labeled "Monkey-eating Eagle," which opened its vast horny beak in a most suggestive fashion and frightened us all to death. Then my cousin Peter, out in Kent, conceived a passion for bird nesting—which boys were still encouraged to do in those days, along with the pinning of butterflies after a quick whiff of chloroform ("which may be had from your local chemist for a few pence"); and I began to perceive that "ordinary birds" included pipits and wagtails, terns and kestrels, Yellowhammers and robins, and a miscellany of crows, not all of which were black.

I went at the age of five to a Church of England primary school. It was my first taste of religion—and a very kindly taste it was, too; the way that religion ought to be: songs and stories and being nice to people. The "nature table" was a shrine: a fir cone, a twig of willow with catkins, a mushroom, and a couple of unidentified rodent skulls from a local bomb site (there were three or four close by) of the kind that nowadays would bring in the disposal squad from Health and Safety, in jumpsuits and welders' masks, to drop them with long tongs into polyethylene bags. But although we all caught measles and sneezles and whooping cough and mumps—"common childhood ailments," as they were called—we did manage to avoid Weil's disease and bubonic plague, although I imagine only by a whisker. There was also a glossy magazine, a huge departure in those austere times, with a picture of shorebirds mysteriously labelled "Oystercatchers and Knots."

I was hooked. It isn't formal teaching that gets you into things, or at least not necessarily. It's the incidentals. I took it to be self-evident from about the age of six that everyone must be obsessed with "nature," and I am still shocked to find how far and how often that is from the case.

The problem, once you are hooked, is how to get close to living creatures, how to engage with them. One way is simply to learn: I was lucky to go to a school where biology was taught brilliantly, and then to an ancient and therefore damp and crumbling university packed with Nobel Prize—winning biologists. I never wanted to be a professional scientist myself. I just liked, and like, being with the creatures themselves, and the ideas, and matching words to the ideas. So I write books about them.

The last such book was on trees—my other life-long indulgence—but as it progressed my friend Barrie Lees said, "I already know about trees," as indeed he does. "What I really want to know about is birds. They keep coming into the garden. They fiddle about. What are they? What are they up to? Why are they doing whatever they do?"

These are good questions. There really are more than five species of birds in London. Indeed, the current list for London stands at 357. They all have names, too: none should be called an "ordinary bird." I am sure that birds do many things for no particular reason, but the more you look, the more you find this isn't necessarily so. The starlings swirling overhead are returning from a hard day's foraging in the surrounding fields. If agricultural practice changes, the starlings suffer. The blackbird that hops from twig to twig as it strips the cotoneaster berries is not a fidget with a low attention span. Its movements are strategic—designed, or evolved, to deny the cats and sparrowhawks a sitting target. The female sparrow with her body close to the ground, shivering her wings, is inviting the male to copulate. The flock of sparrows, twittering and apparently quarreling with no particular aim, are forming social bonds and sizing each other up, showing all the others what they are made of. Among those others are potential mates, and it pays to make a good impression. It is all very serious, because life is serious. Life requires nourishment—which can be hard to arrange; whether you eat berries or blackbirds, you have to be skilled. Life requires mates, or the lineage comes to a stop; and that requires the ability to seduce, and the ability to see off rivals, and—in both cases—the ability to socialize: to know who's who. Then you have to raise the babies, and in most cases you need to do all this before winter sets in. In short, life is complicated and it needs very good timing.

So this is the book that Barrie said should be written. I'm glad he put me up to it. In many ways I'm sorry it's finished.

In Britain, at least, there are three ranks of recognized birdwatchers. First, there are the "twitchers," or "listers" as they are known in the United States. They seek primarily to see as many different species as possible. In these Internet days, twitchers form a network that spreads nationwide, and is even international. If someone in the east of England spots some rarity—invariably a "vagrant"—some poor bewildered and half-starved creature blown off course by a crosswind—then the buzz goes out and hundreds of twitchers in lurid hues of anorak, with cameras and field glasses of all shapes and sizes from pince-nez to howitzer, crowd in around it like a bullfight. In Dorset, in England's southwest, I was recently caught in the frenzy around a Ring Ouzel—like a blackbird but with a white ring around its neck. Listers may be mocked, in some circles, and they sometimes trample on the nests of native birds while they focus on the hapless stray (or so it is alleged). But largely through their efforts we know which birds go where, and to some extent how often. All in all, they are a valuable group.

Then there are the "birders." They are not necessarily pros, but they may, like a fellow I met recently, spend three days waist-deep in an Irish bog for a glimpse of a Bluethroat, or weeks and weeks, their careers in jeopardy, watching the nest of an Osprey. They also help professionals in their national bird counts. Birders, too, are a very useful breed—and one of the most productive examples in all of science of the "pro–am" connection.

Top of the hierarchy are the professional ornithologists. They do the serious research. These days they are astonishing both in their persistence and in their ingenuity. There will be many examples in this book.

For my part, to be truthful, I have stood apart from this hierarchy. I have not risked drowning and hypothermia and duck lice in bogs, or dangled on strings from lonely cliffs to count the parasites of kittiwakes, or led a skein of migrating geese in a microplane. But I have been privileged this past forty years to travel around the world (before the days when air travel was known to be wrecking the climate), and wherever I went I was keen to see what was what. Some of it has been pure indulgence. I saw my first Brown Pelicans from a Jacuzzi in a hotel garden in California (paid for by a kind employer). The closest I have ever been to a kingfisher was on a hotel roof in Panama; it perched about two feet from my tea. I saw my first wild parrots from a hotel pool in Delhi—glowing grass-green in the low evening sun and flying in convoy straight as arrows, as parrots do. But I have also gone out wherever possible with local guides, more or less the world over. All of them were very fine—real enthusiasts with excellent knowledge, who knew where to look. Best of all was John Butler in the Cota de Doñana in southwest Spain—an ex-army man of the finest type; he was tireless, patient, extremely knowledgeable, and endlessly meticulous. As noted in the Acknowledgments, I was shocked to learn of John's recent death. A great loss.

You may think it is foolish of me to acknowledge my own relative lack of direct involvement with the birding scene—you could, after all, be reading books by full-time ornithologists—but here, to a large extent, is my point. Nature is wonderful—it is the center of everything—and if you take a serious interest, it changes your life. The word *jargon*, meaning meaningless jabbering, comes from the French for the twittering of birds. But in truth, the twittering of birds is *never* meaningless. The birds twitter for a reason—and it won't be a frivolous reason. As you become more aware, you start to get a feel for the reasons for things. All nature acquires meaning. You realize then—and it is perhaps the most important thing to realize—that simply to be alive and aware in such a world as this is a privilege. If people in high places felt this, the world would be very different. You don't have to be a pro, or even spend a weekend in a swamp, to see the truth of it. You just have to take an interest and be alert. The point of this book is to nudge people who feel in a general way that birds in particular, and nature in general, are kind of interesting to the point where they start to feel the meaning of it all. After that—well, life can never be the same again.

So, here's the plan. The book is divided into four parts.

Chapter 1, in Part I, points out how very brilliant birds are—an altogether superior class of creature—and also points to a puzzle: that birds do most of the things that mammals do and a few more besides, and yet they seem to differ from mammals in all points of detail in which it is possible to be different. But what's the basis for this claim that birds are "superior"? And why should they be the same as mammals—the same as us—and yet be so different? Chapter 2 provides the answer—or at least as much of an answer as can be provided. It lies, of course, in evolution, for as the great Ukrainian-American biologist Theodosius Dobzhansky famously opined in an essay of 1973, "Nothing makes sense in biology except in light of evolution." Chapter 2 also looks at the still vexed issue of whether birds descended from dinosaurs. If they did, then strictly speaking birds *are* dinosaurs. The dinosaurs *did not disappear* when the putative asteroid struck 65 million years ago. They are calling to us from every hedgerow.

Chapter 3—the first chapter of Part II—asks how many species of birds there really are in the world. To anticipate, it's a lot, but nobody knows how many for sure and never will, for reasons both practical and philosophical. This chapter also asks how we can possibly keep track of so many—to which the answer is: classify. The modern science of taxonomy takes us into some of the most subtle reaches of biology, and as DNA studies come on board, they are producing some big surprises. Chapter 4 is an annotated cast list. It's quite a long cast list, because there are a lot of birds. Without good classification, there would be no sense in such a list.

The five chapters of Part III are the guts of the book. They ask what birds actually *do*—and why they do it, and how. Chapter 5 asks how they get their energy—feeding, in other words—and it's a big question. There's a lot to it. They need the right apparatus, of course (including the great talons and beak of the Monkey-eating Eagle), but they also need the right strategy. Strategy is all. The blackbird's fidgety feeding is part of life's strategy. So, too, are the often vast migrations that a large proportion of birds undertake each year—sometimes from pole to pole—as they strive to take advantage of the tropics and subtropics when the high latitudes are cold, and of the long, long days of the polar summers when they have young to feed. How they do this is discussed in Chapter 6.

Then to sex and reproduction in Chapters 7 and 8—the most absorbing matter of all because it demands such extravagant behavior and because, above all, it requires layer upon layer of sociality. One modern and very convincing theory says that the need to attract mates, to see off rivals, and to raise offspring, with all the social intricacies that this implies, was the prime reason some animals—notably some birds and mammals—evolved such big brains. Reproductive success belongs not simply to the swift and the strong but also to the clever (and amusing and sociable).

Actually, it is not altogether obvious why animals have brains at all—since brains are, after all, such expensive organs to maintain. Trees live immensely complicated lives—it's not just a matter of standing still with their arms out—and they get by without a brain. Even among animals, some have found it expedient to dispense with brains—such as barnacles, which stick themselves to a rock head down, so their head and what they have of a brain are reduced to a blob of glue. It works for them: there are billions of barnacles in the world. Of course, animals that move around—as most do—need brains of a sort to coordinate their movements. But why be as clever as a crow or a parrot? Social complexity may be the answer. But it leaves the highly intriguing question—addressed in Chapter 9—of what actually goes on in the head of a crow or a parrot (or, for that matter, of a sparrow or a chicken)? What do they perceive? What do they think? Do they think like us, or are they as different from us in their cogitations as they are in their anatomy and physiology? In short—what is it like to be a bird?

Part IV is about birds and us. Birds enhance our lives enormously, if we let them. We, in return, are killing them off. We need to understand why we are doing this. The same destructiveness that is killing them is killing us. Here again the birds can help us; by thinking about them, we can gain insight into ourselves, including insight into our own psychology. Again, it seems, the insights that are emerging are not as commonly expected.

Jesus said, "Consider the fowls of the air" (Matthew 6:26).[1] Indeed.

I

A Different
Way of Being

The courtship flight of the Hen Harrier is pure display: each bird shows its mettle as the male passes food to the female.

1

WHAT IT MEANS TO BE A FLIER

"ALL ANIMALS ARE EQUAL," THE RULING PIGS DECLARED IN George Orwell's *Animal Farm*. "But," they added, "some animals are more equal than others."

All animals are equal no doubt in the eyes of God, and all that manage to survive at all in this difficult world are in some sense "equal." But some, by all objective measures, are far more impressive than others; and none, not even the mammals, the group to which we ourselves belong, quite match up to the birds. Birds have their shortcomings, to be sure, as flesh and blood must. But they are, nonetheless, a very superior form of life.

Above all, birds fly.

They are not the only animals that have taken to the air, of course. There are many gliders. Flying fish are remarkably adept, and various frogs and snakes and lizards contrive to parachute from tree to tree; and there is a variety of gliding mammals, including phalangers and squirrels and colugos (sometimes known as flying lemurs). But only four groups have managed powered flight, driving themselves through the air by flapping or whirring their wings. Many insects fly wonderfully. Bats fly well enough to catch insects in the air—and, for good measure, they do it at night. The ancient pterosaurs, contemporaries of the dinosaurs, included some of the biggest powered fliers of all time—and what a sight they must have been! Pelicans, returning home against the evening sun, might give us some idea of what they were like.

But none of these creatures flies as well as the birds. Perhaps this is why birds are still with us and pterosaurs are not. Perhaps this is why bats fly mainly at night; if they are ever forced to fly by day, as they may do in cold weather when there are too few nighttime insects, they quickly get picked off by hawks.

Flight, indeed, is the key to birds. Many have abandoned flight, of course, like penguins and Ostriches, and there are or have been flightless ducks and geese, many flightless rails and auks, at least one flightless ibis, flightless cormorants, and flightless parrots. The famous Dodo was a flightless pigeon, and there was even one flightless passerine (a perching bird)—or so it's said, though it is hard to tell, since the bird is extinct. But all of these flightless types had flying ancestors. Some birds fly but ineptly—including the superficially grouse-like tinamous of South America, which hurtle along with huge bravado but little control, and sometimes end up killing themselves, like twelve-year-old joyriders. On the whole, sensibly, tinamous prefer to stay on the ground.

The fact that birds fly—or at least are descended from ancestors that were adapted to flight—dominates all aspects of their lives. Flight brings huge and obvious advantages, but it is also immensely demanding and so has its downside, too.

WHAT IT MEANS TO BE A FLYING MACHINE

As a mammal, I have often admired and envied birds—as who has not? I remember once, in southern Spain, struggling over the rocks to get to the base of some cliff to catch a glimpse of the Egyptian Vultures that in the evening appear over the edge, riding along the length of it on the up-currents—not for any obvious reason, since it is too late to feed at that time of day but just, it seems, to keep an eye on things, like the squire riding his estate.

After half an hour or so the vultures did turn up. Birds in general have big eyes—their skulls are built around the orbits—and in birds of prey they are particularly big. The eyes of a big eagle are as big as a human's. In birds of prey (and in some other birds such as kingfishers and swallows), the retina has two foveas (particularly sensitive spots), and the eyes as a whole have a greater concentration of light-gathering, color-sensitive cones than any other vertebrate. So the visual acuity of a hawk or a vulture is two or three times as great as ours. Australia's huge Wedge-tailed Eagle has been shown to see rabbits clearly from a kilometer and a half away. So

I daresay that those Egyptian Vultures could see me and my companions more clearly with their unaided eyes than we could see them with all the technology of Zeiss lenses; and it took them less time and almost zero effort to fly the 6 miles or so that they had doubtless traveled than it had taken us to struggle a few hundred yards from the road. As a heavyweight mammal land-bound and land-locked by gravity, I felt inferior. All animals may be equal according to the ideology—but here, truly, birds were the masters.

Yet evolution has not worked entirely in the birds' favor. They pay a price for their magnificence and their skill, and they seem to pay an even higher price than seems strictly necessary. Even the birds that no longer fly—even the ones that have all but abandoned wings altogether, like the kiwi—deep down are built for flight. In general this means they have to be small. If you double the linear dimensions of a bird—or, indeed, of anything: a guinea pig, a brick, a ship—yet retain the same proportions, then you will increase its weight by eight times.

A bird with a body that's 20 centimeters (8 inches) long, like a fairly average thrush, weighs eight times as much as one of 10 centimeters (4 inches), like a small finch or weaver. A 40-centimeter (16-inch) bird like a grouse is eight times heavier again—sixty-four times heavier than the finch, or six football teams' worth of people compared to a single person. Whether or not a bird can fly depends on its wing "loading"—the mass that needs to be lifted per unit area of wing. But the area of a wing increases only four times as the body length is doubled, if the overall proportions remain the same. So a thrush that needs to fly as well as a finch needs wings that are twice as big, relative to its body size, as the finch's— and so on and so on: the bigger the bird, the bigger the wing needs to be, not only in absolute terms but also relative to the body size.

The physics of flight is well understood—or so the textbooks suggest. My own sympathies lie with an airline pilot I once met who regularly took 747s across the Atlantic but told me that he still did not quite believe that such egregious structures could really fly at all, since they are big as a row of small houses and many times heavier than air, and stuffed with people and suitcases for good measure. He did his job and drew his salary in a haze of incredulity.

The point of the scientific theory, though, is that air is not so insubstantial as it seems. A wind of 30 miles an hour is difficult to walk against. A hurricane of 100 miles an hour can flatten entire cities, can pick up roofs and even motor cars and hurl them halfway across a state, as so many people have experienced. The energy thus implied is tremendous. Creatures or machines that can harness this energy can fly.

Two related principles are involved. First, since air does have weight and is fluid, it exerts pressure on whatever is within it. The earth's atmosphere at ground level exerts a pressure of about 15 pounds per square inch. We do not notice this because the pressure inside us—including the pressure of the air within our lungs—inevitably increases or decreases until it equals the pressure that is all around us, and when that is achieved there is no sense of being pressed upon from outside. In the same way, a whale that's a mile beneath the sea does not notice that the water around is pressing down at many tons per square inch. But if the pressure in the atmosphere is reduced in any one place, then the force exerted by the air around it, as it rushes to fill the space, is prodigious. Old-style scientists were fond of demonstrating this by sucking the air from metal containers and then watching them collapse, violently. Tin cans show the principle excellently—cans thicker than a man could readily crush in his hands are demolished by simple air pressure.

So wings, whether of birds or aircraft, are designed so that as they move forward through the air (or the wind moves over them), the pressure above the wing is less than the pressure below. This is achieved by the cross-sectional shape of the wing. In general, the top is more curved than the underside. This means that the air that travels over the top of the wing has farther to travel to get from the front to the back than the air underneath. The air moving over the top is therefore stretched—the air molecules are pulled farther apart—so the pressure above the wing is reduced. Even a slight difference between the pressure on top of the wing and the pressure beneath is enough to provide lift—to raise the flying object bodily. In aircraft, the wing (these days usually of metal; in earlier times made of fabric stretched over a wooden skeleton) is simply shaped to be more curved on top than beneath, to form an "airfoil." In the flight feathers of a bird, the supporting rod, the rachis, is toward the front of the feather, so the front is thicker than the rear and the rear-pointing barbs again are shaped to form an airfoil. In feathers that are not designed for flight—those of the tail, for instance—the rachis runs through the center.

But of course the airfoil does not work unless the air is moving across the wing. So the second requirement is to provide such movement. Many animals (and, of course, some aircraft) achieve this either by starting off from a height and allowing gravity to give them forward speed, or they point themselves into the wind and allow the movement of the air itself to provide the movement. In either case, there are two effects: first, the basic airfoil lift effect; and second, if the wing is simply tilted up at the front, then forward movement into the wind will naturally provide more lift. The same principle is used to provide steerage—with flight feathers (or flaps) on the wing and feathers (or more flaps) on the tail.

The most refined fliers, of course, do not need gravity to provide them with forward movement, and they do not need a pre-existing headwind. They provide their own forward momentum. Aircraft do this with propellers or with jets, and birds (and bats, pteranodons, and insects) do it by flapping their wings. Put the two together—the airfoil wing and a means of creating forward speed—and we have powered flight. Once the basic apparatus is in shape, it can be refined in a hundred ways, as seen both in modern aircraft and in modern birds. Some birds, though, still like a helping hand from headwinds, or from rising currents of air. *Archaeopteryx* probably needed a little help from gravity, too.

If the bird actually wants to flap its wings to power itself along, then it needs muscles for the task; and the bigger the area of wing, the bigger the muscles need to be. The power of the muscle depends (other things being equal) on its cross-sectional area. But when the linear dimensions double and the weight increases eight times, the cross-sectional area of the muscle increases only four times. So the flight muscles of a big bird need to be bigger than a small bird's, not only in absolute size but also proportionately.

Then again, the wing beat is complicated. It is not like rowing, where there is a power stroke through the water and then a recovery stroke through the air that merely returns the oar to its starting point. The lift when a bird flaps its wings comes not merely from increased air pressure below but also from reduced air pressure above—the flying bird is pushed upward, and sucked upward. Because of the way the bird angles its wings as it flaps, the upstroke is a power stroke, too. So it needs not one but two sets of flight muscles—the bigger ones are for the downstroke, but the ones that power the upstroke are big, too.

The wing muscles are on the chest—they are the "breast meat" of the Sunday chicken—and they are anchored to a "keel" or "carina"—a projection of the breastbone. The bigger the flight muscles, the deeper the keel needs to be. So a very big flying bird with proportionately huge wings would need vast breast muscles to power it along—which would necessarily be heavy and make the problem even worse. The great British biologist J. B. S. Haldane pointed out that if human beings wanted to fly like a bird, with flapping flight, in the style of Daedalus and Icarus, we would need a keel that protruded outward from our chests by 2 meters (6 feet) to anchor the necessary muscles—although such muscles would themselves triple our weight so that then we would need even bigger muscles to provide the necessary power; and so on and so on.

Clearly, the business of flying soon gets out of hand once body size starts to increase. As things stand, among present birds, the flight muscles on the breast typically account for about a quarter of the total body weight, although in pigeons it can be more than 40 percent. The biggest flying birds are the Mute and Trumpeter Swans, the Great Bustard, and the American condors—though there was once a South American vulture known as *Tetrornis* that had almost twice the wingspan of today's biggest birds. But although eagles and swans are powerful fliers, they cannot begin to match the swallows or the pigeons for agility; and although they seem vast when they are on the wing, none has a body much bigger than a terrier. There are bigger birds than that, of course, such as the modern Ostriches and cassowaries, and there were many more giants in the past, but the very big birds are all land-bound. They have long since abandoned flight.

Of course, being big is not the be-all and end-all, but survival on this earth is all a matter of niches and large body size is a very handy niche for animals that can hack it. Little animals can duck and dodge and live on scraps, but big animals can eat littler animals, and little animals get eaten. Britain now seems to have eagle-owls, though whether they escaped from aviaries or flew in from their natural breeding grounds in southern or eastern Europe is unclear. But now that they are in Britain, they are happy to feed on hobbies, among many other things—and hobbies, only marginally smaller than kestrels, are formidable birds of prey in their own right. As we will see in Chapter 5, too, big animals find it easier to live on plants—which is a great advantage, since plants are so abundant. On the whole, mammals seem better at being big. The Ostrich has succeeded well enough, but it is the only truly huge terrestrial bird in the whole of Africa, while among the mammals there are dozens of kinds of antelope, plus wild cattle (buffalo), pigs, elephants, hippos, rhinos, and horses (in the form of zebras). Even on the smaller scale, the mammals do better on the ground. Worldwide there are scores of small game birds and rails and others that fly very little or not at all, but there are nearly a thousand kinds of rodents—mice, rats, porcupines, guinea pigs—in the same kinds of niches.

And there is worse. Small bodies mean small heads mean smallish brains. To be sure, intelligence is not just a matter of gross brain size, and some birds are extremely bright (notably crows and parrots). But no bird has quite the physical and physiological wherewithal to develop brains of our size. Birds might continue to evolve for another billion years (we may hope they have that opportunity!), but they surely will never produce great poets, or philosophers, or professors of jurisprudence.

There is also another, more subtle reason bird brains do not quite seem to match those of apes and people. The ancestors of modern birds adapted their forelimbs for flight, and in doing so they lost two of their fingers; the remaining three, at the ends of the wings, just provide a scaffold for feathers, with the feathered thumb forming a separate "bastard" wing at the front to increase maneuverability. Furthermore, birds seem on the whole to be more "modular"[1] in their construction than mammals—and this is another shortcoming. Birds are like several animals in one. Birds are running animals, and/or they are flying animals, but only a few, like baby Hoatzins, use their legs and wings in concert, as climbing mammals routinely do. The birds that have abandoned flight have also more or less given up on wings. Other birds have abandoned their legs; the scientific (generic) name of the Common Swift is *Apus*, which is Greek for "no foot." Largely because their legs are so feeble, some swifts are condemned to fly more or less full time, except when they are raising young. They even copulate on the wing. But no bird ever did what our own, pre-human ancestors were able to do: stop using the forelimbs for locomotion and then use those liberated limbs for new and truly creative purposes.

In monkeys and apes, the arms and hands are dual purpose. They are used for locomotion—running or swinging or both—but also for manipulation: peeling fruit, grooming, and even, in the case of chimps, for using and making tools. Our own ancestors took the process one step further. Once they had come down from the trees and freed their arms and hands from the chores of locomotion, they could use them entirely for manipulation. Thus began an evolutionary, positive feedback loop, an exercise in "co-evolution": the hands at the ends of the long flexible arms could manipulate the environment, and thus gave the brain an incentive for evolving further, to make the manipulations more adept and more rewarding. The evolving brain in turn provided evolutionary incentive for the hands to become more dexterous—and so the two organs worked in concert until we had hands that enable some people to play a Beethhoven violin concerto, and a brain that enabled Beethoven to write it in the first place. By contrast, a bird wing serves little purpose except for flying; it is often used in display and very occasionally in combat, and sometimes to shade the babies or (as in some herons) to provide shade to cut out reflections in the water and make it easier to fish, but all these activities are ad hoc add-ons. Basically wings are for flying and are not much good for anything else—certainly not for making spears and clubs or playing the violin. So birds that gave up flight found themselves with spare forelimbs that were virtually useless for anything else—used by male Ostriches for sexual display but lost almost entirely in the kiwis, the Emu, the extinct moas, and the long-extinct loon-like *Hesperornis*. No positive feedback loop could be set up between the forelimb and brain as in monkeys, and especially as in human beings, so there was never the internal driver to turn clever birds into intellectuals. Storybook owls are wise, but real owls seem rather dim and the Secretary-bird, a long-legged terrestrial hawk from Africa, may look like a bureaucrat in certain lights but it cannot in reality take company minutes. (The question still remains, however, as to whether the brains of birds work in the same way as those of mammals or would demonstrate quite new principles if only we had the means to find out.)

Flying birds can be bigger, though, than a flying mammal can be because they are so beautifully engineered to reduce weight. We see this most obviously in the skeleton. Birds have lost a great many bones that mammals and most reptiles routinely possess, and some of the birds' remaining bones are fused or partially fused together. The most ancient birds, known from their fossils, had jaws with teeth, like reptiles. In all modern birds, the teeth and the hard and weighty bones needed to hold them are replaced by a beak made of horn, which consists of keratin and though often of horrendous strength—big parrots casually crush Brazil nuts—is light. The flying bird is not nose-heavy. (As we will see in Chapter 5, however, there is much more to beaks than meets the eye. Many are intricately engineered.)

In mammals, the backbone with its attendant tendons and muscles is like a great spring that may extend into the tail; we see this in the sinuous bounds of the cheetah or the almost indiscernible flick of the dolphin's tail that shoots it through the water. By contrast, the thorax or trunk of the bird is not a spring but a box: the vertebrae at the back are partially fused; the sternum at the front has a keel for attaching the flight muscles (though some flightless types like the Ostrich have no keel); and the peculiar collarbones, or clavicles, at the front are fused at their ends to form the V-shaped "furcula," or wishbone. The wishbone acts like a spring in flight: the angle between the two arms widens and closes again as the wings beat, maintaining strength and rigidity without compromising flexibility. I know of no other instance in all of nature where a single bone acts as a spring in this way—another example of supreme adaptation and supreme inventiveness in the lineage of birds.

Incidentally, the wishbone was once a prominent feature of the traditional roast chicken. The children vied to pull it apart—little fingers only!—and the one who finished up with the bigger portion got to make a wish. When chickens were raised in the traditional way the wishbone could put up a grand struggle. With only the little fingers, it was often hard to break—and showed what strength and power this modest structure must have in the living bird, like sprung steel. But in modern broiler chickens, raced from the egg to puffed-up oven weight in six weeks, the bones have no time to grow and the wishbone typically breaks and then disintegrates as the bird is handled for slaughter, and many modern children are hardly aware of its existence, except as a few splinters.

Apart from the flexible wishbone, the only movement in the whole trunk, even in flight or when the bird is running its fastest, is in the shoulder and hip joints. The bony tail that serves the dolphin so well, and serves the cheetah as steerage and the spider monkey as an additional limb, is all but lost, reduced to a few fused bones known collectively as the "pygidium." To be sure, many birds rely on their tails for steerage even more than a cheetah does: the long, forked tails of swallows, swifts, frigatebirds, and tropicbirds give them their aerial agility; without them, they could not function. But the tail is not bony. It is made entirely from feathers, that grow from the skin around the pygidium. The tail feathers in effect do the job of a bony tail, but they are much lighter; once grown, they need no blood supply and no internal nerves. To substitute feathers for flesh, blood, and bone is a marvelous piece of delegation. In some birds, including fowl, the tail feathers from the pygidium are enhanced by additional feathers from the back. The "tail" of the peacock, which forms the wondrous fan, is constructed thus; it is not a tail at all.

Overall flexibility comes from the neck, which is always very versatile and sometimes very long. Anatomically speaking, therefore, a bird is like a snake in a box, with two sets of limbs added on (either one of which might at times be all but abandoned), and a rump to wag at the back, with the tail feathers to provide leverage. The beak at the end of the long, flexible neck serves the same function as the human hand at the end of its long, flexible arm. In many birds, from flamingos to shovelers to finches to parrots to hummingbirds to woodpeckers, the tongue serves in many various ways as a limb within a limb, sometimes with wondrous dexterity.

The bones on the whole are hollow and hence extremely light, with great strength relative to their weight. Struts within the long limb bones add to the strength, like the cross-girders in, say, New York's Queensboro Bridge. For good measure, as we will see, some of the bones in most birds are filled with air. All in all, the individual bones and the skeleton as a whole are a supreme exercise in civil engineering.

But the key to the design lies in one of nature's most ingenious inventions: feathers.

THE PROS AND CONS OF FEATHERS

In modern birds, feathers serve three main functions: for insulation, for display, and for flight. Chemically they are similar both to the scales of reptiles and to the hair of mammals—all are made from slight variations of a hardened protein known as keratin—but the structure of feathers is much more refined. All true feathers are branched—a central "rachis" with rows of "barbs" on either side. In feathers intended for insulation, and in some that serve for display (as in the sexual plumes of herons), the barbs remain loose and fluffy. In flight feathers, the barbs hook together like a zipper to form a continuous vane—a strong, light airfoil. Yet another triumph of natural engineering.

Common sense says that these three different functions could not have evolved all together, if only because different functions require different kinds of feathers. Some biologists have suggested that birds first acquired feathers when they began to fly, but since the flight of birds depends absolutely upon feathers, and since flight feathers are the most sophisticated kinds of feathers, fashioned as they are into airfoils, this seems most unlikely. Besides, as described in the next chapter, non-avian, non-flying dinosaurs that also had feathers have now been discovered, and although the fossils of these nonflying dinosaurs are younger than the

oldest known bird, their existence suggests that feathers evolved independently of flight. Indeed, the use of feathers for flying is just another
(very good) example of evolutionary opportunism: natural selection making use of structures that had already evolved for some other purpose.
Most biologists feel intuitively that the very first function of feathers was
for insulation, since some ancient dinosaurs were almost certainly
homeothermic (meaning they maintained a constant temperature) and
needed to keep the warmth in—even in the tropics, deserts at night can be
very cold.

I reckon, though, that feathers could well have evolved first for display,
and mainly for sexual display, for no lineage of creatures that reproduces
sexually can survive for more than a generation if it fails to find mates, and
there is no doubt that in the mating game it pays to strut your stuff. Certainly many modern reptiles go to great lengths to put on displays with
frills and dances—which feathers would have enhanced no end, as many a
modern human dancer in dozens of cultures from Maori to Cherokee to
the Royal Ballet has demonstrated. Feathers add the refinement of dif-
fraction: the light rays broken up into their different component colors so
that the feather shimmers and seems to change color—now you see it,
now you don't. Only a few other creatures can match this wonderfully il-
lusory effect, like the scaly wings of some butterflies and moths. In truth
we will never know for sure how feathers first arose and what first drove
their evolution, but as discussed in Chapter 7, it does not pay to underestimate the driving power of sex.

The feathered, non-avian dinosaurs have all disappeared and so have
the pterosaurs—the ancient reptiles that flew on leathery wings, without
the aid of feathers. The birds survived because they had feathers and they
flew—or this, at least, is how it seems. But flight requires a great deal more
than feathers. Every aspect of a bird's anatomy and physiology is geared
to it.

But nothing in nature is for nothing. Individual feathers are a
metaphor for lightness, but the plumage as a whole typically weighs two or
three times as much as the bones. Thus, though they are the key to flight,
they add to the load. They need serious upkeep, too. Most birds (though
by no means all) have a "preen gland" on the small of the back, near the
tail, that secretes oil, which they spread over the feathers with their beaks
(in other words, "preening"). The feathers are subject to attack by lice and
provide a miniature tropical forest for other small creatures to lurk in.

Birds are beset by parasites, which many contrive to remove in part by dust baths, by regular bathing, and sometimes by "anting"—introducing ants into the plumage to clear out the parasites. My grandfather told me that he and his fellow squaddies adopted the same tactic during the Boer War in South Africa, at the turn of the twentieth century: they would place their clothes over an ant nest, and let the ants go to work on all the exoparasites that lurked within. Those species of swifts that spend months on the wing do not find it easy to preen or groom, and cannot bathe in puddles or in the dust, so they suffer mightily from parasites.

Most birds, too, molt at least once a year (just a few molt less often than this) and replace their entire plumage. Peacocks lose their magnificent trains and grow them again. Guillemots and most geese lose all their primary flight feathers at once and so, for some weeks each year, are flightless—and guillemots migrate, in this flightless state, by swimming. Many birds take the opportunity, while molting, to change their appearance; thus the field guides show the males, in particular, of many shorebirds and songbirds and ducks and others in their brightest mating plumage in the spring, when they are after mates, and in duller "eclipse" plumage, often camouflaged ("cryptic"), in the nonbreeding season. Clearly, birds that change from mating to eclipse plumage and back again must molt twice a year. For all birds, the changing of feathers is a huge physiological burden. So birds pay an enormous price for their feathers. But still they are worth it. Feathers—particularly in the form of flight feathers—are the secret of their great success.

NOBODY'S PERFECT

Beyond doubt, some birds fly better than others. It is also the case that modern birds fly much better than *Archaeopteryx*, generally acknowledged as the first bird of all, which lived in the Jurassic, deep in dinosaur times. So flying has improved over time.

Some modern biologists regard such commonsense observations as heretical. It seems it is politically incorrect to suggest that evolutionary change over time brings about improvement. Such an idea is deemed to be "teleological," implying that evolution has some goal in mind; whereas, so some moderns insist, evolutionary change over time should be seen simply as a random walk, liable at any stage to go this way or that. We cannot speak of "improvement" because there are no objective criteria to measure improvement by. Natural selection does not strive for technical perfection. It merely produces creatures that can survive in their own particular environment, and all creatures that do survive must therefore be considered equally successful. Thus, some moderns agree with Orwell's dictatorial pigs: all animals are indeed equal.

Whether or not the path of evolution leads living creatures toward some prescribed goal is a matter of metaphysics, or indeed of theology, which, alas, is beyond the scope of this book. But improvement, not least in the form of technical proficiency, can be measured objectively, just as the performance of a car or an airplane can be measured. Some birds are more agile than others, or quicker, or can fly farther. Perhaps more to the point, some get more flying hours, or speed, or distance, or aerobatics per unit of energy than others, which means that by the standard definition of physics they are more efficient.

Yet in flying there can be no such thing as perfection. All flight, as an aero-engineer once pointed out to me, is compromise. If you do one thing particularly well, you will do other things less well. High-altitude spy planes can stay aloft for days on very little fuel, but they cannot go fast and are not good for aerobatics. Jumbo jets fly high and fast with a huge payload, but again they are not built for agility and they use prodigious quantities of fuel. Modern fighter planes are wickedly fast and maneuverable, but they are also unstable and without an on-board computer to make rapid microadjustments, they could hardly be flown at all. Planes that are good all-rounders, like the Tiger Moths that still are used for training pilots, cannot do any one thing particularly well. It's the same with birds.

No one was better equipped to discuss the mechanics of bird flight than the great engineer-turned-biologist John Maynard Smith. In a classic essay from the 1950s, "Birds as Aeroplanes," he compared *Archaeopteryx* to a paper airplane. Paper airplanes have more than adequate wings designed for maximum lift—low wing loading—and long tails. They are wonderfully stable, but they don't do much; they travel in a straight line, not very fast, and stop when they run out of momentum. It is clear from its shape that, aerodynamically speaking, *Archaeopteryx* was like a paper airplane: very stable but not maneuverable. Of course, *Archaeopteryx* flapped its wings for powered flight, which paper airplanes cannot, but we can see from its solid bones and its bony jaws and tail that it must have been heavy, and must have expended a lot of energy even to do what it did.

Contrast modern birds. The mighty Peregrine and the even mightier Gyrfalcon can dive or "stoop" on their prey—a pigeon, in the case of the Peregrine—at nearly 300 kilometers per hour (185 miles per hour), or so it's estimated—Formula 1 speed. Sparrows are aerodynamically unstable compared to *Archaeopteryx*, but they make up for this with fabulous coordination, able to respond in microseconds to the smallest fluctuations in the air, like modern fighter jets. Storks, eagles, and vultures have long and broad wings, yet find it difficult to take off from the ground in normal conditions, and a grounded vulture in a northern zoo may not bother to take off at all. Instead, these big birds rely on the up-currents of cliffs or, more generally, on thermals—rising columns of warm air from hot ground that can carry them upward almost out of sight; human sight, that is, for these birds can see all they need to see from far above.

Geese, as heavy as most eagles, are able to power themselves to great heights because their broad, long wings are commensurately muscled. Bar-headed Geese on migration fly over the Himalayas, almost at jumbo-jet height. Swallows, swifts, and nightjars, with their long scimitar wings and long forked tails, are wonderfully agile, able to pick off the swiftest and most elusive insects on the wing. Pheasants and other fowl with their short wide wings and powerful legs take off like Harrier Jump Jets, fast and nearly vertical, though they don't on the whole fly far (although some quail, which are also fowl, migrate). Pigeons fly straight and true for hundreds of miles at 60 to 80 kilometers an hour (40 to 50 miles per hour)—but not nearly fast enough to escape Peregrines. Albatrosses, with their very long narrow wings, are not very maneuverable, but they can fly for thousands of miles almost without effort, like gliders with extremely accomplished pilots, and their wings lock in the stretched-out position; it requires no muscular effort. Since albatrosses fly over the Southern Ocean, there are no thermals to provide lift. But there is constant wind, and the wind high up is faster than the wind lower down, where the waves act as a brake. So albatrosses gain speed at high altitudes and then sink slowly to wave level under gravity, then they turn—still at high speed—to face the slow winds near the waves; and so they find themselves flying into the slow winds at high speed—and this gives them lift, just as a plane is lifted when it rushes along the runway into static air. Far below the soaring albatrosses, in the Southern Ocean itself, penguins "fly" underwater. Their wings are not mere paddles, or oars, just pushing the water backwards; they provide lift—true hydrofoils.

Each of these birds does what it does brilliantly—but, as with airplanes, none can do everything. You cannot be as agile as a swallow or as fast as a Peregrine but also as effortlessly economical as an albatross. Each kind of maneuver requires its own technology, and different technologies are incompatible. In the same way, you can go very fast in a Ferrari, but you cannot use it to take the family on holiday. You just have to choose. Still, though, it remains the case that some machines do what they do better than others that are trying to do the same thing. Some birds are more equal than others.

ENERGY AND POWER

Flight requires tremendous energy—and food is at a premium. As we will see in Chapter 5, there's a lot to eating: the right physical equipment, the right techniques, the right strategy, each and every day and over a whole lifetime. Eating raises physical problems. Birds that eat leaves need to eat a lot and so need big stomachs and diversions of the gut known as ceca (pronounced SE-ka)—and all of this is extra weight. Birds that eat seeds, which contain more energy than leaves, have a special need for gizzards—expansions of the gut commonly filled with gravel for crushing, like a mill; extra weight again. All birds need water and some are very thirsty creatures—more weight again (although some birds, as we will see, like some gazelles, are adapted to live without drinking). Birds go to great lengths to conserve water, so whereas mammals, like us, excrete surplus nitrogen (from protein and amino acids) in the form of urea diluted in great quantities of water, birds excrete surplus nitrogen as uric acid, virtually in crystalline form, with almost no added water. Hence, the characteristic white splashes; and hence the enormous value of guano, the accumulated excreta of nesting birds (usually seabirds), that once was collected for making gunpowder (saltpeter) and then was collected for fertilizer (and in both cases became a significant industry).

But once the fuel is on board, safely packaged in the form of glycogen and particularly of fat, it needs to be "burned" with maximum efficiency to satisfy the bird's enormous appetite for energy, and of the flight muscles in particular. Birds achieve this by ensuring a throughput of oxygen that by mammalian standards is truly staggering. The lungs of mammals are mere bags (or sacs, as zoologists tend to say): the air goes in one way and comes out the same way. With each breath, a little residual air is left in each lung—not very efficient at all. In birds, the lung is a series of parallel tubes like the condenser of a boiler. The air rushes in one end of the lung and rushes out the other end—there is complete clear-out; all the air and the oxygen it contains can be brought into close contact with the blood vessels in the lung walls. From the lungs the air passes to a series of air sacs distributed around the body, varying in number from seven in loons and turkeys to at least twelve in shorebirds and storks. Most birds have nine such sacs: two in the neck, which male frigatebirds dilate into huge red balloons for sexual display, as we will see in Chapter 6; two in the upper thorax; two more in the lower thorax; two in the abdomen. Finally, there is a single sac between the two clavicles (the interclavicular sac), and this one extends into the bones of the wings and sternum. It's the air from this sac that is forced at pressure past the syrinx, with such efficiency. Once the air has been into these air sacs, it does exit the body by the same route it came in—via the trachea, like a mammal. The key difference is that the lungs in mammals are a cul-de-sac, while in birds they are a thoroughfare.

The system that increases efficiency of respiration has another bonus. It makes birds even lighter than they would otherwise be. Real ducks float on the pond for the same reason that plastic ducks float in the bath. They are full of air.

But even here, nothing is for nothing. Birds that hunt underwater need to be heavy. So it is that penguins, loons, grebes, Anhingas, and Cormorants go to various lengths to increase their weight, at least for the purposes of diving. Penguin bones are heavier than the average. Cormorants and Anhingas do not waterproof their wings—and so they become heavy as they swim. Cormorants float low in the water. Penguins and particularly Anhingas float even lower, the latter so low that sometimes only their heads and necks are above water, and so they are also known as "snakebirds." When Cormorants and Anhingas emerge, they have to dry themselves off, holding their wings out like scarecrows. Grebes draw their feathers tight to the body before they dive—the opposite of fluffing out—to squeeze the air from them.

Birds, then, are unique. Indeed, as we will see in the next chapter, they seem so different from other creatures that some leading biologists in the nineteenth century said they could not possibly have evolved from any known creature—and they used the fact that birds exist as evidence that

One of the great sights of the Southern Ocean—diving-petrels flying into waves and out the other side.

Darwin's idea of evolution was wrong. (As we will see, many influential biologists in the nineteenth century were anti-Darwinians.)

And yet, birds in many ways are remarkably similar to mammals. They are designed differently, and to some extent lead different lives, but they are parallel life forms. In their general approach to life they are much the same as us—it is just that, in every way in which it is possible to be different, they are.

THE SAME ONLY DIFFERENT

In almost all general ways birds are just like us—just like mammals, I mean. The biggest birds are nothing like as big as the biggest mammals, but even the smallest are very big, as animals go—for most animals are insects, mites, or nematode worms, and all but a few moths and beetles and megastick insects are tiny. Birds and mammals are both warm-blooded, too. This does not mean that they are gratuitously hot. It does mean that, on the whole (with some exceptions which we will look at as this book unfolds), they contrive to keep the core of the body around the vital organs and the brain at a more or less constant temperature, which except in the tropics in the daytime is higher than their surroundings. The temperatures that both birds and mammals strive to maintain—around 40°C (104°F)—is the best working temperature for most of their enzymes, so this temperature helps them to function at around optimum physiological efficiency. Both mammals and birds maintain this temperature by burning food energy specifically for the purpose of generating heat—indeed, this is how some of them use up to 90 percent of their energy.

But constancy of temperature is just as important as high temperature, so that "homeothermic"—meaning "constant temperature"—is a better term than "warm-blooded." It is largely because they maintain a constant, high temperature that birds and mammals are able to maintain large and complex brains—which one way or another are extremely sensitive to temperature flux. In the mid-nineteenth century, the great French physiologist Claude Bernard wrote of *le milieu intérieur*—the "internal environment"—of animals; and in the 1930s, the American physiologist Walter Bradford Cannon coined the word "homeostasis"—the sum of all attempts to keep the internal environment constant. Homeothermy is the supreme example of homeostasis. External conditions vary far more on land than in water, and of all terrestrial creatures mammals and birds maintain homeostasis most effectively, because in addition to everything else they are homeothermic. To be sure, warm-bloodedness has its downside: since homeotherms burn so much energy just to keep warm, they need to eat far more—sometimes many times more—than "cold-blooded" (or "poikilothermic") animals of the same body weight, such as snakes or lizards. We might say that lizards are low-input, low-output creatures while birds and mammals are high-input, high-output: high maintenance. Both lifestyles have their advantages and disadvantages. In deserts, where there is plenty of free heat from the sun and there is little to eat, reptiles may well have the edge. But most of the time, if the homeotherms can find enough to eat, they tend to outpace the poikilothermic lizards and frogs and outsmart them. In most terrestrial environments, therefore, birds and/or mammals are dominant.

Yet birds and mammals differ in the details of their most basic design. When Charles Darwin first confronted the animals of Australia in the 1830s, during his epic journey around the world on HMS *Beagle*, he commented, "Surely these creatures are the work of a separate Creator!"—and we might say the same about birds in general, as opposed to mammals.

Some differences are obvious.

Mammals have hair while birds have feathers. Many mammals have lost some or almost all of their hair—whales, naked mole rats, and even humans are pretty sparse—but no bird ever loses its feathers entirely (though many have bald patches as in the necks of vultures or in the "brood patches" hidden beneath the feathers of the lower breast, which transmit body heat efficiently to the eggs during incubation—typical of females and sometimes present in both sexes).

All known birds lay eggs with stony shells, while most living mammals bear live young. To be sure, this distinction is not so absolute as it might seem, because some living mammals—the monotremes: the duck-billed platypus and the echidnas—do lay eggs, and most of the very earliest mammals almost certainly were egg-layers. But, as we will explore in Chapter 7, the manner of reproduction has a profound influence on all aspects of life—and especially on family life.

In mammals, the genital and excretory organs share accommodation to some extent—in males, the sperm and urine exit by the same tube (the urethra). But in female mammals, the urethra and the vagina are quite separate, and in both sexes the genitalia are separate from the anus. But in birds, the eggs, sperm, urine, and poop all exit through a single aperture, the cloaca. Birds in general copulate simply by placing the openings of their cloacae together. This seems to work well enough, and most male birds do not have a penis to enable them to penetrate and place the sperm where it is needed. A few, including the Ostrich, do have one; and in some ducks it is a spectacular organ, at least as long as the duck itself—more like a fireman's hose. The distribution of penises among birds seems somewhat arbitrary and hard to explain. Thus, we might argue that a duck needs a penis because it copulates in water—but some birds that copulate in water do so without a penis, and no bird is more land-bound than an Ostrich. Much in nature remains unexplained (and much indeed may be inexplicable).

Most birds are seasonal breeders and, as with many seasonally breeding mammals, the testes and ovaries grow prodigiously before the breeding season but shrink again when breeding is over. Shrinking the organs that are not immediately necessary saves weight, although whatever is shrunk has to be regrown the following season. Nothing is for nothing.

Mammals and birds, both, are noisy creatures. They commonly make their presence felt, and they communicate by sound. But birds are far better at it. Many mammals have an extensive vocabulary—different sounds for different objects and exigencies, like the vervet monkeys' cries of "Snake!" and "Eagle!"; but few can match the range of meaningful sounds that a chicken gives voice to (more on this in later chapters). Apart from (some) human beings, mammals on the whole are not melodious and there is little evidence that they intend to be. Some bellow, like rutting stags, but few sing, apart from human beings, and perhaps whales, and wolves, if we stretch a point (and possibly courting mice, who have now been shown to serenade in ultrasound). Yet many birds are famed for their songs; and, joy of joys, some of the most glorious songsters are the ones we encounter most often.

Here again we find contrast: mammals and birds produce sound in quite different ways—and the avian technology is superior. Both groups have a larynx. In both, the larynx is set in the throat and serves as a valve, helping to ensure that food and drink pass into the esophagus, rather than the trachea (the windpipe). If food and drink do pass into the windpipe, the animal chokes. But in mammals, the larynx is also fitted with two membranes that vibrate like the reed of an oboe, so that the larynx doubles as the voice box. The one exception among mammals—as is so often the case—is the human being. In us, the larynx has dropped down the windpipe and in men, in whom the larynx is especially large, it conspicuously forms the Adam's apple. This arrangement vastly increases the human ability to make and manipulate sounds, since we can use the entire throat and thorax as a resonating chamber, as other mammals cannot, and can articulate sound with wondrous facility, as we do when we speak. All other mammals, with their larynxes high in the throat, produce a limited range of strangulated sounds. Huge efforts have been made to teach chimpanzees to use human language, but only by signing—no one ever supposed that chimps could ever speak, because they don't have the physical apparatus.

But as always there is a downside. In humans, the larynx is too low-set to act as a valve. So we, among all mammals, have a unique propensity to choke. We choke if we try to drink and breathe at the same time. Other mammals do not.[2]

Birds, as so often, have solved this dilemma beautifully—in two ways. First, they have left the larynx in the throat to act as a valve, and have invented a quite new kind of apparatus dedicated entirely to the production of sound. This dedicated soundbox is the syrinx. It is situated low down in the throat, in the angle between the two bronchi, where they first branch away from the trachea to carry air to the lungs. Second, as discussed earlier, birds have a series of air sacs positioned around their body that enable them to breathe with huge efficiency; and these sacs also help to produce sound.

The anatomical details of the syrinx are immensely complicated (and so is the corresponding physics), but the main point is that the syrinx is hard up against the cervical air sac, which is positioned in the neck; and between the syrinx and the cervical air sac is a membrane. Air from the other air sacs around the body is forced through the bronchi and through the syrinx—and this membrane then vibrates. The principle of sound production, then, is not like an oboe reed, as in the mammalian larynx, but rather like the skin of a kettle drum. The pitch of a kettle drum, unlike an ordinary military drum, can be changed by tightening or relaxing the tension rods that are set around the circumference (and are the chief obsession of percussionists in long orchestral works with significant changes of key). The same applies in the syrinx: the tension in the membrane is changed by muscles, and so, correspondingly, is the pitch of the sound it creates. Most syrinxes are fitted with a pair of external muscles. The syrinxes of songbirds such as robins and nightingales are also fitted with six internal muscles that presumably offer greater versatility.

The syrinx is fabulously efficient. Almost 100 percent of the energy in the air that is forced under pressure past the membrane is translated into sound. The efficiency of the mammalian larynx is a mere 3 percent. If Pavarotti had had a syrinx, no opera lover would have been safe. Covent Garden, La Scala, and the Metropolitan Opera House would surely have been shaken to their foundations.

There are further refinements. The depth and resonance of the sound depend in part on the length of the trachea; and in some birds, such as cranes, the trachea is long and coiled, like a euphonium. And so the sound of the crane echoes across the marshes as powerfully and evocatively as the music of a silver band through the valleys of Wales. With typically brilliant avian engineering, the coiled trachea is housed within the bone of the sternum (the breastbone).

Not all birds are sonorous, of course. A grunt and a hiss are the best you can expect from a Mute Swan, or a Turkey Vulture, or the Greater

Rhea. But the sound of the nightingale, though it is a creature of modest size, can fill half a forest, and a skylark can fill the entire landscape. And I once saw a cassowary at London Zoo open its prodigious beak like the door to some hellish tomb, and frighten a little girl so much with his mighty roar that she leapt five feet from a standing start into her father's arms. Lions can shake us up too, of course, but then lions are many times bigger, and they can't do it without preamble. They need to pump themselves up for a good roar, while a cassowary can roar from nowhere. Truly, on the Orwellian scale, birds are more equal than other animals.

But there are many more, far less obvious differences between birds and mammals.

The bones of young mammals grow in three sections—the ends that form the joints (known as the epiphyses) are at first divided from the main shaft by cartilage, and they grow at a different pace from the main bone. In birds, the bones are all of a piece from the outset—although the ends, in young birds, are largely cartilaginous and become bony later. In mammals, the red blood cells (erythrocytes) that carry the oxygen around the body lose their cell nuclei as they develop. Bird erythrocytes retain their nuclei, as do those of reptiles.

Sex is determined in different ways in mammals and birds. In mammals, males have two different sex chromosomes, X and Y; and in this respect they are said to be "heterozygous." The females have two X chromosomes, XX; and are said to be "homozygous." XX and XY are now universally recognized symbols of femaledom and maledom, and have found their way into advertising slogans and onto the doors of washrooms. But in birds, the females are the sex with different sex chromosomes, known as Z and W, while the males have two Z chromosomes. So the females are the heterozygotes, ZW; and the males are the homozygotes, ZZ.

It seems to be the case, like it or not, that in mammals the males are the more flamboyant and the more variable sex. The males are usually bigger and stronger, and take more risks. Among human beings, the most famous poets, composers, and physicists tend to be men, although many of the finest prose writers and biologists are women. But there are also more seriously dim and dangerous men than dim and dangerous women. Some put all this down to expectation and upbringing, but others ascribe the differences to biology. Ninety percent of mammalian species are polygynous—the males have many female mates—so the most productive reproductive strategy for the males is winner-takes-all. Males that behave in a cautious, middle-of-the-road kind of way are not likely to find a mate at all, so they might as well be extremist and go for bust (albeit with a fair chance of getting it completely wrong). Females, on the other hand, will always find reproductive partners in a polygynous society, so their best policy is simply to be sensible. This notion is far from established, but it is plausible.

Some biologists ascribe the extreme qualities of male mammals to their heterozygosity. If their single X chromosome contains genes that make their owners especially clever or especially dim, then those genes will not be tempered by corresponding genes on the matching X chromosome, because there is no matching X chromosome. But female mammals, this argument has it, are likely to be less extreme because all their X genes are doubled up, so that any extreme genes on one X chromosome are likely to be balanced by more moderate genes on the matching X chromosome.

In birds, the males are also—usually—more extreme in their appearance and behavior than the females. But, in birds, the males are the homozygotes. So the theory that links behavior and appearance to heterozygosity or homozygosity seems to go by the wayside (or at least it needs more work). Meanwhile, this all shows rather neatly how the study of birds can throw light on mammals, including human mammals.

I wonder, too, if the catalogue of contrasts between birds and mammals also applies to their brains and hence to their ways of thinking. Clearly, the overall structure of the bird's brain is very different from the mammal's—birds emphasize the cerebellum, concerned with muscular coordination, more than the cerebrum, concerned with what we call thought. But birds clearly do think nonetheless. Crows and parrots in particular are positively bright, often eerily so. But do they think the same way as we do? Or do they have ways of processing information quite different from ours? If they were philosophers, would they invent a different logic? Scientists who have studied computers these past forty years in an attempt to emulate the human brain—artificial intelligence, or AI—have for the most part concluded that human brains and computers are qualitatively different: that we and computers approach the process of making sense of things quite differently (and computers cannot properly be said to "make sense" of anything). What of birds? Are they different again? If they were, then this would be of profound philosophical import, suggesting once more that what we take to be true is just what our particular kinds of brains happen to make of what is out there. In line with Rudyard Kipling, too, "And what can they know of England who only England know?" we would surely discover a great deal about our own psychology if we could compare it with a psychology that was discernibly different. Again, we see that while birds are worth studying for their own sake, they can also tell us a great deal about ourselves. More of the mindfulness of birds in Chapter 9.

So there are huge, general questions to be asked. Are birds really so different from all other animals that their evolution cannot be explained? Were they really just placed on this earth ready-made? And why are birds so similar to mammals in general ways, and yet so different in detail? And then again, why are birds on the whole so similar, one to another, despite the huge discrepancies in size and way of life, while mammals are so immensely diverse?

In truth, bird evolution is now being worked out in impressive detail, and as the story unfolds, the answers to all these puzzles are becoming clear. As Dobzhanksy said, it's evolution that makes sense of biology.

2

How Birds Became

H OW DID BIRDS, THESE MOST BRILLIANT CREATURES, come into being? How come there are so many different kinds, each so beautifully adapted to its own particular niche? How come they are so like mammals, and yet so different?

For many centuries, the answers seemed self-evident. All creatures are as they are because God made them that way. Birds came about as recorded in Genesis 1:20: "Let the water teem with living creatures, and let birds fly above the earth across the expanse of the sky" (New International Version). There was no conflict between the scientists and the theologians because, well into the nineteenth century, science was widely conceived of as a branch of theology. Many of the finest naturalists have been clerics, including perhaps the greatest of all, much admired by Darwin, the Reverend Gilbert White of Selborne in Hampshire (1720–93). White saw things with the naked eye that most of us would miss with binoculars; he kept meticulous notes, he always asked "Why?" and he was not afraid to speculate, most cogently. His book of 1788, *The Natural History and Antiquities of Selborne*, is one of a distinguished few that changed the course of biology. In his introduction, the reverend naturalist makes his priorities clear: "If the writer should at all appear to have induced any of the readers to pay a more ready attention to the wonders of Creation . . . his purpose will be fully answered."

Yet even in White's own time, more and more scientists were wondering whether the account in Genesis should be taken literally. In particular, the new fossils that were coming to light—not least as the civil engineers and miners of the Industrial Revolution dug deeper and more adventur-

Was this the very first bird of all? Archaeopteryx *lived in tropical swamps, deep in dinosaur times, 140 million years ago.*

ously—showed that the creatures that lived in the deep past were often very different from those of today. Some argued that God had simply decided to change the cast every now and again, sweeping away the earlier types in a series of catastrophes. But some began to suspect that, in truth, some of the early extinct types were the ancestors of modern creatures, and that those creatures had changed over time. In other words, some began to think in evolutionary terms.

Yet none could think of a plausible mechanism by which these changes had come about. It was easy enough to see how creatures would vary as the generations passed—because, after all, children do not exactly resemble their parents. The challenge was to explain adaptation. Lineages of creatures over time seemed to become more able to cope with their environment—and it was hard to see how this could happen without the guiding hand of a benevolent creator. So evolutionary musing remained a minority pursuit—not entirely convincing, and potentially blasphemous at that.

Then, in the late 1850s, Alfred Russel Wallace (1823–1913) and Charles Darwin (1809–1882), working independently, outlined a mechanism of evolution that did seem plausible—what Darwin called "natural selection." All creatures produce more offspring than can possibly survive, so the offspring must compete for resources. The ones that survive are the ones that are best adapted to the prevailing conditions. In 1859, Darwin presented this idea at length in his *On the Origin of Species by Means of Natural Selection*. Wallace was and is a very strong player, too, with much to say of great interest, but he acknowledged that Darwin had pondered their shared ideas in greater depth, and natural selection is now commonly called Darwinism.[1] A couple of years after *Origin of Species*, the philosopher Herbert Spencer—who was already an evolutionary thinker—summarized the general notion of natural selection in the phrase "survival of the fittest," where "fit" seems both to imply "healthy" and also "apt" (as in "fit for purpose"). Darwin adopted the phrase "survival of the fittest" in later editions of *Origin of Species*.

Popular histories are wont to imply that the theologians and priests of the mid-nineteenth century were universally hostile to Darwin's ideas, while the biologists welcomed his views with open arms. The truth is far different. Many of Darwin's scientific peers—including some of the most eminent—were outraged; and many theologians took his ideas in their stride, arguing for example that if God chose to perfect His creatures bit by bit and over time, then so be it. Natural selection is, after all, a highly effective way to go about things. When Darwin died in 1882, he was buried in Westminster Abbey at the feet of Isaac Newton, and Dean Farrar said of him in his funeral oration: "This man, on whom for years bigotry and ignorance poured out their scorn, has been called a materialist. I do not see in all his writings one trace of materialism. I read in every line the healthy, noble, well-balanced wonder of a spirit profoundly reverent, kindled into deepest admiration for the works of God."

Why is all this relevant here? Because in the arguments that followed *Origin of Species* birds featured prominently. In 1861, just two years after *Origin of Species*, the world's earliest known bird turned up in a quarry in Germany—forever to be known as *Archaeopteryx*—and the controversy that followed was both scientific and religious and, in both guises, was often fierce.

ARCHAEOPTERYX

Darwin did not derive his idea of evolution from the fossil record, which in his day was still very sparse, but fossils did play some part in Darwin's theorizing. Before they came to light in such numbers, laid out chronologically in successive strata of rock, there was no overwhelming reason to suppose that the deep past was very different from the present. There seemed little reason to doubt that the world's creatures had always been the way they are now, as Genesis implies. When it became clear that the past was very different, some new explanation was required; but because the fossil record was so "spotty," it did not show the kind of changes that Darwin said must have happened. It certainly did not, at that time, show the gradual, step-by-step shift from primitive to modern that he envisaged. Often there were great gaps in time, and sometimes in geography, between successive fossils—too many "missing links." Worse, sometimes it seemed impossible to envisage a link that was halfway plausible—one that really could bridge the gap between the ancient and the extant.

Birds seemed to raise particular problems. Well before Darwin, biologists saw that birds have much in common with modern reptiles. Most obviously, the legs of most birds (or at least the tarsi—the long foot bones that lead into the toes) are scaly—very like, say, the legs of a crocodile. Old-style biologists, steeped in traditional theology, were content simply to see this, like everything else, as part of God's plan. If scaly legs work for crocs, why not use them on birds, too? The fossil record showed that rep-tiles were very ancient—far older than the oldest known birds. Darwin proposed that birds had scaly legs not because God was tidy-minded and economical with his designs but because somewhere among the ancient reptiles was the ancestor of modern birds.

If this were so, then there should be a link between the two: some ancient creature that was half reptile and half bird. But in the late 1850s, when Darwin wrote *Origin of Species*, this putative link was conspicuously missing. Worse, the professor of zoology and geology at Harvard no less—the Swiss-American Louis Agassiz—proclaimed that such a creature that was half bird and half reptile could not possibly exist. Birds are just too peculiar. There was, said Agassiz, no plausible route by which reptiles could evolve into birds. Agassiz was two years older than Darwin, and as a scientist in the 1850s he ranked at least as highly. He established his repu-tation after hiking in the Swiss Alps in the 1840s, when he perceived that the landscape and much of the northern world in general had been shaped by a series of ice ages—which was and remains a truly great insight. (It wasn't floods that caused the greatest dramas of the recent past, but ice!) In 1859, the very year of *Origin of Species*, Agassiz established Harvard's Mu-seum of Comparative Zoology, truly a monument of science. But he did not like Darwin's account of evolution one little bit, and the seemingly in-explicable origin of birds was fuel for his fire.

Darwin had addressed the shortage of "missing links" well before the critics started attacking him. As he put the matter in Chapter 6 of *Origin of Species*: "why, if species have descended from other species by insensibly fine gradations, do we not everywhere find innumerable transitional forms? I believe the answer mainly lies in the record being incomparably less perfect than is generally supposed. . . . The crust of the Earth is a vast museum; but the natural collections have been made only at intervals of time immensely remote."

But then, right on cue, in 1861, the very thing that was most glaringly absent turned up in a quarry in southern Germany: dating from the Late Jurassic, deep in dinosaur times, about 145 million years ago, the fossil of a pigeon-size vertebrate, almost complete apart from the head. At first glance it was just another small reptile. It had a long, whip-like tail with twenty-one or twenty-two vertebrae—very lizardlike. But its forelimbs were long. They were wings.

At first, this seemed nothing too exceptional: winged reptiles—in the form of pterosaurs—were well known by the 1860s. But in pterosaurs, the wing is supported by a hugely elongated fourth finger, while this new creature's wings were supported by the arm bones (the humerus and the radio-ulna)—as in a modern bird. The new creature still had four fingers, one of them long and forming the terminal joint of the wing, while the other three were short and clawed.

What really struck home were the feathers—or at least impressions of them, very clear—forming a perfect halo around the body, the thighs, and, most elaborately, the forelimbs. So its arms weren't simply wings—they were feathered wings; and the feathers on the wings were true flight feathers. Along the creature's breastbone was a smallish but nonetheless convincing keel—anchorage for flight muscles. Here, probably, was a flapper, no mere glider. It also had a furcula—strangely boomerang-shaped to be sure, but a very palpable wishbone nonetheless. By all the definitions of birddom that were then current, it was unmistakably a bird, but a bird that still retained some primitive features of a reptile—most obviously, the lizard-like tail. The fossil was shown to a local paleontologist, Hermann von Meyer, who in 1862 gave it the official scientific name of *Archaeopteryx lithographica*.

By the early twenty-first century, nearly 150 years after the first one came to light, nine *Archaeopteryx* skeletons have been unearthed, and one isolated feather.[2] The first two skeletons to be found, and the sixth, are among the finest fossils ever discovered. The reason lies in the nature of the bedrock. All the fossils have been found in the honey-colored limestone around the Bavarian city of Solnhofen. This stone is of such fine quality, so finely textured, that it is used for lithography—printing with engraved stone plates. This is why the *Archaeopteryx* fossils were found in the first place, and why they were found at the time they were. For the Solnhofen rock had been quarried over many centuries—the Romans used it for pavements. But so long as the stone was simply used for walking on, it could be quarried en masse without close scrutiny. Lithography came into its own in the nineteenth century, but for such a purpose, only the finest stone would do; so, as Pat Shipman records in *Taking Wing* (London: Weidenfeld & Nicolson, 1998), the stones were now quarried by hand, each piece lovingly split to offer the best possible surface, and examined by up to a dozen expert quarrymen before being sold at an enormous premium. Only through such care could such delicate, flattened fossils have come to light. Hence, the name that von Meyer proposed: *Archaeopteryx* means "ancient wing" while *lithographica* refers to the smoothly beautiful stone it was found in. All of the eight *Archaeopteryx* skeletons, and the feather, have lain between two layers of such limestone. The bulk of each fossil has always been in the upper slab while the lower layer (the "counterslab") carried the impression.

Here then, you might suppose, was exactly the link that Agassiz had said could not exist: half reptile, half bird. Yet there were many doubters. Some of the fiercest were in Germany, which is why the Germans did not bag the very first *Archaeopteryx* for themselves. Instead, they allowed it to be sold to the British Museum in London (for £450), and so it has been known ever since as the London *Archaeopteryx*. In London, the fossil fell into the hands of Richard Owen (1804–92), who was Britain's most distinguished vertebrate anatomist but also, yet again, one of Darwin's most vehement detractors. His religion was orthodox; he did not like the idea of evolution, and he did not like Darwin. Owen did not acknowledge *Archaeopteryx* as any kind of "missing link" between reptiles and birds. Instead, in a monograph published in 1863, he simply declared that the London fossil was "unequivocally a bird." It had some characteristics that are seen only in the embryos of modern birds—and embryos are very important in deciding relationships. He also noted that it had some unbird-like features, but these merely suggested "a closer adhesion to the general vertebrate type." He also renamed the fossil *Archaeopteryx macrura*, meaning "long tail." There was no good reason to do this, except that those who propose names for new creatures get their names attached to them in parentheses. Thus it would have become *Archaeopteryx macrura* (Owen) as opposed to *Archaeopteryx lithographica* (von Meyer). Hmm.

On to the scene then came Thomas Henry (T. H.) Huxley (1825–95), who was Darwin's great friend and champion and had his own scores to settle with Owen. In 1868, he published his own account of *Archaeopteryx*. His paper is a master class in putdown. In his title, "Remarks upon *Archaeopteryx lithographica*," he ignored Owen's proposed *A. macrura*. Then, he said, Owen had mixed up the right leg of *Archaeopteryx* with the left, got the pelvis the wrong way round, and misaligned the furcula. Huxley, renowned for his venomous wit almost as much as for his science, could not resist applying the boot: "It is obviously impossible to compare the bones of one animal satisfactorily with those of another, unless it is clearly settled that such is the dorsal and such the ventral aspect of the vertebra, and that such a bone of the limbs belongs to the left and another to the right side" and so on.

Owen predicted that when a head of *Archaeopteryx* came to light it would probably have a beak—because it was, after all, a bird, and birds have beaks. On the contrary, said Huxley, it might well have reptilian teeth to match its reptilian tail. If this did turn out to be the case, then, he said, "*Archaeopteryx* will not the more, to my mind, cease to be a bird, than turtles cease to be reptiles because they have beaks." The real test, he said, was feathers. Some extinct reptiles have a pelvis and/or feet like *Archaeopteryx*. One of the fossils found alongside *Archaeopteryx* at Solnhofen was of small running dinosaur *Compsognathus*, and that had bird-like feet. But no reptile that was known in those days had feathers, and only birds share all three features with *Archaeopteryx*: bird-like pelvis, feet, and feathers. Ergo, whether beaked or toothed, *Archaeopteryx* would remain a bona fide bird. In fact, the same slab that contains the London *Archaeopteryx* also contains a fragment of upper jaw with four teeth, which most paleontologists at first assumed must belong to some other creature. But the four teeth probably belonged to the *Archaeopteryx*. Later, even more complete fossils with skulls have confirmed that *Archaeopteryx* did indeed have teeth—but also that in crucial respects the skull was very birdlike, indeed.

The most recent *Archaeopteryx* to come to light was described in 2005. It is now in the Wyoming Dinosaur Center at Thermopolis and is known, accordingly, as the Thermopolis specimen. Of all the specimens known so far, it has the best-preserved head and feet. The head confirms once more that *Archaeopteryx* had teeth, but the feet reveal new details not shown in the earlier specimens. First, although it has three forward-pointing toes, it lacks a fourth toe pointing backward, as in sparrows or eagles, so it probably was not a proficient percher. This adds weight to the idea, favored by many paleontologists, that *Archaeopteryx* spent most of its time on the ground. Second, and most intriguing, its second toe was "hyperextendible"; it could be turned upward, to point toward the sky. We will see a little later why this is so important.

In general, biologists continue to acknowledge *Archaeopteryx* as the oldest known bird—possibly the direct ancestor of all later birds—and also as the "missing link" between reptiles and birds and hence as evidence for evolutionary theory in general. Huxley was among the first to think along these lines. So, too, was the Scottish biologist Hugh Falconer, who wrote to Darwin: "Had the Solnhofen quarries been commissioned—by august command—to turn out a strange being à la Darwin—it could not have been executed more handsomely—than with *Archaeopteryx*."

So *Archaeopteryx* focuses our search. To trace the deep origins of birddom we need merely to find fossil creatures of a kind that could have given rise to *Archaeopteryx*. This, of course, is much easier said than done. In modern times—since the 1990s, indeed—huge progress has been made. But to appreciate that progress fully, it is worth dwelling briefly on the difficulties. With birds, the problems tend to be writ particularly large.

WHAT'S THE PROBLEM?

Problem one is as already mentioned: the fossil record is wonderful but it is regrettably and inescapably spotty. Fossilization is a rare event. Most of the creatures that are not eaten before their time simply decay. Birds are particularly fragile because their bones are so light and because so many live in forests, where decay is fastest, or along the shore, where everything is liable to be smashed. Entire strata of rocks, bearing witness to goodness knows what, have been eroded away or reburied and recycled deep in the molten magma beneath the ocean bed. Countless fossils have been destroyed forever by the same civil engineers who have brought so many of them to light. Many of the finest fossils are in the most inaccessible places, although this can be to our advantage—finds from China are now transforming our knowledge of bird history. We might be grateful that, until very recently, most of China remained undug, except sometimes at the surface, by a few hundred generations of farmers.

More subtle is the issue of search image. How do we know what we are looking for? To be sure, this may seem obvious enough. In the case of birds we are seeking creatures with bird-like feet, wings, and feathers. Many dinosaurs have very bird-like feet, as Huxley observed, whereas mammoths, say, very clearly do not. We surely can ignore mammoths in our search for ancestral birds. But what of wings? Some ancient reptiles—the pterosaurs—had wings, but in detailed anatomy they were very unlike those of birds. Anyway, we need to discover how wings evolved in the first place—and so we need to find ancient reptiles that did not themselves have wings, but had forelimbs that conceivably could have evolved into wings. What does such a limb look like? How do we know until we find it? But if we don't know in advance what we are looking for, how do we know it when we do find it? If we do not have a clear idea of what we are looking for, then we finish up being totally indiscriminate. But if our search image is too narrow, then we may miss what is essential.

Then again, natural selection does not look ahead. The creatures that survive in any one generation are the ones that do well in the here and now. Indeed, natural selection is decidedly opportunist: it picks up on whatever features are already in place, and exaggerates or abandons them according to circumstance, with occasional genuine novelties. The deep ancestor of birds no doubt was a squat and shambling reptile, and that squat and shambling reptile did not know that its descendants were to become birds. It did not regard itself as a precursor of some other life form. It felt no sense of destiny. It did not set out, in ways that paleontologists could later observe, to hone itself for flight. All in all, evolution does not pursue what an engineer would consider to be a logical path to a prescribed end point (and neither, in reality, do engineers). In short, we cannot necessarily tell, just by looking at it, whether some ancient fossil creature had or did not have the wherewithal to evolve into a bird (especially since the fossils are liable to be fragmented, with a great deal missing).

To understand the evolution of any creature, too, we must address two separate kinds of issues. The first is one of genealogy (more accurately *phylogeny*—a term explained in the next chapter). Who are the creature's ancestors? What lineage does it belong to? The second issue is one of function: by what plausible route could the ancient creatures that we know about have evolved into their modern forms? So in the case of birds we need to know if they really did derive from reptiles, and if so, which ones. But we must also be able to explain how some ancient reptile, which had no wings and must therefore have walked or run or clambered, gave rise to descendants that did have wings and could fly.

Yet despite the difficulties and the caveats, biologists this past two hundred years have put together a convincing account of how birds became. We can never be sure in matters like this (or, indeed, in science as a whole) that we have the truth, the whole truth, and nothing but the truth, but the story that is now widely accepted is coherent and it fits the known facts, and that at least is a good start. The story runs roughly as follows.

IN THE BEGINNING

The first vertebrates came on land about 400 million years ago in the geological period known as the Devonian. They could walk on four legs—that is, they were tetrapods—and they breathed air. But they produced naked eggs that dried out easily, so they had to be laid in water, and they hatched into larvae that breathed through gills. In short, these first tentative terrestrial vertebrates belonged to the grade of creatures known as amphibians, still represented today by frogs and salamanders.

By about 350 million years ago, in the Carboniferous period, some of these amphibians gave rise to a new grade of tetrapods known loosely as reptiles—the first vertebrates that could, if they chose, live their lives entirely on land. They had thick, waterproof skins. They also produced eggs with waterproof ("amniotic") membranes that did not have to be laid in water—and indeed would drown if they were.

Soon after the primitive reptiles first appeared, their lineage divided to form two branches (Figure 1).

One of those branches was, or is, the Synapsida. The synapsids produced some quaint-looking creatures such as *Moschops*, resembling a big lizard with a calf-like head, which became the subject of a children's cartoon in the 1980s, and the "sail-back reptile" with a 2-meter (6-foot)-high crest along its back. We will not find the ancestors of birds among the synapsids. There is nothing birdlike about them.

But the synapsids did give rise to the mammals. The first recognizable mammals arose from among their ranks 220 million years ago. This means that the mammals as a group are roughly as old as the dinosaurs, but they were obliged to live a mainly hole-and-corner existence so long as the dinosaurs were around. Only when the big terrestrial dinosaurs disappeared, around 65 million years ago, were the mammals free to evolve into their present glory—whales, bats, horses, cats, dogs, elephants, squirrels, us. So it is that in whales, bats, human beings, and the rest, the synapsids live on. But the cumbersome synapsids of reptilian grade are long since gone.

The other reptilian branch was, and is, the Eureptilia. The Eureptilia in turn divided into two: the Anapsida and the Diapsida.[3] The Anapsida have enjoyed a minor but honorable presence this past 330 million years or so, and survive in the form of tortoises, turtles, and terrapins. By con-

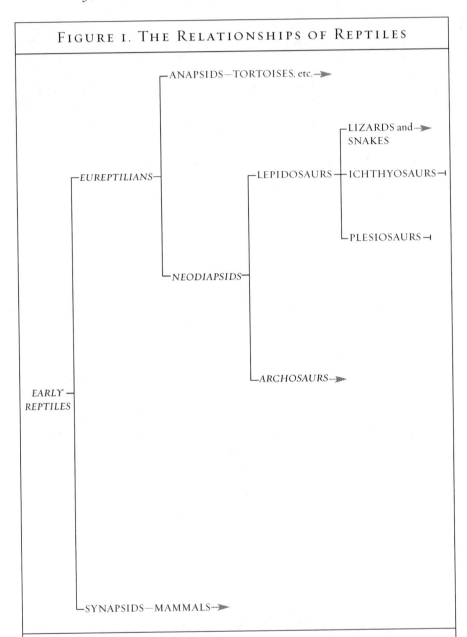

FIGURE I. THE RELATIONSHIPS OF REPTILES

The very first reptiles—the first four-legged vertebrates able to lay eggs on land—divided into two great groups: the eureptilians and the synapsids. The birds belong to the first, and the mammals belong to the second. So the two great groups share a common ancestor, but have evolved along quite separate lines. Specifically, the eureptilians gave rise to the neodiapsids, which gave rise to the archosaurs—and birds are archosaurs.

Source: Adapted from Cracraft and Donoghue, 2004; and Benton, 1997.

trast, the Diapsida divided further into an array of forms, including some that were big and spectacular. Two of those groups are well known to us as fossils, but are long extinct: the fish-like ichthyosaurs of the deep oceans, and the pliosaurs and plesiosaurs, among whose ranks, so some allege, was or is the Loch Ness monster.

But two of the diapsid groups that arose in those early days are still very much with us. One of them is the Lepidosauromorpha, who nowadays are represented by the lizards, snakes, and the lizard-like *Tuatara* that is now being conserved on islands around New Zealand. The other is the Archosauria (Figure 2). The archosaurs in turn diverged to form a whole range of creatures, including the crocodiles and their relatives, very ancient but still very much with us; the extinct flying reptiles, known as the pterosaurs, which include the pterodactyls; and the dinosaurs.

Beyond doubt—everything about them proclaims the fact—the birds also belong among the Diapsida. Specifically, it is clear that birds are archosaurs. Their extinct relatives include the dinosaurs and pterosaurs. Their closest living relatives, their fellow archosaurs, are the crocodiles.

Here, then, we already have the answer to the question posed in the first chapter: How it is that birds and mammals are so similar, and yet are so different? They are similar in basic structure because they share a common ancestor—one or other of the very earliest fully terrestrial vertebrates. They began from the same starting point—a leathery-skinned, squat, and shambling pentadactyl tetrapod ("pentadactyl" meaning that it had five toes) of the kind colloquially known as a reptile. But birds and mammals differ as much as they do because, 330 million years ago, that common reptilian ancestor gave rise to two different lineages. The mammals belong to one lineage, and the birds belong to the other. Mammals are synapsids; birds are diapsids.

The very ancient ancestor shared by the mammals and birds bore very little resemblance to either of its distinguished descendants. It did not have feathers or fur. It had a very small brain and, one may presume, very limited intelligence. It was not warm-blooded. All the special features that make birds and mammals so dashing, and so dominant, have evolved independently within the two groups. In their similarities, therefore, we see one of the most stunning examples of evolutionary convergence in all of nature. But it is hardly surprising, after 330 million years or so of independent evolution, that in detail the two groups should be so different. The ecology of the modern world may be seen, in large part, as the inter-

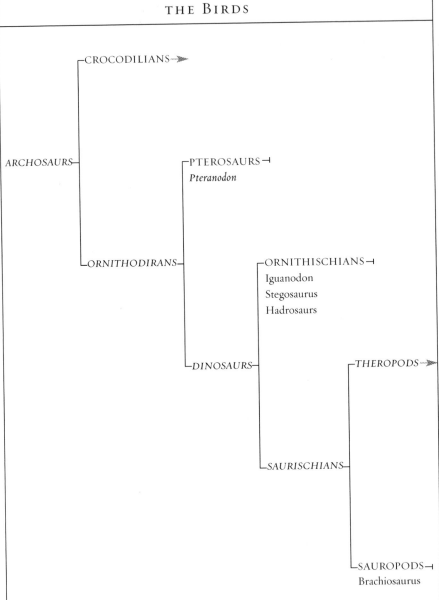

FIGURE 2. HOW THE ARCHOSAURS GAVE RISE TO THE BIRDS

The archosaurs split into two: crocodilians on the one hand, and ornithodirans on the other. Many crocodilians are still with us: crocs, alligators, caymans, and gharials. The only remaining group of the ornithodirans are the birds, which are thought to be theropods.

Source: Adapted from Cracraft and Donoghue, 2004; and Benton, 1997.

play of these two great parallel lineages: the synapsids, now represented by the mammals; and the diapsids, in the form of birds. (Although, as has been pointed out to me, ichthyologists and entomologists are among the biologists who would not agree with this. Fish and insects are very big players, too!)

But how do birds relate to the other archosaurs?

You might reasonably guess, as a first shot, that birds are closest to the pterosaurs. Pterosaurs, too, had long arms and flew. But the structure of the pterosaur wing was nothing like that of a bird's. A pterosaur's wing was constructed from a single, hugely elongated finger (the fourth), while the bird's wing involves the whole arm. Pterosaurs got their lift and thrust as a bat does, from a flap of leathery skin. Birds use feathers. Pterosaur wings and bird wings are yet another example of convergence. But there is no conceivable route whereby one could evolve into the other.

Among living creatures, crocodilians are closest to birds. Those two are the only remaining archosaurs. Sometimes we can see resemblances. Like birds, crocs take good care of their young (and so, it seems, did many a dinosaur). Young crocs chirp to their mothers for all the world like chicks, and sometimes (so I am told by a zoologist chum) they clamber into the lower branches of bushes for good measure. But no modern biologist suggests that crocs are the ancestors of birds. Most—although not quite all—agree that the ancestors of birds are to be found among the greatest diapsid group of all: the dinosaurs.

Once again, it seems that T. H. Huxley was the first to sow the seeds of this idea. In 1868, the same year that he published his paper on *Archaeopteryx*, he wrote: "if the whole hind quarters of a half-hatched chick could be suddenly enlarged, ossified, and fossilized . . . there would be nothing in their characters to prevent us from referring them to the *Dinosauria*." Note the reference to chicks rather than to adult birds: young creatures, and particularly embryos, often give the strongest clues to evolutionary provenance. In America, Othniel Charles (O. C.) Marsh (1831–99), head of the Peabody Museum at Yale, took Huxley's suggestion one step further. In absolute contrast to Agassiz in New York, Marsh was one of Darwin's most vigorous supporters—and in 1872 he seems to have been the first to suggest explicitly that birds not only are related to dinosaurs but are descended from them.

Yet there is more. For, if birds are indeed descended from dinosaurs, then so, modern biologists insist, they are dinosaurs. The textbooks and a century and a half of folklore tell us that dinosaurs went extinct 65 million years ago, but if the modern orthodoxy is right, that just isn't so. Dinosaurs are with us still—on the duckpond, in the farmyard, in the treetops. It is pleasing to reflect that sparrows, in addition to all their other sterling qualities, are miniature dinosaurs. The fact that an idea is pleasing does not make it true. But in this case it probably is.

There is one technical complication. If birds are dinosaurs, then what should we call the creatures who featured in *Jurassic Park?* If we simply call them dinosaurs, how are people supposed to know that we don't mean birds? To get around this problem, some zoologists now refer to traditional dinosaurs as "non-avian dinosaurs." But the easier route is not to acknowledge that this is a problem at all, and just go on calling birds "birds" and the creatures we traditionally call dinosaurs "dinosaurs." Common sense is still a good thing, even in science.

But the dinosaurs were a very mixed bunch. So, which of them gave rise to birds?

WHICH DINOSAUR?

The dinosaurs first arose about 230 million years ago, in the Triassic, and they dominated the world's terrestrial ecosystems for the following 165 million years—an extraordinarily long run, which gave them plenty of time to diversify. Paleontologists have identified many hundreds of kinds from all over the world, and there must be many more to be found in the rocks beneath the Antarctic ice; we surely know only a small fraction of what was once out there. So where among this plethora do we find the ancestor of birds? And how do we find it?

Common sense takes us some way. We are looking for a dinosaur ancestor that *could,* plausibly, have started at some point to use its forelimbs as wings. So we want one that already walked on its two hind legs (bipeds), rather than all four legs (quadrupeds). The dinosaurs as a whole are traditionally[4] divided into two great groups: the Ornithischia, and the Saurischia. Both groups contain some bipeds, and both contain some that have gone back to walking on four legs. So should we be looking for an ornithischian biped or a saurischian biped?

The word *ornithischia* means "bird pelvis," so we might reasonably suppose that the ancestors of birds might be found among their ranks, but this turns out not to be so. Apart from the general shape of the pelvis, nothing else about the ornithischians suggests any close connection with birds. It seems as if the similarity of pelvis is yet another example of convergence. Indeed, taken all in all, the ornithischians look most unbirdlike. The quadrupedal types among their ranks include *Stegosaurus*, with the parallel rows of upright plates along its back, and the famous Late Cretaceous *Triceratops*, like a huge rhino but with a heavy reptilian tail and a bony frill around its neck. The ornithischians also include the herbivorous *Iguanodon*, which was thought to be a quadruped when first discovered and described in the mid-nineteenth century, but now is thought to have been a biped, albeit a seriously heavy one.

Saurischia means "lizard (or reptile) pelvis," which doesn't look too promising. But many other features of saurischians do suggest affinity with birds, so it is here that we should look. The saurischians are yet again divided into two clear groups: the sauropods, which are quadrupedal; and the theropods, which are bipedal. The sauropods include the biggest land animals that have ever lived—by far: herbivores such as *Diplodocus* and *Apatosaurus* (formerly known as *Brontosaurus*); *Brachiosaurus*, which was even bigger; and the absolutely enormous *Supersaurus* and *Ultrasaurus*, which have recently come to light in the United States. But again there is little among such irredeemably land-bound mega-reptilians to suggest any close connection with birds.

By contrast, the bipedal theropods can be very birdlike. They were similar in form to the first-ever dinosaurs of the Triassic, but as the later periods unfolded, first the Jurassic and then the Cretaceous, they became more and more refined and diverse. Many if not most of them were carnivorous—again, like the most primitive kinds, for the very first dinosaurs of all were exclusively meat eaters. Some of the theropods were heavy and mighty, with small forelimbs that could hardly have evolved into wings: like *Megalosaurus*, dating from the Jurassic, which was the one whose hindquarters Huxley compared to a bird; *Allosaurus*, also Jurassic; and best known of all, the highly sophisticated Cretaceous *Tyrannosaurus*, including the one that has become a nursery favorite, the formidable *Tyrannosaurus rex*.

But there was also a group of theropods who were small and slim. In general they had long forelimbs and hands and fingers (although some virtually lost their forelimbs). They are known collectively as "maniraptors," which means "hand seizers" or "hand thieves" (that is, grabbers). In various details—including the mobility of the joints in the shoulders and forelimbs—maniraptors are strikingly similar to birds. So it is generally believed these days that birds arose from among the maniraptors—which means that they *are* maniraptors, a subgroup of the theropods. We have met a maniraptor already in this account: *Compsognathus*, the small running carnivore that shared the limestone of Solnhofen with the first known *Archaeopteryx*, and possibly crunched the odd *Archaeopteryx* in its well-armed jaws. The most famous maniraptor of all was named in 1924: *Velociraptor*, star of Michael Crichton's *Jurassic Park* and hence of Steven Spielberg's film of 1993.

In truth, the *Velociraptors* in Spielberg's film deviate somewhat from the real thing. The real ones have been found only in Mongolia, and although they could be formidable—the most famous skeleton of all was found locked in mortal combat with a *Protoceratops*—they were not particularly big: the biggest was around 1.8 meters (6 feet), but that included the long tail. The man-size beasts of the film were more like *Deinonychus* from southern Montana, which was 2 to 3 meters (7 to 10 feet) long and had a huge, sickle-shaped claw on the second toe of each hind foot; its name means "terrible claw." The *Velociraptor/Deinonychus* dinosaurs of the movie were shown to hunt in packs, which they may have done, although there is no fossil evidence for this. The script told us, too, that they were highly intelligent—comparable with chimps or dolphins—but the size of their fossilized craniums suggests that, although they were no boneheads, they probably had less wit than the average lion—which, if it were not quite so formidable, would be seen as a quaint and rather muddle-headed beast. In the film these superior dinosaurs held board meetings, plotting their hunting strategy like three-star generals, which made for an excellent chase but is highly implausible.

Spielberg also showed *Velociraptor/Deinonychus* with scaly skin. Why would he not? This was 1993, and everyone in 1993 knew that reptiles have scales. But this, it soon turned out, was the most shocking solecism of them all. For by the late 1990s, just a few years after the film was made, biologists were 99 percent certain that *Velociraptor* and *Deinonychus* did not have scales. They had feathers.

Since 1995, a whole series of maniraptors has been unearthed that were not birds and yet had feathers—and in some cases, those feathers were elaborate: not mere down but truly "pennaceous," with a central rachis and barbs along each side. From these discoveries biologists infer that all maniraptors must have had feathers. The discovery of non-avian feathered dinosaurs has been as momentous to the story of bird evolution as the discovery of *Archaeopteryx* itself 150 years earlier.

DINOSAURS WITH FEATHERS

The first three genera of (non-avian) dinosaurs with feathers were found in northeast China, specifically in the Yixian Formation of the Liaoning fossil beds. The rocks that contained them were as creamy smooth as the limestone at Solnhofen; had it not been so, they could not have conserved the fine detail of feathers. But the Yixian rock was formed later than the Solnhofen rock—it is Early Cretaceous rather than Late Jurassic—and by somewhat different means: from volcanic ash, from eruptions in Inner Mongolia. Those feathered dinosaurs were entombed like the people of Pompeii, although not quite so dramatically. They were already dead when the ash covered them, lying at the bottom of shallow pools (the district was riddled with lakes in those days). The ash evidently floated down on them through the water. But the result was much the same. The Yixian dinosaurs were preserved in the finest detail. There are green fields on the surface now, but the rock, with all its wonders, is not far down.

Ji Qiang and Ji Shu-an found the first one in 1995. They described it the following year in *Chinese Geology* and called it *Sinosauropteryx*—roughly translating as "Chinese lizard-wing." It was a small, slim, very dinosaur-shaped bipedal runner up to a meter (3 feet) or so long, including its long lizard-like tail. Its general form tells us that it was a carnivore, and one specimen was found with the jaws of several mammals lying where its stomach would have been. Here was forensic evidence of its last meal: a 100 million-year-old smoking gun. Yet there is nothing outstanding in any of this—except that, as in the first *Archaeopteryx*, there was a faint penumbra of structures around the flattened bones that proved, on close examination, to be the impressions of very palpable feathers. In fact, those feathers resembled those of a modern kiwi—like fur, but with two branches per shaft. But kiwis are modern birds; their primitive-looking feathers have evolved from more complex feathers. The feathers of *Sinosauropteryx* really were primitive. The similarity of feathers between *Sinosauropteryx* from the Cretaceous and kiwis from modern times is yet another example of convergent evolution.

Several specimens of *Sinosauropteryx* are now known, and it is clear that as a whole they were fairly primitive. Most of the many other maniraptors that we know about are more like birds in their details, and are presumably related more closely to birds. Feathers are not the kinds of structures that are liable to have evolved more than once, so if *Sinosauropteryx* had them, and birds have them, then it is reasonable to infer that all the other maniraptors had them, too. The fact that most maniraptor fossils come without accompanying feathers is of very little significance. Fossilization even of the grossest structures is a rare event. The wonder is not that we have so few fossil feathers, but that we have any at all. Even so, at the time of writing, eleven different genera of maniraptors have been found with feathers. Some of them are new to science, but others were already well known, although only in their naked state, with their putative feathers long since obliterated. As we will see, some of the feathers in those maniraptors were elaborate. Indeed, some were apparently evolved for flight.

Radical rethinking is now required. Biologists traditionally have *defined* birds as "animals with feathers." If an animal had or has feathers, then, ipso facto, it was or is a bird. If non-avian maniraptors had feathers, too, then either we must acknowledge them as birds or we must redefine what we mean by *bird*. Most biologists prefer the latter course: feathers are still on the list of diagnostic features of birds, but they no longer stand alone as the sure-fire criterion of birddom. But some suggest that we should stick with the original definition. So, they say, since all the maniraptors had feathers, including those like *Sinosauropteryx* that are not particularly bird-like, then all of them should be ranked as birds. The argument may seem bizarre, but as we will see, there is more to it than meets the eye.

Paleontologists now recognize many different maniraptors—enough to show that they form a coherent group and to map out their family tree, as shown in Figure 3. Some kinds have been known for a long time—like the ornithomimids, which were first described in the nineteenth century by O. C. Marsh. The ornithomimids were ostrichlike—slim, bipedal runners, with long hind limbs that ended in hoof-like claws, but with long, slim forelimbs, as the birds' ancestors must have had. So they do indeed seem birdlike (and their name ornithomimid means "bird mimic"). Indeed, some of them are named after particular modern birds: *Struthiomimus* means "ostrich mimic"; *Anseromimus* was the "goose mimic"; *Gallimimus,* the "fowl mimic"; and so on.

FIGURE 3. HOW BIRDS ARE RELATED TO THE REST OF THE DINOSAURS

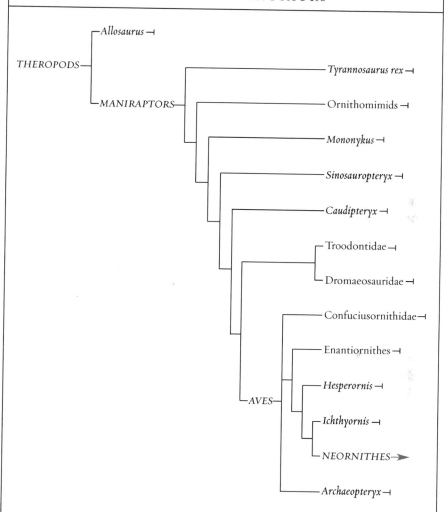

Birds belong—or so it seems—to the particular group of dinosaurs known as the theropods; more particularly, they seem to belong to the specific theropod group known as the maniraptors. The maniraptors once were highly diverse, but all are extinct except for the birds.

Birds as a whole are formally known as the Aves, and the Aves is traditionally ranked as a class. The Aves in past times split to form at least half a dozen different subclasses, and most of them are extinct, too. Indeed, only one of these subclasses remains—the Neornithes. In short, all living birds belong to just one subgroup of Aves, which are just one subgroup of theropods, which are just one subgroup of the dinosaurs, which are just one subgroup of the archosaurs—and so on all the way back to the beginning.

Source: Adapted from Cracraft and Donoghue, 2004; and Benton, 1997.

But in other details they weren't nearly birdlike enough to have been the birds' ancestors. They were more like the birds' third cousins. Closer to the birds, perhaps, were the alvarezsaurids, of whom the best known is *Mononykus*, first discovered in Mongolia in 1993.

Of course, it was generally birdlike—slim and bipedal. But its forelimbs are reduced to tiny stumps—most unbirdlike. Herein, though, lies a lesson. For *Mononykus* was clearly a specialist; its long jaws and pointed teeth suggest that it fed on termites. Specialists are always weird—the weirdnesses of the specialist are just the tools of its trade. Kiwis, after all, have all but lost their wings, but everything else about them proclaims their birdhood. Apart from its jaws and forelimbs, *Mononykus* was so like a bird in other ways that some leading authorities have suggested that it was a bird. For good measure, it almost certainly had feathers; they have never been found on *Mononykus*, but they have been found on a close relative, *Shuvuunia*, also from Mongolia, which was described in 1998. But the latest studies suggest that *Mononykus* and its kind were not birds. Again, at best, they were not too distant cousins.

Even more birdlike were the troodontids (Troodontidae), which again have been known from the nineteenth century (1856). I will say nothing further here except that a species of troodontid quaintly named *Mei long* has recently turned up in China, apparently asleep with its head tucked

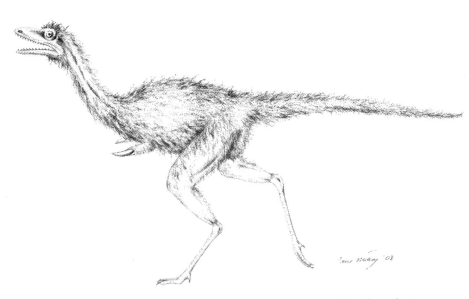

Mononykus at first sight looks nothing like a bird, yet it seems to be a close relative.

under its arm—very birdlike. Again, it is pleasing to think that behavior we assume is exclusively avian in fact was shared by some long-gone creature an almost incomprehensibly distant time ago—not a monster at all, but rather sweet.

Now another shock: *Tyrannosaurus rex*. If you ponder the details of the fossils, in the purist fashion of modern taxonomists, without too many preconceptions, then *T. rex* seems even closer to the birds than the troodontids are. On the face of things *Tyrannosaurus* looks nothing like a bird—it is huge (up to 12.5 meters, or 40 feet, long) and heavy (as big as a fair-sized elephant at around 6 tons), but we must not prejudge these things. Nature does what nature does, and it is the details of the skeleton (and of DNA, when we can recover it, in modern creatures) that reveal who is related to whom, not mere size or way of life. So although *T. rex* seems a most unlikely cousin to the swallow and the sparrow, and although not many biologists believe that it is, it could be.

If *T. rex* was a bona fide maniraptor, then of course it would have had feathers. In a conservative vein, most experts suggest that only the babies would have been feathered—to keep them warm. But if feathers evolved for the reasons I suggest in Chapter 1—for sexual display—then the adults would be even more likely to have been feathered—and extravagantly, at that. In truth it is not possible, and perhaps never will be, to state definitively whether *T. rex* did have feathers—or to confirm or disprove the idea that grown-up *T. rex*'s were turned out like burlesque queens. But I reject that school of biology that feels it is virtuous simply to be dour. If we had only their skeletons to go by, we could not guess that swans could sail and peacocks could dazzle, or indeed that lions have manes or even that camels have humps. Living animals are often far more flamboyant than could be predicted from their bones. Of course, the wildest guesses are the most likely to be wrong. But we can be equally sure that the most cautious interpretations will be wrong a lot of the time. In reconstructions, *T. rex* is generally dressed in scaly leather, like a Gucci handbag. But it might have been decked in gaudy feathers and, in due season, crested and plumed for good measure. It is not good science merely to be puritanical and discard such notions out of hand.

But far more convincingly birdlike were the oviraptors (a term I am using loosely to include the Oviraptoridae and their putative close relatives). They include several highly intriguing types, including some of the principal finds from Yixian.

The one for which the group is named is *Oviraptor*, which again has been known for a long time—since 1924—and again comes from Mongolia. The first bones that ever came to light were on top of a pile of eggs that were thought to be those of a horned dinosaur, *Protoceratops*. Clearly, its discoverers concluded, it was a nest-robber. It had a curious beak, ideally suited, it seems, to seizing and breaking eggs. So it was called *Oviraptor philoceratops*: Greek for "the egg thief with a penchant for ceratops." But more oviraptors turned up that were also sitting on piles of eggs—and inside those eggs were embryos of oviraptors. In short, the oviraptors were not stealing the eggs of other dinosaurs but guarding their own. They died, perhaps, not through miscreancy but from dedication, burned to a heroic frazzle as they shaded their offspring from the Late Cretaceous desert sun. Their powerful beaks, possibly, were not for crushing eggs but for crunching mollusks—which for some reason, unless you are a mollusk, seems far more acceptable. (On the other hand, oviraptors could well have been nest-robbers, too; after all, there are very few mollusks in the average desert.)

The oviraptors are not placed closest to the birds, but they did have some very bird-like features. They had projections on each rib, turning the rib cage into a stiff box, like a bird's. In some oviraptors, too—most obvious so far in *Nomingia*—the tail was reduced to a pygostyle, a base for tail feathers—lightweight, but serving as a bird's tail does for balance and display. All accounts of bird anatomy that I have ever seen state that their stiff thorax and feathery tail, mounted on a pygostyle, are adaptations for flight. But the oviraptors were not fliers; they were runners. Of course, the oviraptors might have *descended* from flying ancestors, as discussed later, in which case the stiffened ribs and pygostyle are not at all surprising. But most biologists agree these days that they probably didn't. So, presumably, the box-like thorax and the pygostyle evolved not as a specific adaptation to flight but simply to reduce body weight, maintaining skeletal strength and maneuverability while reducing bone. Such features can be seen as *preadaptations* for flight—structures that had already evolved for one purpose and then proved useful for something else. But the oviraptors could not have been the ancestors of birds, whatever their anatomy might suggest. They emerged in the Cretaceous, while *Archaeopteryx*, acknowledged as the first true bird, dates from the earlier Jurassic.

Oviraptor itself has not been found with feathers, but some of its close relatives certainly have. In 1997, within a couple of years of *Sinosauropteryx*,

Ji Quiang and Ji Shu-an found *Caudipteryx* and *Protarchaeopteryx* in the ceramic-smooth rocks of Yixian, with impressions showing extensive feathers on the body and arms—which were veritable wings—and a tail like a fan.

Caudipteryx lived in the Early Cretaceous, around 125 million years ago, and was about the size of a peacock. It had a few teeth on the upper jaw—and convincing feathers on its tail and hands. These feathers were long and had a proper "pennaceous" structure, with a central rachis and barbs. They were symmetrical—the rachis ran straight down the middle— so they were no good for flight. Feathers that are aerodynamically worthwhile are asymmetrical, with shorter, curved barbs branching off the rachis near the front or leading edge of the feather and longer, straighter barbs off the back. But, so some biologists say, it is very hard to see how such a feather could have evolved at all unless it originally evolved for flight—and very hard to see why *Caudipteryx* would have such feathers on its arms, unless its arms had previously served as feathered wings. In other words, this argument goes, *Caudipteryx* must have evolved from flying ancestors, just as modern flightless rails evolved from flying rails. But there is an alternative explanation, as we will see.

The name *Protarchaeopteryx* suggests an affinity with *Archaeopteryx* that is not really deserved. Certainly, it was no ancestor. But it was very birdlike: turkey size, and with well-vaned feathers from a short slender tail and similar, symmetrical feathers on its long, slender, clawed, three-fingered hands. Curiously, it had large incisor teeth like its close relative, *Incisivosaurus*—or indeed, like a modern rat.

Even more birdlike, it seems, were the therizinosaurs, who lived throughout the Cretaceous in what is now Mongolia, China, and North America. The claws on their hands could be nearly a meter (3½ feet) long, and they presumably used them as a ground sloth did, for reaching into trees and clawing down the foliage—*therizo* means "to reap," or "cut off." One of the bigger therizinosaurs, *Beipiaosaurus* (found in mid-Cretaceous rocks in Liaoning Province and described in *Nature* in May 1999), was the size of an Emu. Around its fossil bones were the impressions of down-like feathers. But then, biologists nowadays *expect* such dinosaurs to have feathers.

But various features of their fossil skeletons suggest that the closest of all to the birds were the dromaeosaurs. *Dromaeosaurus* was the first to be discovered, in 1922—a slim, running beast. *Dromeus* is Greek for "runner."

By the late 1980s a whole load more had come to light—various, but obviously related, and collectively they were called dromaeosaurs. We have already met two of them: *Velociraptor* and *Deinonychus*. The group as a whole was widespread and clearly highly successful. Their fossils come from North America, Argentina, Europe, North Africa, Madagascar, Japan, China, and Mongolia, and they lasted from the mid-Jurassic—older than *Archaetopteryx*, at around 167 million years old—to the end of the Cretaceous, 65 million years ago.

Among the characteristics of the dromaeosaurs was the curious second toe of the powerful hind legs that could be drawn back to point straight upward and was fitted with a powerful claw. The oviraptors had this, too—and so, too, as we saw earlier, did the *Archaeopteryx* that was found in 2005. This claw was impressive, curved and with a sharp edge like a scimitar. Was it a hook for seizing or a blade for slashing, like the tooth of a sabertooth cat? Did it strike its prey or its adversaries like a modern cassowary, which does such a fine job in disembowelment? Speculation continues.

But although of all the maniraptors, the dromaeosaurs seem closest to birds, there are some curious anomalies. They had long lizard-like tails, rather than a pygostyle. To be sure, *Archaeopteryx* had a lizard-like tail too, and no one doubts that *Archaeopteryx* is a bona fide bird. Yet the oviraptors, which seem far more distantly related to birds, did have a pygostyle. Very confusing. Perhaps pygostyles evolved more than once, or perhaps there is something wrong with the classification. Perhaps more fossils will throw more light. In any case, the dromaeosaur tail was apparently stiffened by ossified tendons, like the tail of a *Diplodocus*, and wagged at the base for balance and steerage.

Most of the dromaeosaurs were runners, like *Dromaeosaurus*, *Velociraptor*, and *Deinonychus*. Bipedal runners don't always make much use of their forelimbs—Ostriches don't, and neither did *Mononykus*. But most of the dromaeosaurs had very mobile shoulders and could hold their arms outward from the body, as very few animals can, and bring them together again in a clapping movement. Presumably this helped them to grasp elusive prey.

Now it really gets interesting. *Dromaeosaurus*, *Velociraptor*, and *Deinonychus* have not been found with feathers—but other dromaeosaurs have. Furthermore, some of them had very well-feathered arms—and not just with downy, warming feathers but with bona fide quill feathers, with a strong central rachis and barbs. If we combine this with the general dromaeosaur ability to bring their arms together—"clap their hands"—we have all the ingredients of powered flight. In short, it seems some dromaeosaurs were fliers.

One of the more impressively feathered dromaeosaurs, with quill-like feathers on its arms, is *Sinornithosaurus*, again from the Yixian rocks. In general, *Sinornithosaurus* was extremely birdlike. But perhaps most birdlike of all was the remarkable *Microraptor gui*, described in *Nature* by Dr. Xu Xing of Beijing in January 2003. Its arms were fitted with quill-like feathers— but so too were its legs. Thus, in effect, it had four wings. (As we will see later, *Microraptor* was not the only feathered creature with four wings.) For good measure, it had a long reptilian tail with a diamond-shaped flourish of feathers at the end that clearly could act both as a stabilizer and as a rudder. In truth, *Microraptor* may not have powered its flight with flapping wings, but it was certainly a glider. Many biologists have argued that birds first developed flight from the ground up—as their feathered dinosaurian ancestors ran faster and faster with their arms extended. But some, notably Alan Feduccia of the University of North Carolina (of whom more later), have long been arguing that birds must have begun up in the trees, as gliders. *Microraptor* could be showing us how this was done.

So it is that the feathered maniraptors have transformed our understanding of bird evolution—and this transformation began only in the mid-1990s. When I began to study zoology formally in the 1950s and 1960s, we were told that bird evolution was a virtual dead duck and would never get any better because birds fossilize so badly. But now, thanks mainly to Yixian, and the Chinese and visiting paleontologists who have worked there, it all seems clear. The maniraptors are a subgroup of the theropod dinosaurs. They were slim, lightweight, and fleet of foot and had feathers—and in some the feathered arms were veritable wings. We can trace a sequence of maniraptors that are progressively more birdlike— ornithomimids, alvarezsaurids, troodontids, oviraptors, therizinosaurs, dromaeosaurs, and then birds. (It turns out that *Tyrannosaurus* probably does not belong in the sequence after all, pleasant though the idea is. But it may still have been feathered.)

But we still have to explain how the long-armed feathered maniraptors came to fly.

HOW DID BIRDS GET TO FLY?

To throw light on the general question—how did birds first begin to fly?—it is sensible to look at *Archaeopteryx*, which, as far as is known, was the first-ever flying bird. Of course, such inquiry cannot answer all our questions. *Archaeopteryx* was already a bird with long feathered wings, while *Archaeopteryx*'s ancestors were reptiles of one kind or another that were on the way to acquiring wings. But we have to start somewhere.

The fossil skeletons of *Archaeopteryx* tell us a great deal—more than we might reasonably have hoped for—but still they leave huge scope for interpretation. The fact that *Archaeopteryx* has a keel on its sternum implies serious breast muscles, which imply powered flight. But its bones and its probable distribution of muscles suggest that it wasn't much of a flier. It could probably manage only a few minutes in the air before it ran out of puff because it probably lacked the respiratory adaptations that provide the energy needed for sustained flight—the throughput lungs and the subsidiary air spaces. It presumably breathed in the relatively inefficient, in-and-out way that reptiles (and mammals) do. Once airborne, it would have flown in a dead straight line like a paper airplane, with very little scope for maneuver, to land in a heap like a modern tinamou, grateful to escape without injury. I have seen drawings of an *Archaeopteryx* landing decorously on a twig like a modern pigeon with perfect control, wings far back and beating fast, legs thrust forward. But I have also seen pictures of Tom from *Tom and Jerry* sleepwalking out of a fifth-floor window. Artists can draw whatever they like, but that doesn't mean it could happen. Still, even to go through its Wright brothers routine, *Archaeopteryx* had first to get airborne. So how did it do it?

There have been two main ideas, and logic suggests that one or the other of them must be true. The first says that *Archaeopteryx* launched itself by running along the ground with its wings outstretched until it was going fast enough to achieve what NASA calls liftoff. Modern swans demonstrate the principle. The other idea says that *Archaeopteryx* climbed to a reasonable height and then threw itself into space. Modern human hang gliders do this. So, too, do the various gliding animals, from flying squirrels to flying frogs. *Archaeopteryx* was not a mere glider, but it might have begun each flight with a glide. At first it would lose height, but it would also gain forward speed. Eventually the air speed under the wings would be enough to provide lift, even though the wings flapped only feebly. Pterodactyls demonstrate the principle precisely. Clearly they had powered flight. But pterodactyls were obviously *not* built for running and must have begun each flight by launching themselves from on high.

So which technique did *Archaeopteryx* adopt? The principles of aeronautics seem to provide the answer, for calculations based on the probable area of its wings and probable body weight suggest that *Archaeopteryx* probably flew at around 12 meters per second, which is about 25 miles per hour. Once in the air, it would just have to keep going. It could not have hovered like a modern kestrel or a tern, or flown backward like a tropicbird or a hummingbird, or changed direction on a metaphorical dime like a martin, or turned somersaults like a Raven. Either it kept going in a straight line or it dropped out of the sky. Aerodynamic theory suggests that if it flew at less than 6 meters per second, which is about 13 miles an hour, it would stall. So if it were to take off from a running start, it would have to have run at at least that speed. But all the calculations suggest that it could not. Two meters per second was probably as fast as it could run—around 5 miles per hour. This may seem horribly slow, although for a creature the size of a pigeon it wasn't bad. But it was not good enough for takeoff.

So to get into the air *Archaeopteryx* would have had to gain height first. Maybe *Archaeopteryx* climbed into the trees, using the clawed toes on its wings. But it could not have been a great climber—it didn't climb well enough to catch the invertebrates it presumably fed on—so it must have fed on the ground. Then it would have been highly vulnerable to running predators such as *Compsognathus*. So we can imagine that it fed as a modern chicken does, pecking at the ground, but then, when threatened, and most unchickenlike, it ran as fast as it could up the nearest slope and launched itself from the top like a hang glider, losing height at first but gaining speed as it fell until it reached the necessary 13 or so miles per hour, when it could gain height again by flapping. Even a few minutes in the air, with a crash landing at the end, would be better than a confrontation with *Compsognathus*. A feeble effort perhaps, but a definite plus. There was enough skill there for natural selection to work upon.

To be sure, this method of takeoff could not have worked unless the Jurassic Solnhofen landscape was laid out like a skateboard park, with artful slopes at convenient intervals. Yet it is possible. But whether as climber or flier, *Archaeopteryx* would surely have been an ungainly beast. Grandly feathered wings, albeit clawed, are not ideal for clambering through the canopy. Monkeys and squirrels would surely have smirked to witness such a scene, had they been around at the time.

But still we have a very neat story. Among the saurischian dinosaurs was a group of two-legged types known as the theropods; some of those theropods were slim-line runners known as maniraptors. All the maniraptors had feathers, and some of them had long arms. One or other of the feathered, long-armed dinosaurs (one that was either a dromaeosaur or closely related to the dromaeosaurs) then turned arms into wings (and sometimes also its legs, as in *Microraptor*). Exactly how this feathered, long-armed maniraptor finally took to the skies is not perfectly clear. It isn't even clear how *Archaeopteryx* did it—and *Archaeopteryx* was already a bona fide bird. But we certainly seem to have the outline of the story.

Or do we?

ARE BIRDS REALLY DINOSAURS?
ARE MANIRAPTORS REALLY BIRDS?

The idea that birds are dinosaurs is immensely attractive. I love it. It has been around for a long time—since the 1860s, indeed—and the growing evidence from the maniraptors, particularly the Yixian maniraptors, seems to sew it up. Game, set, and match.

But nothing is certain in science, except in matters that are entirely trivial. In a field like this, we can hardly even hope for certainty. We have no machines to travel back in time with and see who really descended from whom.

And although "the birds-are-dinosaurs idea" is so beautifully neat, there is room for doubt. The most distinguished of the doubters is the evolutionary biologist mentioned a few pages ago—Alan Feduccia. His career spans nearly forty years. He is a professor at the University of North Carolina, Chapel Hill. He is the author of seven books on the evolution of vertebrates in general and of birds in particular. *The Origin and Evolution of Birds* (1996) was widely acknowledged as "the foundation from which all future investigations of avian relationships will start." Alan Feduccia doesn't like the idea that birds are dinosaurs. Not one little bit.

In truth, he said, the lineage of diapsid reptiles that led to the dinosaurs is, and always has been, quite separate from the lineage that led to the birds, and the two have continued ever since in parallel. The true ancestors of birds lived deep in the Early Jurassic, or the Late Triassic—around 200 million years ago. To be sure, those ancestors have yet to be found—but then, since fossilization happens so rarely, it really is not surprising that any particular fossil, or group of fossils, is missing.

The differences between the dinosaurs and birds are profound, says Professor Feduccia. Huxley and other nineteenth-century biologists concluded that they were related partly because they both have three-toed feet (and "hands"). In each case, the three fingers are all that remains of the five-toed (pentadactyl) hand possessed by the common ancestor of all land vertebrates. But, says Professor Feduccia, birds and dinosaurs did not retain the same three fingers out of the original five. The theropod dinosaurs retained digits 1, 2, and 3 (counting the thumb as 1), while birds have retained digits 2, 3, and 4. The research that led him to this, needless to say, is highly technical, and some distinguished biologists simply ask, "How can you tell?"; and they reject his thesis. But if Professor Feduccia is right, then dinosaurs and birds cannot possibly have any direct connection.

What of the feathered dinosaurs? Again, Professor Feduccia offers a different slant. Some of them, like *Caudipteryx*, clearly do have bona fide feathers—each with a central rachis and barbs along each side. But there is no mystery about this, he says. *Caudipteryx* and others with similar feathers were, quite simply, birds. *Caudipteryx* lived much later than *Archaeopteryx* and was descended from *Archaeopteryx*—or a creature very like *Archaeopteryx*. It had had flying ancestors, but was re-evolving a state of flightlessness. Many modern birds have done this. If an Emu was found in Early Cretaceous rock, says Professor Feduccia, then it too would be seen as a feathered dinosaur.

In some other "feathered dinosaurs" the so-called feathers were more like fur—albeit, as in *Sinosauropteryx*, fur with two prongs. Such fur, he says, is mere "dinofuzz." The impressions could be those of collagen fibers that, in life, were inside the animal. Various other creatures that no one thinks were feathered sometimes leave similar impressions—including the fish-like ichthyosaurs, pterosaurs, and some ornithischians. The creatures with such fuzz were indeed theropod dinosaurs, but they had nothing to do with birds. Overall, birds and maniraptors look very similar. But this—once again—is just convergence.

Then there's the matter of flight.

As Professor Feduccia points out, the idea that the ancestors of birds ran along the ground with their arms out until they took off just doesn't work. As we argued above in the context of *Archaeopteryx*, they could never have got up the speed.

Besides, it's clear that the ancestor of birds was already feathered (even if it wasn't a feathered dinosaur). But why was it feathered? Professor Feduccia does not discuss the notion that I favor—that feathers arose for the serious purpose of showing off—but he does dismiss the other main theory that they arose as insulation. If you want to be insulated, he says, then grow fur. Fur is much simpler than feathers. It isn't conceivable that animals could have evolved such elaborate, branched structures just to keep warm. So, he says, feathers could not have arisen until the earliest birds, or their ancestors, ventured upon flight. For flight they are an aeronaut's dream—light, strong, aerodynamic. So the first truly feathered creatures were fliers, or at least gliders.

But no creature can evolve into a glider if it spends its time on the ground. So, says Professor Feduccia, the true ancestors of birds must have been arboreal (an idea that he shares with the Danish biologist and artist Gerhard Heilmann, whose book *The Origin of Birds* in 1926 is still ranked as a classic). All other animals that have taken to the air use flaps of skin as airfoils. The ancestors of birds developed feathers instead; and of all the vertebrates—indeed, of all creatures bigger than a dragonfly—they became the best fliers by far. *Microraptor* is Professor Feduccia's favorite fossil: a feathered creature that was clearly arboreal, and flew or at least glided, yet lacked the aerial accomplishment of birds. *Microraptor* was not the ancestor of modern birds, for it lived some time after *Archaeopteryx*. (Of course, at least in principle, *Microraptor* could be a descendant of *Archaeopteryx*, but there is no outstanding reason to think that it is. Intuitively it seems more likely that *Archaeopteryx* and *Microraptor* shared a common ancestor, who at present is unknown, but presumably would have lived some time before 140 million years ago.) But it shows what the true ancestor of birds probably looked like.

Alan Feduccia is not the only radical. There are other biologists who acknowledge most of the orthodox story—that birds are indeed maniraptors and that maniraptors are indeed theropod dinosaurs—but they do not see the dromaeosaurs as a whole as the closest relatives of birds. Rather, they say, the dromaeosaurs are not a natural group at all but a mixed bag. Some of them are simply running theropods—closely related to birds to be sure, but not on the main bird lineage. But some of the so-called dromaeosaurs in truth are descendants of *Archaeopteryx*—and so are already bona fide birds. *Caudipteryx* is one such (just as Professor Feduccia argues, though for different reasons). So too is *Protarchaeopteryx*. Or then again, in 1998, Catherine A. Forster, of the State University of New York at Stony Brook, and her colleagues found a creature that they called *Rahonavis* in Late Cretaceous Madagascar. In some ways, *Rahonavis* was more like modern birds than *Archaeopteryx* was, but in some ways was also very reptilian. In modern, orthodox accounts, *Rahonavis* is sometimes shown as a dromaeosaur and sometimes as a bird, a descendant of *Archaeopteryx*. More of *Rohanavis* shortly.

In brief, things are far from settled, and perhaps in a few years' time the picture described in the last few pages will seem very old-fashioned. Perhaps the grouping now known as the dromaeosaurs will have disappeared altogether. Most of its members might be fused with the Archaeopterygidae—the family to which *Archaeopteryx* is deemed to belong. For the rest—who knows?

For my part, I continue to like the idea that birds are dinosaurs, but I have discovered, as the decades have passed, that the ideas that I happen to like are not necessarily true. I also like the idea that science, in the end, is a human pursuit, and omniscience is not in our gift. We cannot hope to be certain about anything—or, if we are, we cannot be certain that our certainties are true. Let us be content simply to acknowledge that birds very possibly are dinosaurs, and that most world authorities think they are, but they might not be; and even if they are, there are still loose ends. But let us also take heart. For although dinosaurs have been known and recognized for nearly 200 years, and *Archaeopteryx* has been on the stocks for 150 years, most of the key information in the modern account is new—it dates only from the mid-1990s, and the first appearance of *Sinosauropteryx*. Even though the study of bird evolution is halfway through its second century, these are early days.

This, then, completes the second part of this narrative: from the first ever tetrapods to *Archaeopteryx*. What happened then?

AFTER *ARCHAEOPTERYX*

As already intimated, the lineage to which *Archaeopteryx* belongs was not the only lineage of feathered maniraptors to take to the air, at least as gliders if not as powered fliers.[5] Indeed, the Mesozoic skies contained a whole suite of flying diapsids: the pterosaurs, which survived to the end of the Cretaceous; various kinds of feathered maniraptors which may not have been birds; and the lineage of the birds themselves. All have died out except the birds. But let us put the rest aside—important though they doubtless were in their day—and focus instead on those ancient, bona fide birds, and see how today's birds emerged from among their ranks.

In truth, the path that has led from *Archaeopteryx* to modern birds is just as complicated, in just the same way, as the path that led from the first slim-line theropods to *Archaeopteryx*. That is, the early descendants of *Archaeopteryx* diverged—radiated—in all kinds of directions, and from that radiation just a few kinds (or, in the case of modern birds, just one kind) of creatures came through.

In fact, the descendants of *Archaeopteryx* evolved along at least five distinct lines—all of them bona fide birds and all of them clearly recognizable as birds. But all the birds that survive belong to just one of those lines. If we rank all birds including *Archaeopteryx* as a class—the class Aves—then we can say that each of the distinct subgroups is a subclass. All modern birds, from Ostriches to wrens, including storks and eagles and penguins and hummingbirds and all the rest, belong to just one of the several subclasses—the one known as the Neornithes, meaning "new birds." All is made clear in Chapter 4. Again we see a parallel with mammals. Now only two subclasses are left to us: the egg-laying monotremes and the therians; and the therians in turn include both the marsupials, like kangaroos and wombats, and the eutherians, which include all the rest, from whales to us. But at least three other distinct subclasses of mammals are known from the fossil record, and are now long gone.

This book looks mainly at the neornithines—the modern birds that play a part in our lives. But we should look briefly at the rest, although they are long gone: true birds, but not modern birds—the kinds that might solemnly be called "non-neornithine avians."

FROM *ARCHAEOPTERYX* TO MODERN BIRDS

O. C. Marsh of Yale, one of Darwin's staunchest American supporters, was a great fossil hunter, and he discovered some very significant birds from the Cretaceous—much younger than *Archaeopteryx*, and much more

specifically avian, but not yet modern. He found his first and most spectacular fossil bird in 1871, from the Smoky Hill Chalk of Trego County,
Kansas. This was *Hesperornis regalis*—*hesperornis* meaning "western bird."
Hesperornis was very much marine. It was cigar-shaped like a modern penguin, but huge—nearly 2 meters (7 feet) long when fully extended. Unlike
a penguin, it had no flippers; its arms were much reduced, like a kiwi's,
surely hidden beneath the feathers or even under the skin. It propelled itself under the water with long powerful legs and huge feet, which it held
out sideways from its body like a frog, driving itself from behind as if with
an outboard motor (ducks do this, too). Clearly it was a formidable predator of fish. It had a long, dagger-like beak, with sharp teeth both on its
lower jaw and to the rear of the upper jaw.

Taken all in all, *Hesperornis* was the most specialist marine bird of all
time—more so than any modern diver, even more than a penguin. It must

Hesperornis, *from the Cretaceous, was the most maritime of all birds.*

have found it hard to come on land at all; surely it would have found it hard even to emulate the penguin waddle (although you can't really tell what animals can or can't do until you watch them in action). Some have suggested that *Hesperornis* produced live young at sea, like a porpoise or an ichthyosaur, but there is no evidence for this. There is no *a priori* reason that birds should not have given rise to avian porpoises, but there is no compelling reason to suppose that they have actually done so. More probably, *Hesperornis* hauled itself ashore to breed, painfully perhaps, like an elephant seal. Later, Marsh discovered another marine contemporary of *Hesperornis*, though smaller, which he called *Baptornis*. Both are now placed in a taxonomic group that in effect has the rank of a subclass, generally known as the Hesperornithiformes.

The following year, in the same site, Professor B. F. Mudge of the Kansas Agricultural College (now Kansas University) found another Cretaceous bird with teeth, and he called it *Ichthyornis*, "fish bird." "Teeth" in this context means true teeth, embedded in bone, not just serrations at the end of a horny bill, of the kind we find in a modern merganser. *Ichthyornis*, in sharp contrast to *Hesperornis*, was a flier. In life *Ichthyornis* must have looked like a small gull (though with teeth). In modern classification, *Ichthyornis* also forms its own subclass, Ichthyornithiformes.

There matters stood, more or less, for a hundred years. Various other fossil birds did turn up, including *Iberomesornis* from Spain, but—until the 1980s—these were assumed to be the remains of modern birds.

But then, in the 1990s, the tide turned. As already intimated, the 1990s outstrip even the 1860s as the greatest decade by far in the whole science of bird evolution. An entire range of ancient but nonetheless unequivocal birds have come to light that to a very significant extent have filled in the yawning gaps between *Archaeopteryx*, *Hesperornis*, *Ichthyornis*, and the moderns. Figure 4 (page 108) summarizes the whole lot (as now understood not least by leading authorities at the American Museum of Natural History, New York).

In Figure 4, the bird shown closest to *Archaeopteryx* is *Rahonavis*—the small slim dinosaur from late Cretaceous Madagascar that Catherine A. Forster described in 1998. Its name—pronounced ra-Hoon-ah-vis—has a fine romantic ring to it: it comes from the Malagasy and translates as "menace from the clouds." In life, *Rahonavis* was a predator about the size of a crow. Its jaws were toothed, like an *Archaeopteryx*, and its tail was reptilian and whiplike—again, like the tail of an *Archaeopteryx*, only shorter, with only thirteen caudal vertebrae rather than twenty-one or twenty-two. Yet again it had the big, sickle-shaped claw on its second toe, shared with other dromaeosaurs and with *Archaeopteryx*. But *Rahonavis* also had several features like modern birds that *Archaeopteryx* didn't have, including spaces in some of its vertebrae for the flowthrough of air. Feathers have not been found from *Rahonavis*, but it had knobs on the forearm bone, the ulna, of the kind that quill feathers would have been attached to—as can be seen in a chicken wing. In truth, there is some doubt whether the ulna with the feather attachments actually belonged to *Rahonavis*. It may have derived from some other ancient bird that died in the vicinity. If it did belong to the rest of the *Rahonavis* skeleton, then *Rahonavis* may well have been a flier. If not, then we just don't know.

All in all, *Rahonavis* remains controversial. Some paleontologists think that it was a true bird and was more modern than *Archaeopteryx*, while others think it was a dromaeosaur; and from its mixture of features, it could have been either. On the other hand, some biologists think that at least some dromaeosaurs, and perhaps others that are now placed in other maniraptor families, should also be seen as birds, more modern than *Archaeopteryx*. The evolutionary history—the phylogeny—as shown in Figure 4 is certainly authoritative but it is not the only version in town.

More modern still was *Confuciusornis*, again from the wonderful, smooth, early Cretaceous Yixian rocks of Liaoning, and first described in 1995 by Hou Lianhai, now at the University of Kansas, and his colleagues. Again, *Confuciusornis* was roughly like a crow in size and outline, primitive to be sure but more definitely birdlike than *Rahonavis*. Like *Rahonavis*, it had pneumatic spaces in its bones, but unlike *Rahonavis*, it had lost the long reptilian tail. The tail bones of *Confuciusornis* were reduced in number and compressed into a bona fide pygostyle.

Next are the Enantiornithes. The name, coined by Hou Lianhai, means "opposite birds": a reference to the shape of their shoulder girdle, which in general form was like a modern bird's but differed in detail. The first enantiornithines came to light in the late 1970s in Argentina. Then, Cyril A. Walker, from the Natural History Museum in the English town of Tring, realized that other fossils already documented were also related to these Argentinian types. In 1988 he named the whole group. *Iberomesornis* was one such—not a modern bird at all, but one of the Enantiornithes. More and more types came to light in the 1980s and 1990s, and now at least forty species have been ascribed to the enantiornithines, from all over the world: Spain, China, North and South America, Madagascar, Australia, and the Middle East. They are extremely various: the smallest were sparrow-size while the biggest known, like *Enantiornis* and the Late Cretaceous *Avisaurus* from South America, had a wingspan as big as a Herring Gull's, at around 1.2 meters (4 feet). They seem to have lived in a correspondingly wide variety of habitats, including lakesides, though only one known type seems to have been marine. Clearly there is a lot still to discover. Recently, an enantiornithine turned up in China with flight feathers on its legs as well its arms. We have already seen this in *Microraptor*. But then, this is a reasonable way to approach the business of flying—so why not?

The enantiornithines looked modern enough, but appearances can be deceptive. Notably, they may not have been fully homeothermic. Animals that are properly warm-blooded, such as modern birds and mammals, grow without interruption through winter and summer, and their bones are smooth when cut across. In contrast, the bones of cold-blooded animals grow in fits and starts according to season, and the interruptions show up in cross-sections like tree rings. Bones from enantiornithines such as *Concornis* (from the Early Cretaceous, whose bones have been found in Spain) show a tree-ring pattern. Modern birds grow quickly, too, while *Concornis* seems to have taken several years to reach fully adult size.

There are still more fossil birds—clearly not modern, but difficult to fit into the family tree. *Patagopteryx* and *Vorona* seem to belong somewhere between the enantiornithines and *Hesperornis*. *Patagopteryx*, found in Patagonia, was like a slim chicken, though with very small wings and flightless (whereas wild jungle fowl, the ancestors of domestic hens, fly quite powerfully). *Vorona*—pronounded voo-roo-na—from Late Cretaceous Madagascar, was first described by Catherine A. Forster and her colleagues in 1996 (from the same deposits that yielded *Rahonavis*). Only the leg bones and pelvis seem to have been found so far, so exactly what *Vorona* looked like in life is hard to judge.

By the time we get to *Hesperornis* and *Icthyornis* and their relatives, we are more or less into modernity. So we seem to have a step-by-step progression from *Archaeopteryx* to modern birds. *Archaeopteryx* had a long whip-like tail—most unbirdlike. *Rahonavis* had a somewhat shorter tail. *Confuciusornis* had a bona fide pygostyle—tail bones squashed into a stump. Later groups are increasingly birdlike. The picture is not as neat as it might seem, however; as we have seen, the tail is also reduced to a pygostyle in some non-avian theropods, including oviraptors. Perhaps the oviraptors are more closely related to modern birds than *Archaeopteryx* is, despite appearances, or perhaps the pygostyle form evolved more than once.

Although the present picture is very pleasing, we cannot be sure that the phylogeny shown here, or anywhere else, is correct. We cannot even be sure that *Archaeopteryx* was the ancestor of all later birds. It may well be that many of the feathered theropods of the Jurassic gave rise to flying forms; that *Archaeopteryx* is just one of the many; and that the true ancestor of all later birds was one of the others. But it is at least a fair bet that the true ancestor of birds was similar, and indeed related, to *Archaeopteryx*. This at least is a good working hypothesis, and in biology a good working hypothesis is the best we can hope for.

Note, finally, that the different fossil birds shown here are all deemed to belong to different subclasses of birds. Indeed, Figure 4 shows nine different subclasses. But modern birds belong to just one of the nine—the Neornithes.

The neornithes clearly began way back in the Cretaceous, so the earliest ones lived alongside *Hesperornis* and *Ichthyornis* and others—and also alongside the small running feathered dinosaurs and the last of the giant dinosaurs. In some fossil mud we can see the tracks of dinosaurs (in the traditional sense) and birds side by side. We can assume that the modern birds evolved from among the ranks of the earlier types, but that does not mean that all the earlier ones disappeared as soon as the new ones came on the scene. To have been an ornithologist in the Late Cretaceous would surely have been heaven.

Only a few of the non-neornithine birds survived the asteroid or whatever it was that brought the Cretaceous (and hence the Mesozoic as a whole) to an end. That event—the "K-T boundary"—really was a watershed. The few non-neornithine birds that did survive the catastrophe died out soon after. So the only birds left to us now all belong to just one avian subclass, the Neornithes.

Yet the neornithines have diverged and radiated so that now they are as varied in their physical form and ways of life as all the rest put together (although no modern bird is quite like a *Hesperornis*). They are so diverse, with so many species, that it is hard to keep track. Indeed, this would be well-nigh impossible were it not for the art, craft, and science of classification, also known as taxonomy.

II

Dramatis
Personae

This perfectly plausible scene includes birds from eight of the thirty or so living orders:
Falconiformes, Charadriiformes; Podicipediformes, Passeriformes (both Passerida and
Corvida), Coraciiformes, Pelecaniformes, Anseriformes, and Columbiformes.

3

KEEPING TRACK: THE ABSOLUTE NEED TO CLASSIFY

IT'S A SIMPLE QUESTION OF THE KIND SIX-YEAR-OLDS ASK: How many kinds of birds are there? But as with most of the questions that six-year-olds ask, the answer is that nobody knows, and nobody can ever know—at least not exactly. The reasons are both practical and philosophical and lead us into issues of relevance to all biology.

To begin with, what does "kind" mean in this context? Generally it means "species." But species is a very tricky concept—and can be especially tricky when it comes to birds.

WHAT IS A SPECIES?

If scientists don't define things carefully, they cannot communicate their ideas precisely, and without precision, science falls apart. But the "things" of which nature is composed feel no obligation to be easily definable. Even very obvious entities such as "leg" and "stomach" are hard to pin down in practice. Is the leg of an insect really comparable with the leg of an Ostrich? Does an octopus have "legs"? More abstract things—or perhaps they should be called concepts—are even harder to define. Richard Dawkins has pointed out that the word *gene* is used in at least half a dozen different ways.

Species is even more abstract and even harder to pin down. Always the dilemma is the same. We can define the thing we are interested in very tightly, like lawyers, but then our definition may not correspond very closely to nature. Or we may define the thing in question generously, to account for the realities of nature, and finish up not with a definition but with an endlessly expandable descriptive essay—more like a blog. So the question "How many different kinds are there?" has no meaning unless we define *kinds*, and if we stipulate that *kinds* means "species," we then have to ask, "But what is a species?" And the answer is not easy. As with *gene* (or *leg*, or *stomach*), *species* means different things in different contexts.

How, first of all, do we decide in practice that bird X belongs to species A, and bird Y belongs to species B? By appearance is the short, practical answer (more broadly, by their measurable attributes, physical and other-wise; but appearance is what it boils down to most of the time). This raises all kinds of problems. First, many groups of species look identical, or almost so, and yet have nothing much to do with each other. In Europe, the Willow Warbler and the chiffchaff (of the family Sylviidae) are very hard to tell apart. Both are small and yellowish-greenish, flicking their wings as they fossick in the foliage for tiny insects. Both are spectacular migrators, lodging in Britain only in the summer. The good reverend nat-uralist Gilbert White was the first to distinguish clearly between them, in the late eighteenth century—and he told them apart not by their appear-ance but by their very different voices. The Willow Warbler is a beautiful songster—"a liquid fall of notes" it's been said—while the chiffchaff merely calls, "Chiff-chaff!" America has its own wood-warblers—very similar to the Old World warblers, although of a quite different family (Parulidae). Anyone who can distinguish in the field between the Bay-breasted Warbler, the Blackpoll Warbler, the Pine Warbler, and the Palm Warbler, all in the genus *Dendroica*, is a better birder than I—indeed, they are known even in the United States as "the confusing warblers." They are all greenish or yellowish with dark streaks and, just to stir the pot a little more, several different *Dendroica* warblers may forage together in the same tree, often a tall tree where they are hard to see, each exploiting a slightly different niche. Owls raise the same kind of problem. Most owls are noc-turnal most of the time and do not distinguish their own species by ap-pearance but by their hoots and screeches and other eldritch calls, but human beings are visual creatures, and faced with two owls that are silent, or indeed dead, we often cannot tell them apart. So it is that the world's species list of owls continues to grow as ornithologists look more closely, both in the field and in the laboratory, at their DNA.

Then there is the opposite problem: that there may be enormous variation in plumage, and sometimes in other features, among members of the same species. Often there is a sharp distinction between males and females—the phenomenon of "sexual dimorphism." Usually (though by no means always!), the males are brighter and sometimes positively flamboyant, while the females are duller and often downright "cryptic," meaning they are camouflaged. The general theory has it that the males are bright for sexual reasons—their priority is to attract mates—and the females are dull for pragmatic reasons—they often spend more time incubating eggs and feeding young, often in the open, and need to be inconspicuous. In Mallards, the dandified drakes have green iridescent heads while the females are decked out in autumnal tweed. But both sexes have a flash of purple-blue on the wings, known as a speculum. The speculum is a color code among ducks—not to enable birders to tell one from another, but to help the ducks and drakes themselves to see who's who. (Even so, as we will shortly see, ducks often get it wrong.)

We can see that male and female Mallards belong together because we can observe them any time we like on the local pond, and watch them knocking about together. But life is not always that easy. For a long time the males and females of the Eclectus Parrot were taken as separate species: the males are emerald-green on the back with scarlet underwings and flanks, while the females are crimson with violet-blue belly and underwings. And since the Eclectus Parrot lives in the forests of New Guinea and Australia, their social to-ings and fro-ings are not easily observed (at least, not in the wild). In many birds the juveniles look quite different from the adults—and behave differently. The adult male American Redstart (another from the American wood warbler family, Parulidae) is very dark with orange bars on its wings and tail, while the females and juveniles are greenish and grayish and altogether duller. But just to keep birders on their toes, the males retain their juvenile plumage during their first year of breeding. Why? Even more to the point, adult male plumage is supposed to attract the females, but if male American Redstarts can attract females even though they still look adolescent, why do the adult males bother with their customary finery, which surely increases the risk of predation and is physiologically costly? Why, for that matter, does any bird bother? I do not presume to know and neither, I venture, does anybody else.

Some birds are polymorphic: different individuals of the same species may adopt any one of two or more distinct colors or patterns. The Ruff is a European sandpiper (a shorebird) in which a breeding male develops a ruff of feathers round its neck like an Elizabethan dandy—white in some individuals (generally the subordinates) and brown in others (the dominants). The Gouldian Finch of Australia, one of the very colorful grassfinches, usually has a black head, but one in four has a red head and one in a thousand or so has a yellow head. The genetics of this are actually quite simple, but the result is striking. (Incidentally, the Gouldian Finch is not a "true" finch at all, but a waxbill. More in Chapter 4.)

Many species of birds (and other creatures) are spread over large areas and vary continuously along the length of their range. The individuals in place A are very slightly different from those at place B, but similar enough to interbreed freely; and the individuals in place B are very similar to those in place C, but very slightly different again; and so on. Such a population—spread out geographically and varying continuously from one end to the other—is called a "cline." Occasionally, the individuals at either end of the cline have become so different that they *cannot* breed with each other and so they form distinct species, even though each can and does interbreed with the population next to it, which in turn breeds with the next, and so on all the way down the line. One group of gulls forms a cline that runs right around the Arctic Circle, and their color varies continuously from a silvery gray to dark slaty gray. All along the cline the gulls in any one place breed perfectly happily with their neighbors on either side. But the gulls at the two extreme ends of the cline are so different that they cannot, or at least do not, breed with each other. In fact, both ends of the circular cline meet around Britain and are regarded as separate species. The silvery-gray one is the Herring Gull, *Larus argentatus*, and the dark one is the Lesser Black-backed Gull, *Larus fuscus*.

Sometimes populations of creatures that all breed together, and are clearly a single species, get broken up. Clines get broken very easily, as they may extend for hundreds of miles and there is plenty of scope for discontinuity. Populations may be fragmented for all kinds of reasons. Sometimes the landscape changes around the animal: arms of the sea invade and isolate part of the range; the climate dries and the forest is reduced to scattered woods, and many woodland creatures (including some birds, even some that fly perfectly well) are reluctant to cross the open spaces in between. Thus it seems that the grouse and ptarmigans of North America and Eurasia are as various as they are—eighteen species of them, each adapted to a slightly different habitat—because the initial population has been divided in this way.

Sometimes one group of the parent population, often a very small group, will wander by chance into another territory and become cut off from the parent population. We know from the way that vagrants keep turning up that birds are often swept away by winds. Sometimes they finish up on the remotest islands—and then, if more than one turns up, they may start to breed. Creatures that live on continents have to compete with hundreds of other creatures, notably with predators. They need to be tough, swift, well-armored, elusive, or very clever. Animals that fetch up on remote islands are often spared such inconveniences: remote islands are most unlikely to contain predatory mammals unless they have been imported by human beings, because the only way a predatory mammal could get there is by "rafting" (floating across on weed or debris)—which does happen, but not often. When isolated animals have no predators to cope with, their genes begin to show us what they can really do. Creatures that on continents stay small, the better to hide from predators, may become large. They also tend to lose their instinctual fear of predators, since there aren't any. Birds tend to lose the power of flight because, after all, flight takes a great deal of energy and is potentially dangerous (not least because of crosswinds). So, apart from the Ostrich, cassowaries, and so on, which are very large and swift or fierce and therefore can live on continents, most of the flightless birds are islanders. The Dodo is the classic: a flightless giant pigeon from Mauritius, way down in the Indian Ocean. Or, rather, it was the classic because flightless island birds tend to go extinct as soon as human beings turn up with their entourage of rats and cats, and so it was with the Dodo by the end of the seventeenth century.

On islands, animals often diverge spectacularly to form several or many different species in double-quick time, mainly because when they first arrive they find many new niches with no resident rivals. Then we find various groups of species which in fact are all closely related. On the Galápagos Islands, a very small group of tanagers, which arrived less than 5 million years ago from the South American mainland, evolved in all kinds of directions to become the fourteen different "Darwin's finches." On Hawaii, also about 5 million years ago, a small group of chaffinch-like birds flew in from mainland South America and evolved to form the even more astonishing variety of Hawaiian honeycreepers.

There are now twenty-two different species of these honeycreepers, and it seems that another twenty or so have been driven to extinction in the past few hundred years by the many exotic animals and diseases brought in by humans.

In general, when a single species of bird (or any creature) becomes fragmented into different populations, and the different populations become isolated from each other, each group starts to evolve along slightly different lines. The different subpopulations—but still not different enough to qualify as separate species—are then called "subspecies" or "races." Sometimes it is hard to decide whether two similar-but-different populations really are races of one species or are different species. In various ways, this can matter. In the 1970s and 1980s, American conservationists tried heroically to rescue the Dusky Seaside Sparrow of Florida from extinction, and alas they failed. In the end the wild population was reduced to one female and seven males, and the species faded away. But later it transpired that the Dusky Seaside Sparrow was not a distinct species of seaside sparrow but merely a subspecies, a race. Its loss was still a pity, but not quite as bad as the loss of an entire species (although Seaside Sparrows as a whole are now in trouble). Contrariwise, many populations that we believe to be a single species sometimes turn out to be several. Sometimes only their DNA shows the difference—and DNA studies are new. Species of owls are constantly being split into several species under closer examination.

Breeders of animals produce different subpopulations artificially—and these artificial creations are called "breeds." (Plant breeders call their artificial subpopulations "varieties.") Different breeds within the same species may differ far more in appearance than entire suites of wild species. Indeed, in the early nineteenth century, the distinction between race and species was by no means clear, and when Charles Darwin was working on his magnum opus, *On the Origin of Species by Means of Natural Selection*, he spent many happy if confusing hours talking to pigeon fanciers, trying to work out what the difference really was. By then the fanciers had bred many, wildly different types—fantails and tumblers, pouters and croppers, frillbacks, nuns, monks and mokees, and many more. In general, the breeders assumed that each kind had descended from some different wild species of pigeon, but Darwin thought otherwise. All, he concluded, were descended from the wild Rock Dove, *Columba livia*. Yet there is far more visible difference between, say, a Silesian Pouter and a Schmalkalden Moorhead than there is even between a kestrel and a sparrowhawk—two birds that, as we will see, may hardly be related at all. So why do we say that all the domestic pigeons are simply different "breeds" of the same species, while those two middle-size birds of prey are of different species? Why, more broadly, should we decide that two owls or warblers that look identical are in fact different species—or that two populations of Seaside Sparrow that look different are really the same species? What makes them different if not their appearance? Why won't taxonomists simply accept the commonsensical notion that "If it looks like a duck and quacks like a duck, then it is a duck"?

Intuitively we feel that female and male peafowl or Mallards belong in the same group—species—because they breed together, and that two identical-looking owls that do not breed together are really different. This has led to formal definitions of *species* based not on appearance but on reproductive potential—who breeds with whom. Thus, at university in the early 1960s, I learned to define *species* roughly as follows: "A population of creatures may be considered to be of the same species if males and females can breed together to produce fully viable offspring, or may be presumed capable of doing so."

This is a neat definition that has stood me in good stead this past forty-five years, but as the professor himself went on to discuss, it opens some very large cans of worms. What, first of all, does "fully viable" mean? Well, we can easily see what it does *not* mean. When two creatures of different species mate, the usual result is that nothing happens. If a dog ever has his way with a cat, no kittens or puppies or any kind of offspring result. Cats and dogs are just too different. Similarly, in the United States, a Rusty-Margined Flycatcher was once observed to copulate with a Ruddy Ground-Dove, two birds that seem even further apart, genetically, than cats and dogs. We can presume no progeny resulted.

But if two different yet similar species mate, then offspring may sometimes be produced and these are called "*inter*specific hybrids" (and if two different races, subspecies, varieties, or breeds of the same species are crossed, the result is an *intra*specific hybrid). Interspecific hybrids may look good, and may be very strong, but often they have some defect that makes them unable to compete, at least in the wild, and so they are not "fully viable." The best-known example—so well-known that it is part of common lore—is the mating of horses with donkeys to produce mules. Mules are very tough, but they are sexually sterile. An animal that is sterile cannot be said to be "fully viable." The fact that their hybrid offspring are not fully viable confirms that horses and donkeys are indeed different species. Similarly, when the different species of grouse and ptarmigan overlap in the wild, they may interbreed, but the hybrid offspring are usually infertile.

But all is not so simple. Sometimes the presumed lack of viability is not so easy to see, but is there nonetheless. So it is that Ireland, much of Europe, and the west of Scotland have Hooded Crows, while in England and the south of Scotland the "Hoodies" are replaced by Carrion Crows (which the English tend simply to call "crows"). Along a line that runs through western Scotland and the middle of Ireland the two kinds meet, and there they interbreed to form hybrids. There is nothing obviously wrong with the hybrids: they are not sterile. But the hybrids do not spread north to invade the territory of the Hoodies, or into crow territory in the south. For some reason, they cannot compete well with either of the two parent species. So they remain confined to a narrow hybrid zone, hemmed in by their parent species on either side.

There are many examples of such hybrid zones in nature in all kinds of creatures. In the eastern United States, the Blue-winged Warbler interbreeds with the Golden-winged Warbler where the two overlap, to produce a distinct hybrid known as Brewster's Warbler. Brewster's Warbler does not spread dramatically into either of its parent species' territory, but it may occasionally mate successfully with one or other of the parent species to produce the rare Lawrence's Warbler. (Again, whoever can spot a Lawrence's in the field, or indeed a Brewster's, deserves a Golden Twitcher's Award.)

Sometimes, however, hybrids thrive perfectly well—and sometimes too well. Ducks are great hybridizers: it seems that just about any species of duck can and on occasion does hybridize with any other species of duck. On the canal and rivers near my house, the ducks often seem to be a fine old mixture. Geese, too: near me, hybrids between Canada Geese (which have invaded Britain in force) and Greylag Geese (chief of the native species) are common. Sometimes the hybrids are so "viable" that they oust one or other and sometimes both of the parent species. So it was that in the 1940s Sir Peter Scott, a pioneer conservationist and an outstanding wildfowl expert, decided to introduce Ruddy Ducks from North America into his newly established wildfowl center (now the Wetland Centre) at Slimbridge in Gloucestershire, in the west of England. He was very aware of the dangers of introduced species—no one was more aware—but he argued that there are no British ducks even remotely like Ruddies, so why not? Unfortunately some Ruddy Ducks found their way to Spain (it's not far, for a duck) and met the White-headed Ducks, which look very different (they have white heads) but are from the same "stiff-tailed" group of ducks. The two species interbred and soon there were fears that the White-headed Ducks would disappear altogether, "swamped" by the dominant genes of the Ruddies. The situation is now closely monitored. I would certainly support those conservationists who believe that, where necessary, the Ruddies in Spain should be culled (although the culling is very expensive).

You might well feel, very sensibly, that sexual compatibility depends simply on genetic distance—that the more different two creatures are, the less likely they are to be able to interbreed. You might also suppose that in this sophisticated age we could measure the genetic distance by directly analyzing the DNA, the stuff of which genes are made. Up to a point these eminently sensible generalizations apply, but still nature pulls surprises. In the 1980s, a Pine Siskin mated with a Red Crossbill and produced at least one daughter. No one witnessed the mating, but the offspring turned up in the backyard of an ornithologist, Dr. Dan Tallman of Northern State College, South Dakota; and he passed it on to Dr. Richard Zusi of the Smithsonian's National Museum of Natural History in Washington, D.C., who traced its ancestry through its DNA. The hybrid was feeding in a flock of Pine Siskins, though it had the look of an outsider. In truth, a great deal of what happens in nature is known only through such chance encounters—the right expert in the right place at the right time. Thus, Dr. George McGavin of the Oxford Natural History Museum recently found a new species of spider in Borneo: it lowered itself on a thread of silk in front of his nose as he was having lunch. I wonder how many new creatures any of us might have discovered if only we had known what we were looking at.

The story of the siskin and the crossbill shows that it isn't simply the overall genetic distance that determines who can breed with whom. There seem to be specific genes at work; and if the specific genes within each partner are compatible, then mating may be successful even if the overall genetic distance seems great. Again, the opposite also applies, at least among plants. So it is that most varieties of apples or plums cannot breed with members of their own variety because specific "incompatibility genes" prevent such inbreeding. I know of no comparable examples among animals, but many do seem to have a psychological aversion to incest (although many apparently do not).

Such difficulties need not defeat the taxonomist. Usually things are more straightforward. It is clear, for example, that the Ostrich is just one species, divided into several races (which are color-coded, some with blue necks and some with red or pink), and that Ostriches cannot mate successfully with any other living bird, so we know where the species *Struthio camelus* begins and ends. Even so, the overall picture is blurred. Biologists might choose to devise the specific concept of species and to a large extent this concept does correspond with what we see in nature. Ostriches are Ostriches and Ospreys are Ospreys, and never the twain shall meet, with each other or with anything else. But nature does not always recognize our neat definitions—which, in the end, are our own inventions. Since nature defies our attempts to define the concept of species, we cannot say how many there really are. We are reduced to saying, "It depends what definition of species you have in mind."

There is one further complication. It makes perfect sense to base our definitions on reproductive habits and genetic compatibility if we confine our studies to living species, whose breeding habits we can observe and even manipulate. But such definitions are of much less use (not totally useless, but largely so) when applied to extinct birds that are known only from stuffed specimens in glass cases or—if they died a long time ago—from fossils in various states of petrification. When all we have of a creature is its bones, with all its DNA long faded away, we have only appearance to go on. Modern classification, as we will see, is based on the genealogy of creatures—who is literally related to whom and who descended from whom—and to work out genealogy properly, we need to incorporate the extinct types, including those known only from fossils. Yet it can be hard to reconcile different sets of information when one set is based on reproductive habits and genetics as well as the appearance of skeleton, feathers, and all the rest; and the other is based only on the appearance of the bones. Again, these problems need not cause us to despair, but they certainly help to remind us that in the end, when all the work is done, omniscience is not in our gift, and nature feels no obligation to be easily comprehensible.

And even if we could be absolutely sure where one species ends and another begins, and could identify them all in the field from a single glimpse, we still could not say how many there are.

STAND STILL AND BE COUNTED

If all species really were distinct from all others—no hybrids, no halfway subspecies—and if we could identify them all instantly on sight or at least by sound, then, surely, we could just go out and count them. Of course, life is never so simple.

Many birds are extremely elusive and some, including some of the elusive kinds, are also extremely rare. Few people in Britain have seen a Corncrake or a bittern, yet there are bitterns now in London, though mostly they hide among reeds and feed at night (but I have seen them flying by day in East Anglia). Nightingales may fill half of Europe with their song in due season, but you rarely see them because in the main they sing from the heart of bushes and small trees. My best audience with a nightingale was by chance, not in Britain but in southern Spain, when one came to serenade us or at least to announce his presence to other nightingales from a low branch right by our picnic. This was lunchtime. Nightingales sing at least as much by day as by night.

Britain is full of people, the landscape is generally accessible, and it has more birders per head of population than almost anywhere. But most species of most terrestrial creatures live in tropical forest, some of which is all but impossible to get to—it takes enormous fortitude and a great deal of time and therefore money to explore the steep and forested slopes of Venezuela or Indonesia—and some forests you can't get to at all for political reasons or because, horribly, so much tropical land has been mined. Even when a tropical forest is accessible, you find that you can spend an enormous amount of time in it without seeing any animals at all. Only in Hollywood are the various pith-helmeted heroines beset by snakes and giant centipedes at five-minute intervals when, for whatever implausible reason, they fetch up in the jungle. A botanist in Costa Rica told me he has worked for several days a week in the forest over the past twelve years and has only once seen a snake. The animals are all around, but they are wonderful at hiding. In India and Africa, even elephants disappear into the shadows. Small birds that live in the canopy, 30 meters (100 feet) up, in what from the ground seems gloom (although in truth when seen from overhead is in bright and merciless sunlight) are all but impossible to see. Many tropical forest species, too, from trees to insects, are surprisingly local in their distribution, so you won't see them unless you happen to be in exactly the right area.

So it is that new species of birds turn up all the time. Some, hitherto, were entirely unknown to science and some of them are large and spectacular. The beautiful Congo Peafowl is one such, decked out in various shades of glossy green and chestnut, but it was first made known to science only in 1936. Though it is called a peafowl it is also very like a guineafowl; perhaps it represents some kind of evolutionary link between the two. Another member of the pheasant family, one of the francolins in the genus *Francolinus* was discovered in the mountains of Tanzania only in the 1990s; it is popularly known as the Udjunga Forest Partridge. Three new species of grasswrens—pleasant little perching birds from Australia—have been identified within the past fifty years: the Gray, the Kalkadoon, and the Short-tailed.

Sometimes—surprisingly often—birds that are thought to be extinct turn up again, sometimes after decades. So it was that the Eyrean Grasswren was first reported in 1874 from northwest of Lake Eyre in South Australia, but then it was not seen for the better part of a century and ornithologists feared it was gone forever. But a few were spotted in the 1960s and finally, in 1976, one was captured—definitive proof of continued existence.

Australia is big and grasswrens are small, and although some are doing well in suburbs, others are scattered far and wide in the remotest places, so it really isn't surprising that they duck out of sight. Bigger birds get lost, too. Most famously, the spectacular Ivory-billed Woodpecker was last photographed on the American mainland in 1938, and in Cuba in 1948. But there have been many reported sightings since, including many in the 1990s, and ornithologists from several states, including Arkansas and South Carolina, are keeping up a concerted search, sometimes with good grant money behind them and always in high hopes of finding them.

The world is also constantly losing species. Most vulnerable are island species, partly because their populations tend to be small to begin with; partly because they have nowhere to run if and when their islands are invaded by human beings and by the dogs, cats, pigs, rats, goats, monkeys, exotic birds, and diseases they tend to bring with them; and partly because island birds have evolved in the absence of predators, are often far too tame for their own good, and as we have seen are often flightless. Nowadays tropical forests are threatened, too. Dr. Andrew Mitchell of the Global Canopy Programme, Oxford, estimates that an area the size of Belgium is currently being lost from the Amazon every year. In Indonesia, it is estimated that an area the size of a football field is lost every second. This matters hugely. An estimated 74 percent of all bird species rely on forests, and most of these—60 percent of the whole—need tropical forest; many of them, almost certainly, live only in relatively small areas, so that if their particular patch of forest is wiped out, then they are wiped out along with it. Putting everything together, BirdLife International now suggests that about one in nine of all bird species are in danger of extinction within the next few decades. In some groups the proportion is far higher. About one in four members of the pheasant family (Phasianidae) are now thought to be threatened, if not on their beam ends.

So for all kinds of reasons we cannot really say how many different birds there are in the world. Even if the problem were in practice easier, we would still confront the general philosophical issue identified by John Stuart Mill in the nineteenth century: that however much we know, or think we know, we can never be sure that we know all there is to know. However many kinds of birds we count, we can never be sure we haven't missed a few (or a lot). Furthermore, we can never know how much we don't know, except in the broadest terms, for it is logically impossible to gauge the extent of our own ignorance unless we are already omniscient.

You may feel that all this is of marginal importance. People have been looking seriously at birds over most of the world for many hundreds of years, and now there are formal, well-equipped university teams on the job and battalions of freelance professionals and expert amateurs, so that by now, obviously, we must pretty well have tracked them all. But in 1961, in modern times—I was just about to go to university—Oliver Austin declared in his marvelous *Birds of the World* (every birder should have a copy) that the world contains around 8,650 species of birds and added, somewhat incautiously, "It is unlikely that [this] round figure . . . will change by more than 1 or 2 per cent in the near future." But the excellent *New Encyclopedia of Birds*, edited by Chris Perrins and published in 2003, lists 9,850 species. If you think that this really *must* be the limit, then remember the cautionary tale of England's Martin Lister, who in his *Tractacus de Araneis*

(Treatise on Spiders) of 1678 described thirty-four spiders and three har-vestmen, and while he issued a modest disclaimer, "I would not wish any-one to suppose that I have described every single type," he also asserted "that it is not easy to find in this island any new species that I have failed to describe." Now the British species list of spiders stands at around 650. I do not mock Dr. Lister; he was a fine physician and a far better natural-ist than I could aspire to be. But nature is difficult.

So most authorities these days would not claim (Perrins certainly does not) that 9,850 must be the limit. Indeed, most when pressed for a figure these days tend to say, "Well, there are probably around ten and a half thousand." Given that this allows quite a lot of leeway (and extinctions are running apace), I will take this as a rough ballpark figure: 10,500 it is. Or thereabouts.

How can anyone get a grip on 10,500 different species? How can we possibly keep track of them all? "Classify" is the answer.

THE CRAFT AND SCIENCE OF CLASSIFICATION

Classification is a large part of thinking. We say things or ideas have "meaning" when we can place them into a category. Everything is in some category or other, and every category is part of some larger category. That is classification.

A propensity to classify is built into all sentient creatures. Even a tree can distinguish the category of pollen that can fertilize its flowers from the (much larger) category of pollen that can't. Many laboratory experi-ments have shown that animals, even the kind that don't seem particularly bright, can do much more than that. Pigeons can distinguish paintings by Monet from paintings by Picasso, not just by recognizing individual paint-ings but also by their style. This was shown in laboratory experiments, and is the more remarkable because fine art is not part of the pigeon's normal environment, even though the occasional one from London's Trafalgar Square may make its way into the National Gallery.

Human beings have made many different stabs at classifying the crea-tures around us. We need to do this. We rely entirely on other creatures for our own survival, and we need some way of keeping track—of distin-guishing dangerous from not dangerous, edible from toxic, potential mates from potential trouble. Some systems of classification, in some folklore, have been frankly whimsical—or so at least they seem to outsiders, who don't understand the underlying beliefs and thought processes. Many have been entirely pragmatic; thus chefs and fishmongers speak of "shellfish"—anything crusty on the outside, and soft and maritime within—and so they lump crabs, which are crustaceans, with oysters, which are mollusks.

The fishmonger approach, and folk taxonomies in general, work perfectly well for ad hoc purposes within their own societies. But many have felt, these past few thousand years, that nature itself is not ad hoc—not just a random collection of creatures; that there is some reason they cluster into various fairly obvious groups, like birds and mushrooms and monkeys. Many have felt there is, in fact, a natural order and any taxonomy should reflect that order. So the search has been on over the centuries for a natural classification—one that does not simply help us to keep track but also reflects how nature really is.

The first person to seek such a natural classification formally, or at least the first that we know much about, was Aristotle. He didn't get very far, but he did pin down some of the main problems, which is a vital beginning. Without any preconceptions about who really belongs with whom, he tried out various criteria by which to group similar creatures to see how far he got. What about the number of legs? Why not? They are conspicuous and easy to count, which is always a good start. But humans have two legs and so do birds, whereas cats and dogs have four. Intuitively, one feels that people are more like cats than they are like chickens. If we want to be more formal about it, we can point out that chickens have feathers and lay eggs, like eagles and sparrows, whereas humans and cats have hair and bear live young, which for the first few weeks or months of life are suckled. Somehow—it isn't easy to say why, but somehow—hair and suckling seem more important than number of legs. So, at least in this context, number of legs is not a satisfactory criterion—although it does get us a long way if we want to distinguish spiders from insects. More generally, we see that creature A may have some things in common with creature B, and other things in common with creature C—so then we need ways of deciding whether and why the similarities with B are more or less important than the similarities with C. Intuition takes us a long way if we are dealing with familiar creatures like humans and cats, but not very far if we are dealing with creatures out of the plankton or even, as we will often see, with birds.

Overall, Aristotle's approach was commonsensical, and up to a point he was successful. But his approach in the end was mainly pragmatic. He devised sensible ways of putting living creatures into categories, but he did not provide any convincing basis for those categories. He did not show why his groupings should be considered "natural."

Aristotle is still a huge presence in Western thinking, but his views on the natural world did not survive the rise of recognizably modern science in the seventeenth century. Seventeenth-century natural philosophers, in the form of physicists, conceived the idea of scientific "laws": the deep principles on which the universe is run. Biologists caught the mood, and although the concept of a law is harder to apply to living creatures than it is to the stars, they began to seek an underlying pattern. Several seventeenth-century naturalists tried to classify living creatures based on what they perceived to be the natural order of things. In particular, John Ray—a great naturalist who was much admired by Gilbert White a century later—attempted a comprehensive taxonomy of all living things. Of course, no taxonomy can be considered natural unless it does indeed conform to the innate order of nature—and why should there be such order? Why, for that matter, should the stars, and light, and cannonballs falling from towers conform to "laws"? But to seventeenth-century scientists the answer was obvious. To a man—Galileo, Newton, Descartes, Leibniz, Kepler, Robert Boyle, and of course John Ray—they were deeply devout. The laws of nature were the laws of God. Nature is orderly because God has a tidy mind. All science was embedded in a theological framework.

This was how things stood, more or less, until the mid-eighteenth century. On to the scene, then, came Carolus Linnaeus. He, too, was devout—he took it to be self-evident, at least through most of his life, that the order he sought in nature was the order of God's own mind—but nonetheless he transformed the practice of taxonomy. In fact, he made two separate, huge contributions. First, he showed the world how to name species economically and consistently—devising a method that is still the standard. Second, he spelled out, formally, a system of classification that is absolutely to the point: a nesting series of categories, small categories in larger categories that are contained in still larger and all-embracing categories. His system is still with us, in essence, although now somewhat modified. We should look at each of his two contributions in turn.

LINNAEUS AND THE NAMING OF NAMES

Linnaeus wrote about all the creatures that were known in the eighteenth century, but he was primarily a botanist, and an exploratory botanist at that, who did much to increase the world's knowledge of plants, especially in the wild and woolly northern latitudes where few others cared to venture. In those days (indeed, even now), communication was a huge problem. There was no photography. Botanists and apothecaries knew how to preserve plants in dried form, but until well into the nineteenth century, when the chemists got seriously on the case, museum curators invariably lost most of their collection to insects, mites, and fungi within a few years. From the early Middle Ages onwards, there are herbals with beautiful drawings and paintings of plants and other creatures, but none that I know would find their way into a modern, refereed journal. They did not really serve for detailed identification.

Instead, the early botanists and herbalists, including Linnaeus in his early days, wrote long descriptions of each plant, with details of hairiness and glossiness and where they grew and whether or not they had bulbs and were or were not edible, and so on and so on; and at least until the nineteenth century, internationalists that they were, all naturalists wrote in Latin, the *lingua franca* of science, though not necessarily as spoken by the Romans.

The early herbalists were astute and truly expert. They knew, for example, that the huge diversity of plants, and indeed of all living things, is all variation on a far smaller number of themes. Thus, many flowers were buttercup-like, and buttercups in general were called *Ranunculus*. But there were many kinds of buttercups that lived in different ways in different places and with discernibly different physical features. So, for example, there was one that crept along the ground and their long descriptions of it would include the word *Ranunculus* but also the word *repens*, "creeping." Descriptions of an aquatic variant, the water crowfoot, included the term *aquatalis*. Another had a bulb and was referred to as *bulbosus*. And so on.

Linnaeus could hardly be called a lazy man—he did set out to classify all living creatures, plus minerals; he was tireless at self-promotion; and he led vigorous botanical expeditions for local amateurs complete with brass bands. But he did not like scribbling for scribbling's sake. He grew tired of the long Latin descriptions. But then he noticed (and surely he was not the first to notice, but he seemed to be the first to draw formal attention to the fact) that long descriptions were unnecessary. At least, a long description was needed when a plant was first described, but after that, all that was needed for reference was the general name—the *generic* name—plus one key word from the description. Thus, the creeping buttercup could simply be called *Ranunculus repens*; the water crowfoot, *Ranunculus aquatalis*; the bulbous buttercup, *Ranunculus bulbosus*—and so on through all the scores of different buttercups throughout the world. Two words each would do the trick.

Indeed, two words were all that was needed to refer to each and every kind of creature in the world. The nomenclature was minimalist, yet there need be no ambiguity. In the same way, there are 6.5 billion people in the present world and countless institutions, but all of them could in principle be reached with a one-line e-mail address. It is remarkable, but it is the case.

Linnaeus's two-word summaries form the "binomial" system of naming. Sometimes a third name is added these days to denote a particular subspecies, but all living creatures, including all the smallest microbes, are given Linnean binomial names: *Ranunculus repens*, the creeping buttercup; *Homo sapiens*, the human being; *Larus argentatus*, the Herring Gull; *Phylloscopus trochilus*, the Willow Warbler; *Phylloscopus collybita*, the chiffchaff. There need be no mistake, ever. And because the names continue to be basically in Latin (though much modified and abused for all kinds of ad hoc purposes), they are universal. Britons, North and South Americans, Germans, Russians, Chinese, Malaysians—all, in this unique and circumscribed context, speak the same language. Indeed, the binomial nomenclature of Linnaeus has provided humanity with the only universal language it has ever possessed, since the mythical Tower of Babel.

Some Linnean conventions are worth nothing before we move on. The first of the two names is the "generic" name, the name of the genus, as in *Larus*—gull, *Phylloscopus*—warbler, *Falco*—falcon, and so on. The second name is the specific, or species name.

The complete scientific, or Latin, name is always written in italics. The generic name always begins with an uppercase letter, and the specific name always begins with a lowercase letter, even when the specific name is based on a proper name—as in *scotica,* meaning Scottish. Newspapers seem to delight in getting this wrong.

If you are referring to lots of gulls in one paragraph or page, then after spelling out the generic name once, it is legitimate from then on to reduce it to one letter, so you can say "Sometimes among mixed flocks of the Herring Gull, *Larus argentatus*, and the Lesser Black-backed Gull, *L. fuscus*, you may also see Common Gulls, *L. canus*, or even the odd Little Gull, *L. minutus*—so always look carefully." But you have to avoid ambiguity. It would not be intelligent to say, "The Herring Gull, *Larus argentatus*, is often seen with the Bar-tailed Godwit, *L. lapponica*," because the generic name of the Bar-tailed Godwit is not *Larus*, but *Limosa*.

Having observed that the various buttercups are variations on a theme of *Ranunculus* and that the gulls (or at least, the biggest of the ten genera of gulls) might all be classed together in the genus *Larus*, Linnaeus went on to suggest that genera could then be grouped into bigger groups and so on upward to create a complete, eminently satisfying hierarchy.

THE HIERARCHICAL CLASSIFICATION OF LINNAEUS

With the concept of genus formally in place, Linnaeus then suggested that genera could be grouped into orders, orders could be grouped together to form classes, and classes could be combined within kingdoms. Thus, his hierarchy contained five ranks that, starting from the bottom, were: species, genus, order, class, kingdom. As far as he was concerned, there were and are only two kingdoms—animals and plants (in which he included fungi, though he really should have known better). Within this simple, five-ranked hierarchy all living creatures could be enfolded and readily accounted for. Any new creature that heaved into sight could simply be plugged into the overall structure.

Linnaeus's basic framework remains, although since his time, as the world's species lists and general biological knowledge have grown, taxonomists have found it necessary to add a few more ranks. Between genus and order they have inserted the rank of family. Between order and class, zoological taxonomists have added the rank of phylum (while botanists have invoked the equivalent "division"). Linnaeus's two kingdoms have now been expanded into at least half a dozen. Fungi have their own kingdom, quite separate from the plants (and, in truth, fungi are much closer to animals). The red seaweeds form a fourth kingdom, the brown seaweeds a fifth, and the various small creatures that used to be classed as protozoa now form a series of kingdoms of their own, although exactly how many seems to depend on the taste of the particular taxonomist. (Previously they were shoved in with the animals, or more generally were called "protista" and often divided somewhat arbitrarily between the animals and plants.) Now, too, there is a ranking even higher than kingdom—the domain. All the kingdoms mentioned so far (animals, plants, fungi, and the various seaweeds and protists) have similar cells. All cocoon the bulk of their DNA in a discrete nucleus, and accordingly are grouped together to form the domain of the Eucarya (which means "good kernel" or "good nucleus"). But there are also two domains of prokaryotes—creatures that do not parcel their DNA into nuclei: the Bacteria and the Archaea. Until the 1970s, no one realized that the Archaea and the Bacteria should be classed entirely separately, yet at the molecular level they are far more profoundly different from each other than, say, animals are from plants. And so we have, in order of increasing specificity: domain, kingdom, class, phylum/division, order, family, genus, species.

So, taking the complete classification, we can say that the Herring Gull is of the species *Larus argentatus*, in the genus *Larus*, in the family Laridae, in the order Charadriiformes, in the class Aves, in the phylum Chordata, in the kingdom Animalia, in the domain Eucaryota. Since this system contains eight ranks while Linnaeus's original version had only five, I feel we should say it is neo-Linnean rather than Linnean. But nobody else that I know of uses the term neo-Linnean, so I cannot claim that it is the convention.

Modern taxonomists have also found it convenient to introduce intermediate rankings—an indefinite number of them—between the eight formal rankings. We have already met "subspecies," which essentially means "race." Sometimes big families are divided into subfamilies, which are sometimes called "tribes" (although sometimes tribes are nested within subfamilies). Orders may be divided into "suborders" (which stand between order and family) or joined together into small groups called "superorders" (which stand between order and class). Some taxonomists are even more persnickety, and invent even more intermediate rankings. But while the eight basic rankings are obviously useful and indeed necessary, too many names applied to too many intermediate groupings can drive you mad. That is counterproductive.

Just a few more conventions are worth noting. First, the names of animal families always end in *-idae*, as in Laridae (gulls), Parulidae (American wood-warblers), Paridae (tits and chickadees), and so on. Informally, zoologists commonly spell the families in lower case, and drop the *-ae*. So members of the Laridae can be referred to informally as "larids" while members of the Paridae are "parids." This isn't just insider talk. It is sometimes useful.

The names of bird orders—the ordinal names—always end in *-iformes*. All three syllables should be pronounced: eye-FORM-eeze. Again, too, zoologists are content to express the names of orders informally, and to drop superfluous syllables. So Charadriiformes can be written "charadriiforms" and pronounced kar-ad-RI-forms. Passeriformes is rendered informally either as "passeriforms" or, easier still, as "passerines." Similarly, groups such as the Neornithines may be presented formally, with a capital N, or informally, "neornithines."

One final irritation: in some accounts you find the common names of birds spelled in lowercase, as in "green woodpecker." In others you find the first name begins in uppercase and the second name in lowercase, as in "Green woodpecker." In yet others you find both names begin in uppercase, "Green Woodpecker." Intuitively I prefer the first option—green woodpecker. But problems arise when you want to distinguish between the specific species of Green Woodpecker and some other woodpecker that just happens to be green; or between a European Blackbird and a "black bird"; and so on. If you use upper case for both names, all such problems are avoided.

Here, in this brief account of Linnaeus, we see one of the prime functions of classification: it helps us to keep track. So it is that nearly 10,500 species of birds are known, still rising, and 10,000 is far too many for all but the most dedicated or obsessive to keep track of. But those 10,000 are grouped within about 170 families; and it is surprising how many you can become familiar with just by osmosis—noting the family names in passing in the guidebook. In turn, those 170 or so are grouped within thirty-two orders (including the orders of the moas and the elephantbirds, which are extinct but only recently)—and anyone can remember thirty-two. Thus, with a little orderly classification, any of us can come to grips with, or at least get a handle on, a plethora of creatures that otherwise would be quite beyond our compass.

The Linnaean system (in its neo-Linnean form) still prevails, at least as a practical device. Considered purely as a system of classification, its logical structure is hard to improve upon. But the conceptual basis of classification has shifted absolutely since Linnaeus's day. For in 1859, Charles

Darwin published *On the Origin of Species by Means of Natural Selection*; after Darwin, most taxonomists agreed that nature seems orderly simply because all creatures evolved from common ancestors. Ever since Darwin, the basis of classification was not God's presumed tidy-mindedness, but the historical fact of evolution.

CLASSIFICATION RECONCEIVED: DARWIN AND THE PHYLOGENETIC TREE

Darwin did not suggest simply that birds had descended from more primitive birds, or that land animals had descended from fish. Far more than that, he suggested that all the creatures that have ever lived on earth were descended from a single ancestor that must have lived—well, a very long time ago, although no one in the mid-nineteenth century knew quite how long ago that was. If all creatures are descended from the same common ancestor, then all are literally related, one to another. Human beings are related to apes—many cultures have noticed the similarity. But apes in turn are related to all other mammals. And if you go back far enough, then you find that mammals are related to birds, and to reptiles and fish; and all those vertebrates are related more distantly to insects and spiders and worms; and all those animals are related to mushrooms and molds and to seaweeds and plants; and more distantly still they are all related to bacteria and archaeans.

Suddenly, after Darwin, the craft and science of taxonomy became less arbitrary. It is presumptuous, after all, to try to read the mind of God—which, in effect, pre-Darwinians had been trying to do. But if everything is related to everything else, then the classification seems to define itself. Different species in the same genus are like brothers and sisters. Different genera in the same family are like cousins. Human beings and Herring Gulls are, say, fifth cousins. Animals and buttercups are twenty-seventh cousins. There is nothing arbitrary about this. Classification is based on relationships that in turn reflect evolutionary history. History is reality—not necessarily traceable, but real nonetheless.

Suddenly, too, taxonomists knew what they were trying to do. No longer was it enough simply to decide who looked more similar to whom. The task after Darwin was to discover who was closely related to whom, and who was more distantly related. Thus, taxonomy after Darwin became a matter of genealogy—although genealogy on this vast scale is called "phylogeny," from the Greek *phylos*, meaning "race." The family tree that lies at the basis of modern taxonomy is better known as a "phylogenetic" tree. The branches of the tree correspond to the Linnean rankings. The mighty boughs that spring from the main trunk are the domains. The smaller but still mighty branches that spring from them are the kingdoms—and so on all the way down to the branchlets, which are the genera, and the twigs, which are the individual species.

How do we work out the literal relationships between different creatures that would enable us to build the phylogenetic tree?

CONVERGENCE, DIVERGENCE, HOMOLOGY, AND CLADISTICS

Post-Darwinian taxonomists who are trying to work out who is literally related to whom in general must operate in much the same way as taxonomists always have: they must look for similarities, primarily in appearance. Here the post- and the pre-Darwinians face the same problems. To begin with, the problem that Aristotle encountered: that creature A may have some features in common with creature B, and others in common with C—so which features are the more significant? Which (the post-Darwinian would ask) indicate a true relationship and which do not?

Two further phenomena compound the problem.

The first is *convergence*: often two or more creatures—birds, plants, or whatever—of quite different lineages adapt to the same general way of life in the same kinds of ways and finish up looking much the same. The other is *divergence*: two or more creatures who are closely related, and did descend from a common ancestor who was exclusive to them, may adapt to very different environments and finish up looking very different from each other. Often we see them both at the same time. Thus, some of the auks from the Northern Hemisphere—especially the murres and the big, extinct flightless Great Auk of North Atlantic islands—look very much like penguins. But while penguins are at least loosely related to loons and albatrosses, auks are closer to gulls. So penguins and albatrosses, and auks and gulls, present us with classic cases of divergence. But penguins and auks are also a classic case of convergence.

In the light of all this, how can we work out who is really related to whom, and who merely looks as if they are related? Here we encounter the vital concept of *homology*. The idea of homology dates from the eighteenth century and the word was coined by the great French biologist Jean-Baptiste Lamarck (1744–1829). Intuitively we can see that although the wing of a bird differs in form and function from a human arm, they are essentially the same organ: wings and arms are both forelimbs. A little embryology (of the kind that was fashionable in the early nineteenth century) would show bird wings and human arms developing in exactly the same way from "limb buds" at shoulder level. Accordingly, bird wings and human arms are said to be "homologous." By contrast, the wings of flies serve the same purpose as the wings of birds, but we can see even without being experts that they are quite different: extensions of the cuticle of the thorax. Accordingly, birds' wings and flies' wings are said merely to be "analogous." Why should two different creatures share homologous features? Before Darwin, biologists would simply have said that this was part of God's plan. After Darwin, they took it as evidence that they must have shared a common ancestor.

In short, if we want to distinguish true relationships between creatures, so as to work out their phylogeny, we should not look simply for features that they have in common ("shared characters"); we should be looking for shared characters that are also homologous. If we can look at the embryos, then this can be easy enough to do. But there are also various rules of thumb by which to decide whether the features of adult animals are homologous, and these rules are all that we have to go on when looking at fossils. In the end, none of the rules gives sure-fire answers. In deciding homologies, and hence relationships, there is no escaping the need for common sense and experience—expertise, in fact—and when experts rely on their expertise, they inevitably disagree.

So phylogenetic trees are put together on the basis of shared, homologous characters—yet even this, as it stands, is not quite enough. For example, pigeons, macaws, Budgerigars, and pheasants all have wings made from forelimbs, and all have feathers. Feathers and wings are shared, homologous features, and taken together they tell us that all four of those creatures are birds. But what we really want to know is which of those birds is more closely related to which. The mere possession of wings and feathers does not tell us this, because they all have wings and feathers.

Therefore, we need a way of distinguishing between shared, homologus features that tell us only that pigeons, macaws, Budgerigars (Budgies), and pheasants are all birds, and are not related to flies; and shared, homologous features that will tell us which of those four birds is more closely related to which.

This problem was obvious at least by the early twentieth century, but the taxonomist who really pinned it down was a German entomologist called Willi Hennig (1913–76). As far as pigeons, macaws, Budgerigars, and pheasants are concerned, wings and feathers are merely "primitive" characters. They tell us that all four are broadly related, but do not help us to sort out the more detailed relationships. For fine-tuning, we need to find homologous characters that some of them have and some do not— homologous characters that are *not* shared by all of them. These special features Hennig called "derived" characters. In the case of this little quartet, the emphatically hooked bill of the macaw and the Budgie, with their nostrils placed high on the nose in a discrete little mound known as the "cere," plus a few other features, puts them clearly together within the order of the parrots (known as the Psittaciformes). The hooked bill is the "derived" character that shows that macaws and Budgies are more closely related to each other than either is to pigeons or pheasants.

So it is that modern taxonomy has several origins: first, the general attempt to find tidiness in nature, which culminated in Linnaeus; second, the idea that the perceived tidiness was not imposed ad hoc by a tidy-minded God, but reflects ancestry, which we owe to Darwin. The idea of homology (from Lamarck) helps taxonomists to decide in a general way who is related to whom; and Hennig's adjustment—distinguish primitive homologous from derived homologous characters—provides the final refinement.

Hennig called his approach "cladistics." The term derives from the Greek *clados*, meaning "branch." Each branch on the phylogenetic tree that is built by cladistic techniques is called a "clade." There are small clades— the twigs—corresponding to individual species, and bigger branches that correspond to families, orders, and so on. The rules of cladistics decree that a clade is not a true clade unless it contains the common ancestor of the creatures on the particular phylogenetic branch, plus all of that ancestor's descendants. This also gives us a proper definition of *taxon*. A "taxon" (like a clade) may be small (just one species) or middle-size (family or order) or very big (class or phylum or kingdom or domain), but it cannot qualify as a true taxon unless it is a true clade. To be a true clade, as opposed to an ad hoc grouping, it must contain the common ancestor of the creatures in the group, plus all of that ancestor's descendants.

Just a few more technical terms are worth taking on board. A true clade—a single ancestor and all its descendants—is said to be "monophyletic," Greek for "of one race." As we will shortly see, many of the bird orders and families that are now commonly recognized probably are not monophyletic. Many of them contain birds that are not closely related because they don't share their most recent ancestor, and then they are said to be "polyphyletic." Some others are incomplete—some of their members have been placed in other groups where they do not belong, or have given rise to other groups that ought to be included in the parent group. Such incomplete groups are said to be "paraphyletic." We will come back to this.

Ancestors that are successful may give rise to many thousands of different descendants—just as *Archaeopteryx* has given rise to at least 10,500 species of modern birds. But we know that the original population of *Archaeopteryx* did not suddenly divide into 10,000 daughter species all at once. Indeed, said Hennig, it is probable that ancestral species generally give rise to daughter species simply by splitting into two. Then each of the two daughter species splits into two and so on and on until we have many thousands (with plenty of dead ends—extinctions—along the way).

If we really understood the phylogeny of a lineage, then this is what we would show: a long series of bifurcations. But usually we do not know the phylogeny in such detail. We are forced simply to show the various daughter species that we know about springing from the ancestor all at once, like the prongs of a fork. When we do think we know enough to show all the bifurcations, we say that the tree is "resolved." When we can't, and are obliged to show a toasting fork instead, we say it is "unresolved." The phylogenetic trees of birds, shown in the next chapter, are largely unresolved. Taxonomists have made enormous strides with bird classification, but there is still a long way to go.

When an ancestor splits to give rise to two daughter species, the two are called "sister species." But the expression sister species—or, more generally, "sister group"—is also used in a slightly subtler way. Thus, although taxonomists may have good reason to believe that some ancient beast was indeed the ancestor of later types, they can never be sure. So instead of showing *Archaeopteryx* as the direct ancestor of all subsequent birds,

Hesperornis and *Iberomesornis* and eagles and chickens and ducks, they show *Archaeopteryx* as the sister group of all subsequent birds. The implication is meant to be that *Archaeopteryx* may have been the direct ancestor of all subsequent birds, but even if it wasn't, then it was very like the creature that was the direct ancestor.

Cladistics, and all the terms and technicalities that come with it, now run through all truly modern taxonomy. Hennig probably would not be remembered except by professional entomologists were it not for cladistics. As it is, he is one of the seminal figures in the all-pervasive science of taxonomy.

There has been one other huge advance since Darwin. Modern taxonomists don't simply rely on studies of anatomy and embryology to tell them who is related to whom. In the early twentieth century, taxonomists began to judge the relationships by immunological means—by seeing how violently different animals reacted to each others' antigens. Then they began to compare the proteins of different creatures more directly. Proteins are made by genes; and since the 1960s and especially since the 1980s, taxonomists have found better and better means directly to compare the DNA from different creatures, the stuff of which genes are made. DNA studies do not, as at first was hoped, offer a royal road to unequivocal truth. But they are proving revelatory nonetheless.

DNA

At the end of the 1950s, at Harvard, Dr. Charles Sibley was comparing proteins from various birds. But in the early 1950s—the announcement was made in 1953—James Watson and Francis Crick at Cambridge University described the three-dimensional structure of DNA. They found that each DNA molecule consists of two strands wound around each other, and this simple fact led to the first of a whole new series of techniques by which to explore the relationships between different creatures.

For soon it was discovered that if DNA was heated almost to boiling point, the two strands would come apart; and in 1959, Julius Marmur and Paul Doty, who like Sibley were at Harvard, found that if the DNA was cooled slowly after the strands had separated, then they recombined. Then, quite soon, it became clear that if single strands from different kinds of creature were mixed, the two different kinds would recombine one with another, to form DNA—DNA hybrid molecules. The more closely the two creatures were related, the more eagerly their DNA would recombine. These recombinations were known as "DNA hybrids." The first such hybrids were announced in 1962, formed using DNA from various viruses and bacteria. By the mid-1960s, after much refinement, DNA—DNA hybrids were made between "higher organisms." (But let me emphasize that it was only the DNA that was being hybridized—isolated molecules, *not* the creatures that provided those DNA molecules.)

This suggested a whole new approach to taxonomy. The degree of relationship between two different creatures could be gauged by mixing single strands of their DNA and seeing how closely they recombined. If the two creatures were closely related, they should recombine firmly. If they were distant, then their DNA strands would not fit each other well. The approach is roughly analogous to the immunologists' technique—seeing how eagerly particular molecules from different species reacted. But it is far easier to quantify the degree of recombination between different strands of DNA than it is to quantify the reaction between antigens and antibodies.

Charles Sibley moved to Yale in 1965, taking his bird-protein work with him. A year later, Jon Ahlquist joined him as a graduate student. In October 1973, Sibley learned about DNA—DNA hybridization—and this, he realized, should give far more insight into the relationships between animals than protein analysis could. So in January 1974, he and Ahlquist started making DNA—DNA hybrids from pairs of birds and they continued the work without interruption until July 1986. By then they had produced 30,054 DNA—DNA hybrids—26,064 from birds and 3,150 from mammals. In 1990, Sibley and Ahlquist published their great work, *Phylogeny and Classification of Birds: A Study of Molecular Evolution.* In it they describe the relationships between 1,700 different bird species from which they constructed a complete phylogenetic tree. Sibley and Ahlquist showed extraordinary relationships between birds that were never hitherto suspected and proposed that others that were thought to be similar were in fact far apart.

Many of their initial findings have now been questioned, but many of the later studies would not have been done at all if Sibley and Ahlquist had not prepared the path. Truly this is a classic book, from a classic study.

Of course, DNA—DNA hybridization raises problems of its own. DNA always changes over time as new mutations come on board and genetic drift and natural selection take their toll, which is why DNA differs from creature to creature. But DNA may change at different rates in different creatures. If the DNA of one or the other of two creatures that are being compared has changed especially rapidly, then they will appear to be very different even though they may have shared a common ancestor only recently. So it became necessary not only to look simply at the affinity, or lack of it, between different pairs of species but also to explore the fine structure of the DNA itself. Techniques for doing this rapidly came on line in the 1980s, and so began the age of direct DNA analysis—an age that is now in full swing.

But still there are snags. One of them, at least in principle, is the same as Aristotle noted: that some sections of DNA from creature A may resemble those of creature B, while other sections are more like creature C. Taxonomists can get round this to some extent by statistical means—but statistics offer only probabilities, not the cast-iron certainties that some were hoping for. Or then again, just half a dozen species taken at random could, in theory, be related one to another in any one of many millions of different ways. Again, statistics help to show which of the theoretically possible relationships is or are the most likely—but, again, the statistics deal only in probabilities. Besides, different statistical approaches, applied by different computer programs, tend to give different results.

So, as we will see in the next chapter, there are still many problems with bird taxonomy despite the continuing, intensive explorations of DNA. Even so, we should not be gloomy. Ornithologists have been formally classifying birds since the seventeenth century, and DNA studies are still new, but already they have sorted out a great many problems. And sometimes they have come up with solutions that are highly counterintuitive—the kind of thing no one could possibly have thought of otherwise.

One spectacular study will make the point and close this chapter on a pleasing note. It concerns one of the biggest birds of prey that ever flew—and perhaps the biggest of all the raptors that actually killed its own prey. This was Haast's Eagle, of New Zealand, *Harpagornis moorei*.

Haast's Eagle was first discovered in a former swamp, in 1871, by Julius Haast (1822–87), who founded New Zealand's Canterbury Museum. Seventy Haast's Eagle skeletons have now been found, and the biggest have a wingspan of more than 3 meters (well over 9 feet). The living birds must have weighed 10 to 14 kilograms (22 to 30 pounds). By comparison, Golden Eagle females (the larger sex, as is usual with raptors) rarely reach 2.4 meters (8 feet) from wingtip to wingtip, or weigh more than 7 kilograms (15 pounds).

What could such a mighty bird of prey have eaten, in a land that lacked rabbits and hares and the other kinds of creatures that eagles tend to rely upon? The answer, it seems, is that they preyed upon moas—flightless birds that included the tallest birds that have ever lived, taller (and heavier) even than a modern Ostrich. Many moa skeletons have been found and many a moa pelvis is punctured with holes that correspond precisely, in size and spacing, with the talons of Haast's Eagle. Evidently the eagle threaded its way through the forest like a hawk, hit the moa at great speed from the side (the force of impact is of prime importance in making a kill), seized the back of the moa in one great foot, and struck at its head with the other. Haast's Eagle disappeared in the period that Europeans call the Late Middle Ages, which is when the moas themselves were also wiped out.

But who exactly was Haast's Eagle? Who were its relatives? Where did it come from? How and when did it get to New Zealand? Many of New Zealand's birds have relatives in Australia, so that is where late nineteenth-century and early twentieth-century ornithologists looked for the relatives of Haast's. There was a ready candidate: the mighty Wedge-tailed Eagle, *Aquila audax*. Comparison of their bones suggested that the two great eagles were sister species. Case closed.

But then, a few years ago, Michael Bunce from Canada and Richard Holdaway in New Zealand looked at the DNA in Haast's Eagle bones from the Museum of New Zealand, Te Papa Tongawera. The bones were about 3,000 years old, but New Zealand has a cool climate and the DNA was still in good condition. The DNA bore little resemblance to that of the Wedge-tailed Eagle, but it was remarkably similar to that of the Little Eagle, *Hieraaetus morphoides*, from Australia and New Guinea, and also to the Eurasian Booted Eagle, *Hieraaetus pennatus*. For aficionados, the difference between the mitochondrial cytochrome-b gene from Haast's Eagle and the two small *Hieraaetus* eagles was a mere 1.25 percent.

Such a small difference suggests that the lineage of Haast's Eagle and that of the *Hieraaetus* eagles must have divided just 1 million years ago. But the *Hieraaetus* eagles weigh only about 1 kilogram (2.2 pounds). This suggests that the ancestor of Haast's Eagle was small, that it arrived in New Zealand only about a million years ago, and yet in that brief time—brief by biological standards—it increased in size by ten to fifteen times.

As Bunce and Holdaway comment, animals—even big, specialist, vertebrate animals—can undergo spectacular evolutionary change in double-quick time if the circumstances are right.[1]

Bunce and Holdaway also looked at DNA from sixteen other living eagles and concluded that eagle classification is in a fine old mess, especially that of the "booted" eagles—the ones with feathered tarsi, to which Haast's Eagle seems to belong. In the next chapter, where I discuss birds of prey, I am sticking to the present-day conventional classification, but it comes with significant caveats. To start with, it seems clear that Haast's Eagle needs a new Latin name: not *Harpagornis moorei* but *Hieraaetus moorei*. The world's mightiest eagle belongs among the little ones.

So DNA studies are raising a great many puzzles, just as they are solving others. Many traditional groupings of birds—genera, families, orders—are proving to be far less clean and tidy than taxonomists would like: some polyphyletic, some paraphyletic, and some, frankly, a mess. Again, we find that birds are raising particular problems. Most of the thirty or so orders of modern birds arose within just a few million years of each other, mostly between 70 and 80 million years ago. When different lineages have radiated more or less all at once, a long time ago, it is hard to say which arose first and which arose later.

Without such knowledge it is impossible to build a neatly resolved phylogenetic tree, with a neat series of bifurcations, in the way that Willi Hennig envisaged. Even so, there has been immense progress in the past twenty years and we can hope for much more in the next few decades. The next chapter looks at progress so far.

*All the main groups of ratites, living and dead. Top row: moa, cassowary, elephantbird,
Middle: Ostrich. Bottom: kiwi, rhea, Emu. Bottom right: the closely related,
but non-ratite, tinamou.*

4

ALL THE BIRDS IN THE WORLD:
AN ANNOTATED CAST LIST

IN ONE WAY IT'S EASY TO CLASSIFY BIRDS. AT LEAST WE CAN
say what a bird is. At least among modern creatures, we can say with
absolute certainly that anything with feathers is a bird and anything
that doesn't have feathers is not (assuming it's old enough to grow them).
All the creatures with feathers also have beaks and two legs that end in
scaly toes with claws, and although these features are not unique to birds,
they certainly help to clinch the diagnosis. No one, in short, can mistake a
bird. This is more of a bonus than you might suppose. With many animals, particularly some of those that float or crawl in the sea, it can be very
hard indeed to say, even roughly, what group they belong to.

But there the simplicity ends. We can easily see which living creatures
are birds and which are not, but to say which bird is related to which—to
put them into natural groups, and to arrange those groups in the phylogenetic or genealogical tree that is every modern taxonomist's dream—is
very hard indeed.

To begin with, the very qualities that make birds as a whole so easy to
distinguish from other creatures also make them hard to tell apart. Mammals vary from whales to bats to monkeys to naked mole rats to kangaroos
to platypuses, but all birds are much of a muchness. All are descended
from an animal that flew (*Archaeopteryx* being the best candidate), and
flight imposes tremendous restraints on body form. Modern birds are
even more constrained: they all belong to the subclass Neornithines—
only one of several that sprang from *Archaeopteryx*. We don't know what the
first ever neornithine was, or what it looked like, but we can be sure that

it, too, was a flier. Modern birds fly in many different ways, and although some have given up flying altogether, they still carry the legacy of that demanding and anatomically restrictive ancestry.

There are a great many species of birds—roughly 10,500—but there are only a limited number of ways to live on this earth: only a limited number of ecological niches. So we find that many different birds, from many different groups, may all finish up doing much the same sorts of things, which in turn means they finish up with similar anatomical adaptations. In other words, among birds, the phenomenon of convergence looms very large—and enormously complicates the attempts to classify. Contrariwise, birds from any one group may essay many different ways of life, and this divergence can be just as confusing.

Owls, raptors, and storks illustrate the point. At first sight you might suppose that owls and raptors—hawks, eagles, falcons, vultures, and so on—belong together. Both are predators with hooked beaks and they catch their prey in their talons. Sometimes indeed they have been lumped together as "birds of prey." Look inside, however, at the bones in the neck and the differences become clear; the two have nothing to do with each other. For many decades the raptors have been placed in one order (the Falconiformes) and the owls have had an order all to themselves (Strigiformes). Their resemblance is yet another case of convergence. They are both predators, and hooked beaks and talons are good for killing, seizing, and tearing flesh.

You might suppose, though, that the raptors at least form a coherent group. But, in truth, the New World (American) vultures and condors are very different from the Old World (Eurasian, African, and Australasian) vultures and condors. The Old World vultures are closely related to eagles and hawks, but the New World types—as was first pointed out in the nineteenth century—in some ways are more like storks. For instance, storks and New World vultures both lack a septum behind the nostril, so you can see in one nostril and out the other (as can be checked in your nearest aviary). Both excrete over their own legs—which is said to be a cooling device, although I am not sure what the evidence is for this. Their relationship, you may feel, could be sorted out instantly by looking at the birds' DNA. But people have tried, and so far it doesn't. Some studies suggest that New World vultures are indeed closer to storks while others—as shown on the phylogenetic tree in Figure 4—leave them grouped with the other raptors. The clue to the confusion is also to be seen in Figure 4.1: the raptors as a whole seem to share a deep ancestry with the storks. More DNA studies may sort this out once and for all—or they may not.

At least, though, you might think, that if we take out the New World vultures, the rest of the raptors form a proper group—a bona fide clade, a "true" taxon. But, actually, not so. Amateurs may sometimes confuse sparrowhawks (which are hawks) with kestrels (which are falcons), but in fact, it seems, the hawks and falcons are very different. For one thing, hawks tend to have brow ridges—eyes hooded with bone, seen most clearly in eagles, which are close relatives of hawks—while falcons do not. In fact, in some ways falcons seem so different from hawks and eagles and so on that some feel they should be placed in their own order. As you can see in Figure 4, a consensus of modern studies of many kinds suggests that falcons do share their ancestry with hawks, and that they can legitimately be placed together—but Figure 4 is not the last word. There can be no last word.

In fact, in this tale of raptors and storks we seem to see both areas of confusion—convergence and divergence—in full spate. If the New World vultures really are storks, then this tells us that the storks are astonishingly diverse. And if the New World vultures are storks, and the falcons are not related to hawks, then the Falconiformes as a whole are a remarkable example of convergence. We also see that DNA studies do not provide instant solutions. Problems that puzzled traditional taxonomists often confound the modern molecular taxonomists, too.

But as you will see as this chapter unfolds (and as is intimated in Figure 5), the picture gets worse. Modern studies (not just of DNA, though this plays a large part) suggest that the order in which the storks are traditionally placed, the Ciconiiformes, is a fine old mess. In fact, storks may have little to do with the other members of their alleged grouping. By the same token, the pelicans may be quite distinct from the order to which they gave their name, the Pelecaniformes. Possibly the ciconiiforms and the pelecaniforms should be combined into one, big heterogenous order. But the flamingos, which are traditionally bundled with the storks, seem to be have nothing to do with either the storks or the pelicans. Some modern studies suggest that their closest relatives are the grebes. On the other hand (albeit based on much less evidence), flamingos have sometimes been linked with ducks. This remains a serious mystery—which again may be sorted in time but there can be no guarantees.

Equally problematic are the traditional orders of the nightjars and their relatives, within the Caprimulgiformes, and the swifts and their relatives, traditionally placed in the Apodiformes. Again, it seems, the two orders should be either combined or else redivided along new lines.

There are still problems with many of the families, too—particularly, perhaps, those of the Passeriformes (passerines). Some of the conventional families seem to be polyphyletic (containing species from several different lineages) while many others are surely paraphyletic (incomplete). I will not discuss the details, but merely note that in the coming years the family names of passerines could be revised wholesale. I have given the family names of the passerines just as Perrins describes them in his *New Encyclopedia*, but I have grouped them into what might be called suborders or parvorders according to the classification suggested by Joel Cracraft of the American Museum of Natural History and his colleagues in 2004.

Why is it all so confusing? "History" is the general answer. All living creatures represent the outcome of many millions of years of evolutionary history. All have been shoved this way and that in the past by their ancestors' needs to adapt first to one set of conditions and then to another. Sometimes they have been separated from their relatives and perhaps been introduced to brand-new territories with no competitors, where they are free to diverge as wildly as they choose. Sometimes they have been thrown into contact with other and perhaps fiercer creatures who have pushed out the more extravagant types among them and forced the rest to concentrate on doing a few things well. All the histories of all creatures have been further complicated both by continental drift and by wildly changing climate that has sometimes made all the world tropical but in the past 2 million years has produced a succession of at least a dozen ice ages.

In the years just before *Archaeopteryx* came on the scene, around 140 million years ago, all the land in the world was roughly divided between two supercontinents: Laurasia in the north, including the land that now forms North America and Eurasia; and Gondwana in the south, including present-day Antarctica, Africa, Madagascar, Australia, New Zealand, and India. Madagascar was already separated from Africa (and its creatures, especially its plants, really are very strange). By the time *Archaeopteryx* appeared these great land masses were already breaking up—but still the world was nothing like it is today. The mighty South Atlantic that now divides Africa from South America was the merest channel, and South America was still some thousands of miles from North America. Australia was still in the deep south, almost joined to Antarctica. India was still an island, floating north, but still miles from any continent. In the north, North America was still joined to Eurasia, or very nearly.

By the end of the Cretaceous, 65 million years ago, several and perhaps many orders of modern birds had evolved, including ratites, rails, and ducks. They shared the world with long-gone types like *Heperornis*, and pterosaurs, and some of the mightiest dinosaurs, including *T. rex*. North America was still joined at its northern tip to Eurasia, so that the Atlantic did not open into the polar regions. South America was still very much an island. So too was India, which did not make contact with Asia until around 55 to 50 million years ago; and as it did so, it scrunched the Asian landmass and created the Himalayas (which would still be rising, except that erosion stops them from getting much taller). The final step toward geographical modernity was taken only about 3 million years ago, when South America finally established firm contact with North America via what is now the Isthmus of Panama. Continental drift continues, of course, as is made evident in the continuing earthquakes and their resulting tsunamis.

That is another story. What matters here is that any group of birds that first appeared anywhere over the past 140 million years could have been carried spectacularly across the world by the movement of continents alone. Solnhofen, now in temperate Germany, was on the Equator when *Archaeopteryx* lived there. Single populations of birds could have been, and clearly were, divided by, for example, the birth of the Atlantic Ocean. Other populations have been brought together as formerly separate landmasses collided. On the other hand, Madagascar has been an island since well before the birth of *Archaeopteryx*, New Zealand has been isolated for about 80 million years, and Australia finally broke with Antarctica around 35 million years ago. All the continents have been affected by the ice ages of the past 2 million years. Much of North America and Eurasia was under ice, which drove out the birds and divided their populations. The tropics were affected, too, because they became both cooler and drier—so that continuous tropical rainforest was reduced over large areas to scattered woods, and within those isolated or semi-isolated woodland habitats new species could evolve, not just of birds but of all creatures. When the climate warmed again the forests grew back and the newly formed species were thrown together to come to terms as best they could. Many surely went extinct, but the rest formed new kinds of ecosystems—of immense complexity.

Add to all this the fact that birds can cover huge distances quickly because they can fly—even the nonflying kinds had flying ancestors—and even if they were obliged to walk and had been as slow as snails, which they very obviously are not, they could still have shuffled themselves around the world a thousand times over in the time that has been available to them.

Today we find that although many groups of birds remain highly localized for all kinds of reasons, many others—like the ducks and geese, gulls and shorebirds, raptors, passerines, and so on and so on—have spread themselves all around the world. But at the same time, many groups of closely related species have become divided into subgroups for one reason or another, and then each subgroup has evolved along different lines. Sometimes the separated groups become more different from each other; sometimes, perhaps having become different, some at least of them start to evolve along similar lines—and so we see convergence again. In this way, for example, we find wonderful parallels between many New World groups and Old World groups. The astonishingly various American family of the Icteridae shows such effects brilliantly. Icterids are passerines—perching birds. Some resemble the Old World orioles—and indeed Americans call them "orioles." At least some of the icterid "blackbirds" resemble the European Blackbird (which is in the thrush family, Turdidae). The meadowlarks are indeed larklike. The grackles look for all the world like crows, but they are not—they are icterids. And so on. It is all very confusing. It took many decades of serious anatomical study, abetted these days by molecular studies, to sort it all out (and surely the sorting is far from over).

Finally, it is always intriguing to ask where in the world particular groups of birds first arose—and the answers are rarely simple. We can't judge simply by where they live now. All modern ratites now live in the southern continents—and surely the first ones were Gondwanan—but very ancient Ostrich fossils have been reported from central Asia, very much in the Northern Hemisphere. Albatrosses now belong mainly to the south, but again there are ancient types from the North Atlantic. However, since fossilization is a rare event, and new groups of creatures are bound to be rare when they first appear, it is extremely unlikely that the oldest fossil we happen to know about is actually the first ever. So just because we find fossils of some bird somewhere, that doesn't mean that that is where that bird first arose. We get a better clue by looking at modern birds and making commonsense assumptions. Thus, if a particular region contains the most primitive-looking species of the group, and has the most diversity, then it is reasonable to assume that that region was, indeed, the place of origin. For these reasons it is assumed that both the anseriforms—the ducks and geese—and the passerines arose in the Southern Hemisphere. So, too, probably, despite the misplaced Ostriches of Asia, did the ratites.

Nowadays, most taxonomists arrange all the birds in about thirty orders, give or take a few. Thirty is a convenient number. No one can remember 10,500 species, but thirty orders are easy to get your head around. So any of us—in rough terms—can readily get a grip on all of the birds in the world (taking the rest of our life to fill in the details). The following account—in line with the excellent, modern *New Encyclopedia of Birds* edited by Chris Perrins—describes thirty-one orders.

But whether or not these orders are really valid—whether all the birds that are placed within them really share a common ancestor—and what the relationships between those orders is, as we have seen, often uncertain. Figure 4 summarizes both the present state of knowledge and the confusion. The right-hand side of the figure simply lists the modern orders; the left-hand side shows the genealogic (phylogenetic) tree of birds as now put together—from all the world's evidence—by Joel Cracraft and his colleagues (2004). The caption to Figure 4 gives the details. Ideally, the family tree and the standard list of orders should match each other precisely. But as you can see, the standard listing sometimes matches the phylogenetic tree and sometimes does not. Nature is difficult, and birds can be especially difficult.

As you can also see from Figure 4, the thirty-one orders of birds form two main groups. The Palaeognathae include five orders of ratites, plus the tinamous. The remaining orders—all twenty-five of them—are classed as Neognathaes.

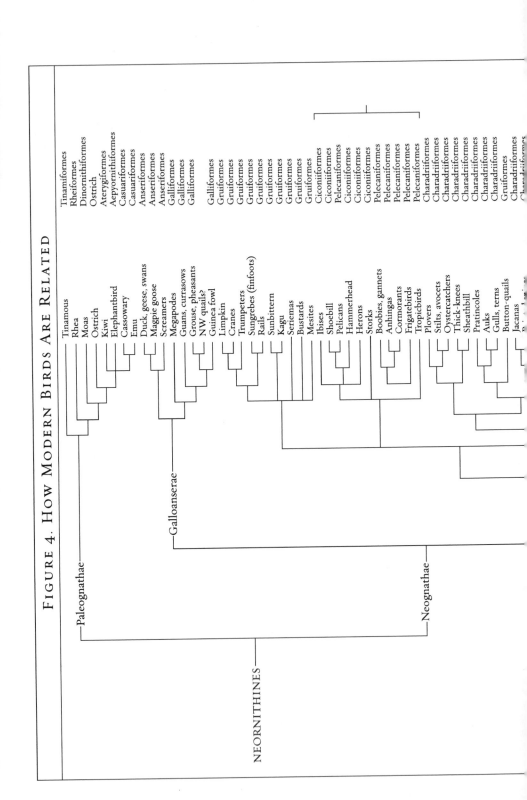

FIGURE 4. HOW MODERN BIRDS ARE RELATED

Tinamous — Tinamiformes
Rhea — Rheiformes
Moas — Dinornithiformes
Ostrich — Ostrich
Kiwi — Aterygiformes
Elephantbird — Aepyornithiformes
Cassowary — Casuariformes
Emu — Casuariformes
Duck, geese, swans — Anseriformes
Magpie goose — Anseriformes
Screamers — Anseriformes
Megapodes — Galliformes
Guans, currasows — Galliformes
Grouse, pheasants — Galliformes
NW quails? — Galliformes
Guinea fowl — Galliformes
Limpkin — Gruiformes
Cranes — Gruiformes
Trumpeters — Gruiformes
Sungrebes (finfoots) — Gruiformes
Rails — Gruiformes
Sunbittern — Gruiformes
Kagu — Gruiformes
Seriemas — Gruiformes
Bustards — Gruiformes
Mesites — Gruiformes
Ibises — Ciconiiformes
Shoebill — Ciconiiformes
Pelicans — Pelecaniformes
Hammerhead — Ciconiiformes
Herons — Ciconiiformes
Storks — Ciconiiformes
Boobies, gannets — Pelecaniformes
Anhingas — Pelecaniformes
Cormorants — Pelecaniformes
Frigatebirds — Pelecaniformes
Tropicbirds — Pelecaniformes
Plovers — Charadriiformes
Stilts, avocets — Charadriiformes
Oystercatchers — Charadriiformes
Thick-knees — Charadriiformes
Sheathbill — Charadriiformes
Pratincoles — Charadriiformes
Auks — Charadriiformes
Gulls, terns — Charadriiformes
Button-quails — Gruiformes
Jacanas — Charadriiformes
Painted — Charadriiformes

Paleognathae

Galloanserae

Neognathae

NEORNITHINES

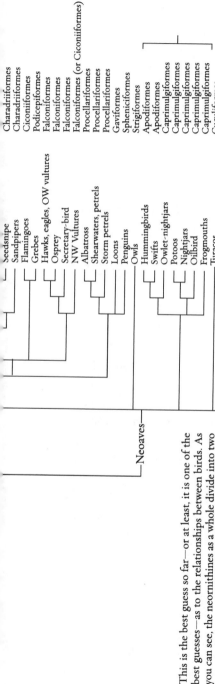
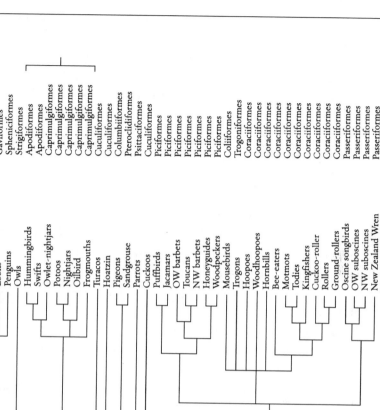

Charadriiformes
Charadriiformes
Ciconiiformes
Podicepiformes
Falconiformes
Falconiformes
Falconiformes
Falconiformes (or Ciconiiformes)
Procellariiformes
Procellariiformes
Procellariiformes
Gaviiformes
Sphenciiformes
Strigiformes
Apodiformes
Apodiformes
Caprimulgiformes
Caprimulgiformes
Caprimulgiformes
Caprimulgiformes
Cuculiformes
Cuculiformes
Columbiiformes
Pterocidiformes
Psittaciformes
Cuculiformes
Piciformes
Piciformes
Piciformes
Piciformes
Piciformes
Piciformes
Coliiformes
Trogoniformes
Coraciiformes
Coraciiformes
Coraciiformes
Coraciiformes
Coraciiformes
Coraciiformes
Coraciiformes
Coraciiformes
Coraciiformes
Coraciiformes
Passeriformes
Passeriformes
Passeriformes

Seedsnipe
Sandpipers
Flamingoes
Grebes
Hawks, eagles, OW vultures
Osprey
Secretary-bird
NW Vultures
Albatross
Shearwaters, petrels
Storm petrels
Loons
Penguins
Owls
Hummingbirds
Swifts
Owlet-nightjars
Potoos
Nightjars
Oilbird
Frogmouths
Turacos
Hoatzin
Pigeons
Sandgrouse
Parrots
Cuckoos
Puffbirds
Jacamars
OW barbets
Toucans
NW barbets
Honeyguides
Woodpeckers
Mousebirds
Trogons
Hoopoes
Woodhoopoes
Hornbills
Bee-eaters
Motmots
Todies
Kingfishers
Cuckoo-roller
Rollers
Ground-rollers
Oscine songbirds
OW suboscines
NW suboscines
New Zealand Wren

Neoaves

This is the best guess so far—or at least, it is one of the best guesses—as to the relationships between birds. As you can see, the neornithines as a whole divide into two great groups: the paleognaths and the neognaths. And the neognaths again divide into two great branches: the Galloanserae and the Neoaves. Most living birds are Neoaves.

The right-hand column shows the orders in which the various birds are traditionally placed. In most cases, most of the time, the traditional, defined orders do seem to summarize the true evolutionary relationships pretty well. But as the diagram shows, there is confusion and disagreement in many areas, including the relationships between the traditional members of the stork order, Ciconiiformes, and the pelican order, Pelecaniformes; and between the traditional Apodiformes, which contained the swifts, and the Caprimulgiformes, the nightjars. The group that includes the kingfishers, woodhoopoes, hoopoes, trogons, and mousebirds also needs more sorting. Other examples are described in Chapter 4.

Most species of living birds belong to a single order, the Passeriformes. This requires its own chart—Figure 5.

Source: Adapted from Cracraft and Donoghue, 2004.

THE RATITES AND tinamous—strange birds with primitive-looking palates:

THE PALAEOGNATHAE

The palaeognaths, taken all in all, are extraordinary. Six of the seven palaeognath orders are known as "ratites": the Ostrich, the kiwis, the cassowaries and Emu, and the rheas, which are still with us; the moas and the elephantbirds, which are now extinct, alas, but only recently. The biggest ratites—the Ostrich, Emu, cassowaries—remind us that birds really are dinosaurs, or probably so. If birds hadn't had to compete with mammals over this past 65 million years, there might be many more ratite-like birds. But although the existing types are very successful, the more versatile form of the mammals seems to lend itself more readily to life on the ground.

The seventh palaeognath order, less obviously peculiar than the ratites but strange nonetheless, is that of the tinamous.

Palaeognathae means "ancient jaws," and refers to the bones of the palate: in general the layout is more complex than in most modern birds. The term "ratite" (including six of the seven palaeognath orders) relates to the shape of the sternum. The breastbone does not have a deep keel (a carina), as in most birds, but is flat; the word *ratite* derives from the Latin for "raft" (contrasting with the keel, but continuing the maritime theme). The ratite breastbone has no keel because keels are for attaching flight muscles to, and since ratites do not fly, their breast muscles are feeble. The tinamous do have a keel on their breastbone—with wing muscles to go with them. So tinamous are not ratites. But they do have a primitive-looking palate, which means they are palaeognaths. Tinamous, in short, are non-ratite palaeognaths.

But although ratites and tinamous have similar palates, does this really mean they are related? Might this be convergence again? T. H. Huxley was among the first to suggest this grouping way back in the 1860s, and taxonomists have been arguing ever since. Ratites might all look very similar to the casual observer—the rhea, for example, is commonly called "the South American Ostrich"—but underneath, they are all very different. If they differed because they adapted to different ways of life—different-shaped beaks, for example—then we could just put this down to the kind of variation we might find among any group of birds—finches or crows, for example. But they differ in profound ways, which to some at least suggests that different ratites have different ancestors. Thus in 1980, in *The Age of Birds*, Alan Feduccia pointed out that they have very different pelvises. In all of them the pelvis is basically basketlike to hold their enormous guts, but this is a functional matter, which could just reflect convergence. It's the details that vary—the kind that suggest different ancestry.

Neither was Professor Feduccia particularly convinced by the palates of the ratites (and tinamous). It is simply a neotenous feature. Perhaps they are not so much primitive as neotenous. Neoteny is common in evolution; it is the phenomenon whereby new groups of creatures may arise, not by adding more features to those that were there before, but simply by continuing childhood features into adulthood. Thus, domestic dogs can be seen as neotenous wolves—as adults they retain the physical features and the trusting character of puppyhood. The palates of embryonic birds may be more complicated than in adults, and perhaps the ratites never shake off this infant complexity. In short, says Professor Feduccia, Ostriches and their ilk may be seen as giant chicks. But each ratite is a different giant chick.

Professor Feduccia, too—like Richard Owen in the nineteenth century, who never lost the opportunity to take a pop at T. H. Huxley—suggested that particular ratites may have descended from various, different neognath orders. In particular, the extinct Ostriches from Asia (if Ostriches they are) first appeared in the fossil record at the time when giant flightless rails from Mongolia had disappeared. Perhaps the Ostriches descended from the rails.

Despite the caveats, modern DNA studies do seem to support Huxley's original idea. The palaeognaths as a whole seem to emerge as a true clade. Within that clade, the tinamous emerge as the sister group to all the ratites. Within the ratites, rheas are the most primitive and are the sister group of the other ratites. Then come the moas; then the Ostrich; and then the remaining three groupings that so far cannot be resolved and are presented as a trident—the kiwis, the elephantbirds, and the cassowaries-plus-Emu. This at least is what has emerged from by DNA studies by Professor Alan Cooper and his Oxford colleagues. This arrangement cannot be taken as gospel, but is certainly as good as any in this uncertain field.

So here are all the palaeognath orders, as they appear in Figure 4.

THE TINAMOUS: ORDER TINAMIFORMES

Tinamous—about 45 to 50 species of them, depending on who's counting, all in the single family TINAMIDAE—look superficially like guineafowl, though with longer beaks. They have egg-shaped bodies and smooth feathers, softly colored apparently for camouflage but very pleasing to look upon; and live on the ground in forest, woods, and savannah from southern Mexico deep down into Chile. They are elusive—more heard (a mellow, flute-like call) than seen. When threatened they stay still, with neck extended, or just creep away, or hide in holes. They break cover and take flight only as a last resort. They have good running legs, but they fly badly and in either case, they quickly run out of puff—perhaps because their hearts and blood vessels are surprisingly small. Guineafowl, by contrast—like all fowl—typically take off like a Harrier Jump Jet in one powerful leap and then fly powerfully. Even so, the pheasant of the Americas (which, of course, are fowl) fail to penetrate far into tinamou country, suggesting that the tinamous may outcompete them. Tinamous, incidentally, have lovely glossy eggs. Those of the Chilean Tinamou are glossy purplish chocolate. In other species they may be blue, green, or yellow.

THE RHEAS: ORDER RHEIFORMES

The rheas look roughly like Ostriches, although much smaller: long legs, long necks, a small head with an all-purpose beak, and a big shaggy body. Yet the differences are clear. Most obviously the rheas, like most birds, have three toes while the highly specialized Ostrich has one big toe and one small one.

There is just one family—the RHEIDAE—but with two species: the Greater (or Grey, or Common) Rhea, which stands about 1.5 meters (5 feet); and the Lesser (or Darwin's) Rhea, at less than a meter (3 feet). Darwin identified the latter during his world trip aboard HMS *Beagle* in the 1830s. He noted that the bones of the rhea he was eating for dinner one night were different from the ones he had gnawed on other nights. So he rushed to the galley to rescue what was left and behold: a new species! More of rheas in later chapters.

THE MOAS: ORDER DINORNITHIFORMES

There were no terrestrial mammals in New Zealand until the Europeans brought them in; the only native kinds were a few bats in the air and a few seals, fur seals, and sea lions around the edge. The biggest terrestrial herbivores were a suite of about thirteen species of birds that Alan Feduccia has likened to "huge, flightless geese": the moas. Among the moas were the tallest birds of all time: *Dinornis* (meaning "terrible bird") stood 3.5 meters tall—almost 12 feet, half as big again as a big modern Ostrich. The moas had lost all trace of wings—not even a vestige, as kiwis have. While the Ostrich is "cursorial" (a supreme runner), the moas were irredeemably "graviportal" (weight-bearing), with three great plonking toes on sturdy legs that in the biggest kinds were like a rhino's.

But although the moas were New Zealand's top herbivores, they were not the top birds. Even the biggest were preyed upon by the mighty Haast's Eagle. Yet it wasn't the eagles that finished them off. It was people—and for once, it seems, we cannot blame Europeans. The Maoris arrived in New Zealand from the South Seas about a thousand years ago and established a moa culture—eating their meat, taking their eggs, making extravagant ceremonial cloaks from their long, Emu-like feathers. By about the fifteenth century, the moas were all gone.

THE OSTRICH: ORDER STRUTHIONIFORMES

There is only one family of Struthioniformes—the STRUTHIONIDAE—with just one living species: *Struthio camelus*, the Ostrich. But that one species, even though now sadly depleted, still ranges throughout Africa. It varies significantly along the way and so is divided into several subspecies—two with blue necks and two with pink necks, neatly color-coded.

Ostriches are by far the largest of living birds. The males stand up to 2.5 meters (8 feet) tall and weigh up to 115 kilos (250 pounds). The females are a little smaller. Yet they seem comical to human eyes, designed to be bizarre. They have big, round fluffy bodies—fluffy because their feathers are without barbs. They have long legs and a commensurately long neck, as mobile as an octopus's tentacle, though usually held vertically. The tiny head is perched atop like a pearl on a pin—and set with huge eyes, perhaps the biggest eyes relative to body size of any bird or mammal, fringed with luxurious lashes.

Yet there is nothing frivolous about an Ostrich. The legs are long, so the animal can take great strides, but the muscles are bunched near the top so that the tibiofibular (drumstick) bulges like a football player's thighs, while the tarsometatarsus (the fused foot) is a skinny and scaly stick. So the leg has minimal weight and minimum inertia. With each stride an animal must first thrust each foot forward and then drag it back again. If the extremities were heavy, this would impose enormous stress. To reduce the weight of the extremities even further without losing strength, the Ostrich has unique feet: just one big toe, and one small one. Thus the overall arrangement is very like a horse, in whom the driving muscles form the buttocks and the toes are reduced to a single hoof.

Thus, like horses or desert antelopes, Ostriches can walk for mile after mile with no problem at all—and when occasion demands (although not unless), they can run at 50 kilometers per hour (30 miles per hour), although some pundits claim that they are at least twice as fast as that. Recent studies[1] show that of all the many gaits an animal might adopt, walking and running are the most efficient; and walking is most efficient when slow, while running is best done fast, when for much of the stride both feet are off the ground and each foot just flicks the ground. The long legs of the Ostrich with their one and a half toes are built for the fast route.

As for the silly neck and head and great wide eyes—well, they offer a wonderful view. Ostriches have marvelous vision, and the neck serves as a periscope. It is hard to creep up on an Ostrich. If you do get too close, their great clawed toe can do a great job of disembowelment—kicking forwards, as birds tend to do, not backwards, like a horse. Ostriches do not, when danger threatens, bury their heads in the sand. With their speed and aggression they have no need for such nonsense. Indeed, before human beings took away their territory and started shooting them, they were common throughout the entire African continent, and well beyond. The Arabian subspecies disappeared only recently. More on their eating and their wonderful, rich social and family life in Chapters 5, 6, and 7.

THE KIWIS: *ORDER APTERYGIFORMES*

Kiwis are archetypal island animals—weird, specialist, and extremely vulnerable to any immigrant beast that has evolved street wisdom on some other continent. There are just three species in a single family, the APTERYGIDAE, which all live in New Zealand and on some surrounding islands. They are nothing like the other, towering ratites to whom they seem to be related. The two larger kinds—the Great Spotted Kiwi and the Brown—are only 35 centimeters tall (14 inches). The Little Spotted Kiwi is a mere 25 centimeters (10 inches). In all kiwis—as in Emus but in contrast to Ostriches—the females are larger, by up to 20 percent. Kiwis are seriously land-bound. Their bones are heavier than in most birds—only partly hollowed out. Their generic name, *Apteryx*, means "without wings," though they do have tiny, vestigial wings hidden among their feathers. Their virtual winglessness presumably helps them to maneuver, and indeed they run at speed through dense vegetation, where they hunt for invertebrates by night (hunting by scent, more like a mammal than a bird). They match the European Robin for aggression, chasing other kiwis from their extensive territories, where they live in pairs and hide by day within a series of burrows. Their breeding is tempestuous, with much snorting and chasing. They lay only a single egg, but it is the biggest relative to body size of any bird: a quarter of the weight of the female. Incubation, safe within the breeding burrow, is hugely drawn out—for up to three months. Kiwis are New Zealand's national bird, but they have been given a bad time nonetheless and now are declining by 6 percent per year—halving every decade—though there are now heroic attempts to save them.

THE ELEPHANTBIRDS: *ORDER AEPYORNITHIFORMES*

The heaviest bird of all was not *Dinornis*, the mighty moa. It was *Aepyornis maximus*—the largest of the elephantbirds of Madgasascar. It was only a little taller than an Ostrich, at 2.7 to 3 meters (9 feet). But at 450 kilos, getting on for half a ton, it was three times as heavy. Elephantbirds seem to have arisen late in evolution; at least, no fossils are known before the Pleistocene, which began only 2 million years ago, and since Madagascar had already been an island for at least 150 million years, it isn't at all clear how they got there. Earlier fossils from other places including Egypt have been claimed as elephantbirds, but all are highly dubious. In the Pleistocene the only big predator on Madagascar was a crocodile, which did not seem to trouble the elephantbirds. The initial population split to form seven known species of different sizes, and all were still going strong when human beings first arrived on Madagascar at about the time of

Christ, about 1,000 years before the Maoris hit New Zealand. Radiocarbon dating of the eggshell fragments that are still found all around shows that all seven species were still widespread as late as the tenth century AD. Indeed, elephantbirds seem to have given rise to the legend of the Roc, the huge bird that bore Sinbad to safety in the *Tales of the Arabian Nights*—although the Roc was a fabulous flier while elephantbirds decidedly were not. As late as the thirteenth century, the great Italian explorer and merchant Marco Polo surmised that the Roc came from Madagascar. But some time after the tenth century the elephantbirds were all gone. Humans achieved very easily what the crocodile did not.

Many whole eggs have been found, as well as fragments, and among them are the largest ever produced by any creature—bigger even than any dinosaur egg. Some are more than 30 centimeters (12 inches) long and when fresh must have weighed more than 10 kilos (22 pounds): as big as seven Ostrich eggs, equivalent to nearly 200 eggs from a chicken or 10,000 from a hummingbird. Some X-rays have revealed that one of the recovered eggs at least has an embryo inside.

THE CASSOWARIES AND THE EMU: ORDER CASUARIFORMES

The order Casuariformes divides neatly into two families.

Family CASUARIIDAE are the cassowaries from New Guinea and northern Australia—the most colorful of the ratites with glossy black plumage and bright-colored wattles around the beak, and a huge "casque" along the head like the crest on a conquistador's helmet, which is used, so it seems, to butt through the dense vegetation and sometimes, apparently, for digging. The casque is not made of horn (keratin) as in a hornbill, but of calcified cartilage, quasi-bone, fused with the skull and covered in tough skin. There are three species; the biggest is the Northern, at about 1.8 meters (6 feet), and is the biggest land vertebrate of New Guinea (and also lives in Cape York, which juts north from Queensland). All three sustain their considerable bulk by gobbling forest fruits, whole, with supplements of fungi and small animals. They are notoriously aggressive. If two of them meet, it usually ends in a fight. The inner toe has a long sharp claw with which many a hapless native and probably not a few visitors have been summarily disemboweled. The people of Papua use these claws as spear points. The wings of cassowaries, though much reduced, remind us that although the ratites have long been flightless, they did have flying ancestors. They retain the quills of flight feathers—not as a mere vestige but for serious purpose. They are sharp, like spikes, and are used both for defense and attack. Again, depressingly, cassowaries are suffering in the wild. The Northern and the Southern kinds are officially classed as "Vulnerable," while the smallest, the Moruk, is "Lower Risk/Near Threatened." Because they eat fruit they are great scatterers of seed. If they go, their forests will suffer, too.

There is only one species of Emu, in the family DROMAIIDAE: *Dromaius novaehollandiae*, reflecting the days when Australia was called New Holland. Emus stand around 1.75 meters (nearly 6 feet), with the females slightly heavier than the males. Emus have remarkable, forked, "double" feathers that hang shaggily; each feather has an aftershaft, a secondary feather that branches from the base of the main feather. Like Ostriches, rheas, and cassowaries, Emus do have wings, but they are scarcely 20 centimeters (8 inches) long. Both sexes grunt and hiss, and the females boom: an aperture between the windpipe and a special air sac in their necks improves the resonance. Their wonderful migrations in search of food and their bohemian family life are described later.

And that winds up the palaeognaths. Now we come to the second great division of the modern birds.

THE BIRDS WITH modern-looking palates:

THE NEOGNATHAE

The neognaths include all the modern birds—all twenty-five orders of them—that are not palaeognaths. Their name means "new palate," implying that the layout of the palate bones is simpler than in the palaeognaths. Although the neognaths are extremely various—everything from storks and eagles to rails and hummingbirds—it seems that they did indeed arise from a common ancestor. Various as they are, then, they do form a true clade.

Joel Cracraft (and many other modern taxonomists) further divide the neognaths into two. The Galloanserae include just two orders—those of the fowl and of the ducks, geese, and swans—which are commonly called "waterfowl." The Neoaves include all the rest.

FOWL AND WATERFOWL: THE GALLOANSERAE

It may seem strange to link the fowl and the waterfowl. Farmyard hens and farmyard ducks do not seem particularly similar. Some still conventional classifications do not show this specific linking. But beneath the feathers and skin the similarities abound, and although none of the fowl look particularly like waterfowl, some of the waterfowl do seem very like fowl—notably the screamers of South America. So the two orders between them really do form a true clade.

CHICKENS, PHEASANTS, QUAILS, PARTRIDGES, GROUSE, AND THE MEGAPODES: ORDER GALLIFORMES

The fowl range from quails to turkeys, but they all look much the same when you boil them down: plump bodies; short wide wings for rapid take-off; strong legs; and either short tails, as in quails, or sometimes extraordinarily long tails (or trains), as in pheasants—which include Common Peafowl, whose train is legendary. Many have ornaments, including the combs and wattles of farmyard chickens, especially the cockerels. There are six families:

Family PHASIANIDAE, 187 species, cover most of the world except for Antarctica and southern South America (where the tinamous roughly fill their niche). The family is traditionally divided into four subfamilies, which probably are not true taxa, though they are convenient. They are:

First, the thirty-two species of New World quails, which do indeed live in the Americas, from northern Argentina to southern Canada. Many are crested. The Bobwhite Quail is the best known for the dubious reason that it is the most widely hunted. It was the one that former U.S. Vice President Dick Cheney was after when he decided to shoot his lawyer instead. Lawyers are easier to hit.

Then there are fourteen species of Old World quails from Africa, Asia, and Australia. They don't look like great fliers, but the Common Quail migrates between Asia and Europe while the Harlequin Quail of Africa and the Rain Quail of Asia are nomadic—they follow the rains and hence follow the food supply, as Emus do.

The partridges, from Africa, Eurasia, and Australia, form the third group. More than forty of the ninety-two species are commonly called "partridge"; forty-one are called "francolins," all native to Africa; and seven species are called "snowcocks." The Giant Snowcock of central Asia weighs in at more than 6 kilograms (13 pounds)—as big as an oven-ready Christmas turkey. The Udzungwa Forest Partridge was discovered in the mountains of Tanzania only in 1992.

The fourth subgroup includes the forty-eight species known as "pheasants"—a loose grouping of big fowl with bright-colored males and that includes such garden wonders as Lady Amherst's Pheasant, the Golden Pheasant, the three species of peafowl, and the five species of tragopan—named after *tragus*, the billy goat, and *Pan*, the half-goat semi-deity, because they have brightly colored fleshy horns on their heads that they erect during courtship. Completing the subgroup are the Himalayan Monal and the Jungle Fowl, the ancestor of the domestic chicken.

Second of the six galliform families is the TETRAONIDAE: eighteen species of grouse, sage grouse, prairie chickens, and ptarmigans. The biggest, the Capercaillie, at 6.5 kilograms (14 pounds) is even bigger than the Giant Snowcock. Grouse and their kind are adapted to the north, with feathered nostrils to filter the cold air and feathered feet that act as snowshoes. Many kinds burrow in the snow to roost, while Red Grouse feed under the snow and thus are hidden from eagles (though big owls can surely hear them burrowing). Ptarmigans are white in winter and blend in with the snow (but the Red Grouse stays red, although it is really a ptarmigan). Overall, in the tundra, grouse account for a fair proportion of the total vertebrate biomass, but then they are preyed upon by lynx, martens, foxes, raptors, and of course humans. Some do come farther south, however; the Greater Prairie Chicken ranges right down to the Gulf of Mexico. Some of the different species may hybridize where their ranges overlap, although the hybrids seem to be infertile.

The turkeys, just two species, form the MELEAGRIDIDAE and are American. The Wild Turkey ranges from Canada to Mexico and is the bird that Benjamin Franklin wanted to adopt as the symbol of the United States (to be usurped, inevitably, by the more obviously imperious Bald Eagle). Native Americans domesticated the species several times, for they are easy to raise on berries and seeds, including acorns, which they crush in a muscular gizzard packed with grit, plus invertebrates and the odd snake. The Wild Turkey was once endangered but may now be saved—

still hunted, but with laws to prevent over-exploitation. The smaller Ocellated Turkey of Central America is still in trouble.

Guineafowl—six species of noisy, social birds from Africa—form the family NUMIDIDAE. They include the Black, the White-breasted, the Crested, the Plumed, and the Vulturine, which is the biggest. But the one that is best known—because it has been domesticated for eating and ornament the world over—is the Helmeted. In the wild, at least outside the breeding season, Helmeted Guineafowl move to the waterhole in single file in flocks of 2,000 or more, with the dominant males in the lead as scouts. Once watered, they advance across the plain line-abreast like a police cordon—an efficient way to seek out bulbs, seeds, and insects. When danger threatens—jackals, baboons, snakes—the adults form a protective "swarm" around the youngsters, like musk ox defending against wolves. Helmeted Guineafowl feed only on the ground, but always roost in trees; so if you keep them for eggs in more northern climes, they are less vulnerable to foxes. But while the Helmeted builds its life around the waterhole, the Vulturine is adapted to semi-desert and does not seem to need to drink at all. Physiologically it matches the desert antelopes, such as addax.

The MEGAPODIIDAE—twenty-two species in seven genera—include the megapodes, or "big feet." They also include the brush-turkeys, including *Talegallas*; Malleefowl; and the Maleo. All are basically turkeylike to look at but in truth are very different. They live in remote places within Oceania: Australia, New Guinea, and many an island. Yet they are not remote enough. Nine of the twenty-two—almost half—are to some extent endangered, one critically so; and the archaeological record shows that in the past few thousand years throughout Oceania, human beings have killed off least another thirty-three species. So the twenty-two are just a shadow even of recent glories. What makes the megapodes so vulnerable is their method, unique among birds, of raising their eggs: in custom-built compost heaps or in holes in the ground. More in Chapter 7.

Finally, the family CRACIDAE includes the guans, curassows, and chachalacas: fifty species in Central and South America. They have big bodies and small heads on thin necks, short rounded wings, and long broad tails, and—unusual among game birds—they live mainly in trees—although, like all other game birds, they mostly come to the ground to feed. The modified windpipes of guans in particular amplify their calls into some of the loudest and most penetrating of all birds. They have also modified some of their main flight feathers (the outer primaries) into spines, which drum when they shake their wings (like the outer tail feathers of the snipe). The four species of piping-guans make spectacular drumming flights, with deep, raucous cackles for good measure. Biggest of the cracids are the curassows, with wattles and knobs on their faces. Cracids as a whole are good to eat and easy to hunt, and several are threatened, although some are now bred in captivity. The White-winged Guan was thought extinct in 1870 but turned up again in 1977.

DUCKS, GEESE, SWANS, AND SCREAMERS: ORDER ANSERIFORMES

The waterfowl form the order Anseriformes, in which two families are commonly recognized: the ANHIMIDAE, which are the screamers of South America; and the ANATIDAE, which include all the rest—ducks, geese, and swans.

The screamers, in the family ANHIMIDAE, seem to be the sister group of all the other anseriforms. They demonstrate most obligingly why the anseriforms as a whole seem to be closely related to the galliforms. They look at first sight like pheasants or turkeys, though their long toes are partially webbed; and although they sometimes stride across open plain, they generally prefer the water's edge and often wade and sometimes swim, searching for lush vegetation and the odd insect—but the youngsters swim more than the adults, in pursuit of their wading parents. Large though screamers are, their big splayed feet allow them to walk on floating vegetation, almost like jacanas. Oddly, they have gas-filled spaces beneath the skin, like bubblewrap, and crackle when handled. They really do scream, too. There are three species, all from South America. The Horned Screamer has a long (up to 15-centimeter, or 6-inch) whippy rod of cartilage that grows from the middle of the forehead and curves forward like a fishing rod—presumably for sexual display.

The family ANATIDAE, containing all the rest of the anseriforms, are very diverse and difficult to classify. On the whole, swans are swans, but ducks are immensely various, and some waterfowl that are commonly called "goose" are closer to ducks than to most other geese. Commonly the ANATIDAE as a whole are divided into fifteen tribes, though not all of them are true taxa. Some beyond doubt are polyphyletic and some seem to be paraphyletic. Classification is especially difficult because different species readily hybridize—sometimes in the wild, often in captivity. Indeed, 400 different kinds of interspecies duck hybrids are known. Sometimes members of different tribes hybridize, again suggesting that the present classification is more ad hoc than is desirable. Also, the Musk Duck from Australia was once bundled with the stiff-tailed ducks, but it does not seem to be related to them and for the time being it has no tribe to go to. For what the tribes are worth, here is a brief overview.

The first tribe, the Anseranatini, contains just one species, the Magpie Goose from Australia, shown here as the sister group of all the rest and so different from all the rest that it is sometimes placed in its own subfamily. They look primitive, and are. Like the screamers, they have very little webbing between their long splayed toes, and big though they are, they

The most primitive of all the living anseriforms—Australia's Magpie Goose, whose feet are only partly webbed.

can perch perfectly happily on branches (as some ducks can). Almost all of the other anseriforms molt their wing feathers all at once so that for a time they cannot fly, but Magpie Geese molt them one or a few at a time and are never flightless (and indeed are good fliers). For good measure, and again in contrast to most anseriforms, the males are bigamous—the two "wives" usually are related to each other. All three build the nest together and incubate the combined clutch of eggs.

The true geese are the Anserini, with fifteen species in two genera. *Branta* includes such luminaries as the Brant, Barnacle Goose, and Hawaiian Goose, while *Anser* includes the Snow Goose, White-fronted Goose, and so on. They turn up a lot in the rest of the book.

Eight species of swan form the Cygnini. Seven of the eight are in the genus *Cygnus*: the Tundra or Whistling Swan, which includes the subspecies known as Bewick's Swan, plus the Black, Black-necked, Mute, Trumpeter, and Whooper. The most primitive swan is the relatively small and short-necked Coscoroba Swan from South America, with its purewhite plumage and its rounded and almost luminously pink bill. It has its own genus, *Coscoroba*.

The remaining twelve tribes are best thought of as ducks, although some of them are commonly referred to as geese (like the sheldgeese).

The biggest of all the tribes by far—fifty-seven species—is the Anatini, the dabbling ducks. Mallards are Anatini—very common both in the Old World and in the New and the parent species of most (but not quite all) farmyard ducks. Teals, wigeons, pintails, shovelers, and the Mandarin and Wood Duck are also in this group. The dabbling ducks feed from the surface, either on whatever floats or by reaching down into the waters below. Shoveler feed like baleen whales: they take in a beakful of water and then strain it through combs on the tongue and the edge of mandibles to filter out the nutritious plankton.

The whistling-ducks or tree-ducks—eight species, widely spread through the tropics and subtropics—form the tribe Dendrocygnini (which means "tree, or branch, swans"). Their call is indeed a whistle; and they do perch happily in trees, although they also swim and sometimes dive.

The stifftails are the Oxyurini and they number seven species. Five come from the Southern Hemisphere and two from the Northern—the latter are the Ruddy and White-headed Ducks who are famously hy-

Wood Ducks, like these, can live only where there are woodpeckers to make the holes in trees that they need to nest in. See Chapter 8.

bridizing, as described in Chapter 3. The stifftails are smallish and dumpy and excellent divers; their tail feathers are indeed stiff, used as rudders. These ducks have immense penises that are withdrawn into the body when not in use and account for a large proportion of the overall mass.

The steamerducks plus the various torrent ducks and the Blue Duck—six species in all—are lumped together in the Merganettini, but may not be close relatives at all. The steamers have very short wings and cannot fly, but when they need to escape their enemies in the seas around southern South America, they churn the water with legs and wings like a paddle wheeler. The torrent ducks catch aquatic invertebrates in the torrid white waters of the Andes, gripping the slippery rocks with their sharp claws and steering through the rapids with their long stiff tails. The very attractive Blue Duck does much the same in the rushing mountain streams of New Zealand.

The Spur-winged Goose and the comb ducks from Africa and South America, just three species, form the Plectropterini. They are mostly black and white and look in general like geese. The males are much bigger than the females, and—unlike "true" geese—they form only weak pair-bonds, with the males generally taking little interest in child care.

The shelducks and sheldgeese—fifteen species in all—form the Tadorini. The seven species of shelducks are beautiful, big, bright, fond of the sea, and occur just about everywhere. The eight species of sheldgeese (in four genera) are again from the Southern Hemisphere—apart from the Blue-winged and Egyptian Geese of Africa, which have wandered a little way north. Unlike most waterfowl, but like the Magpie Goose, the sheldgeese molt their wing feathers in sequence and do not go through a flightless phase. They may not molt every year, but every other year.

The Pink-eared and Salvadori's Ducks together form the Malacorhynchini—though they may not be closely related. The Pink-eared of Australia—small, zebra-striped, with a pink patch on either side of the head—has a huge bill with lamellae around the edge like the baleen of a blue whale, and with the same function: to filter small invertebrates from the water. In this it is like a shoveler, only more so. Salvadori's Duck has a bright yellow bill and is also zebra-striped and favors fast-flowing streams in the mountains of New Guinea.

The pochards and the scaup make up the Aythyini: seventeen species in four genera, occurring worldwide. They feed by diving, typically in fresh water but sometimes in the sea. Scaup and Tufted Ducks with their little plump round bodies, predominantly black, stand out very clearly. Red-crested Pochards are conspicuous for their big gingery heads. India's Pink-headed Duck was probably a pochard, too. In the nineteenth century it was one of India's favorite game birds (meaning that people shot it), but now it seems to be extinct—eliminated, apparently, not so much by the "sportsmen" as by the draining of wetlands. The Pink-headed Duck was often painted, when it generally looks very handsome, though photographs show a somewhat scrawny bird, with a stiff neck and a big head. But photographs can lie, too. In truth, Indian ornithologists have not quite given up on the Pink-headed Duck, any more than the Americans have given up on the Ivory-billed Woodpecker. There have been many reported sightings in the past half-century, but most of them were probably pochards; if a Pink-headed Duck does ever turn up for certain, that will very likely prove to be a pochard, too. The two are placed in the same genus (*Netta*). If it really is a pochard, then we would expect it to be handsome, like its portraits, not scrawny, as the photos show.

Twenty species of mergansers, the Smew, the scoters, the Long-tailed Duck, the goldeneyes, and others form the Mergini. Mostly they are sea ducks, although some breed around freshwater. All feed by diving; the bulkier kinds live on invertebrates, and the more slim and agile saw-toothed kinds—including the mergansers—feed on fish. The eiders are sometimes placed in the Mergini, but sometimes given their own tribe, the Somaterini.

The three remaining tribes contain only one species each: one for the White-backed Duck of Africa (tribe Thalassornithini), one for the Cape Barren Goose from Australia (Cereopsini), and one for the Freckled Duck (Stictonettini)—yet another Australian.

As is obvious from these descriptions, the anseriforms greatly favor the Southern Hemisphere. All the tribes are found in the south and some are exclusive to it: those of the Magpie Goose, the Musk Duck, the White-backed Duck, the Cape Barren Goose, the Freckled Duck, the torrent and steamerducks, the Spur-winged Goose and the comb ducks, the Pink-eared and Salvadori's. Some groups that are found worldwide are most various in the south, including the sheldgeese and mergansers. No tribe is exclusive to the Northern Hemisphere. All the conspicuously primitive types are southerners—the screamers, the Magpie Goose, and the Coscoroba Swan. All in all, it's a fair bet that the anseriforms as a whole arose in the south. Australia is a strong candidate—although perhaps it's just that a great many have survived there because they were spared the attentions of foxes, cats, and egg-stealing rats (until the Europeans brought them in, that is).

THOROUGHLY MODERN BIRDS: THE NEOAVES

Most living birds—all except the fowl, waterfowl, ratites, and tinamous—belong to the huge grouping known as the Neoaves. *Neoaves* is Greco-Latin for "new birds": and *Neornithines*, the subclass to which all living birds belong, is Greco-Greek for "new birds." Taxonomists are nothing if not economical.

Twenty-three orders of neoavians are commonly acknowledged, and are described here, but some of the present ones need serious rethinking. I will discuss them as we go.

Note, if you would, how in Figure 4 the various neoavian orders are grouped into nine subdivisions. The first (which is further subdivided into four big groupings) includes a huge range of birds from the rails and cranes to the penguins. The next includes the owls, which on present evidence do not seem to be very particularly related to any other order. Then comes a group that includes the nightjars, swifts, and hummingbirds. Then the turacos, and then the unique Hoatzin—both without obvious close relatives. Then come the pigeons and sandgrouse, which seem to be sister groups. Then the parrots and then the cuckoos, both without obvious close relatives. Then, finally, comes a big grouping that includes the puffbirds and jacamars, the woodpeckers and toucans, the trogons, the kingfishers and hornbills, and then the huge order of the passerines, which includes more species than all the other neornithine orders put together. The genealogical tree of birds is not yet worked out fully, but the taxonomists are getting there.

CRANES, RAILS, CRAKES, COOTS AND GALLINULES,
THE LIMPKIN, TRUMPETERS, BUSTARDS, BUTTONQUAILS, MESITES,
THE KAGU, THE SUN BITTERN, SERIEMAS, FINFOOTS, AND
SOME HUGE AND ANCIENT PREDATORS: ORDER GRUIFORMES

DNA and other studies suggest that the order Gruiformes is a true taxon, a true clade, even though its members are so immensely various (and one of them really doesn't seem to belong and should surely be placed elsewhere). They are an ancient group: the oldest known gruiform fossils—of cranes—date from the Paleocene, around 60 million years ago. Presumably the very first gruiforms lived long before that—deep in dinosaur times. Now there are no fewer than eleven families.

Family ARAMIDAE contains just one species, the long-beaked, long-legged Limpkin, which catches aquatic invertebrates, especially snails, in the American tropics and subtropics.

The GRUIDAE are the cranes, which include the tallest of all flying birds. They are fabulous dancers and migrators, and are much admired and loved worldwide—but nine of the fifteen species are endangered.

Closely related to both the Limpkin and the cranes, so DNA evidence suggests, is the family PSOPHIIDAE: the three species of trumpeters. They are elegant birds from tropical South America, like long-legged, chicken-size coots with dark feathers soft as fur on their heads and necks and long, draping, plume-like feathers on their wings and backs in subtle shades of green, gray, or off-white.

Cousins to all the above is the family HELIORNITHIDAE, with two species of finfoot—one in Africa and one scattered throughout Asia—plus the Sungrebe from tropical America, all of them swimming on rivers and lakes like long-legged grebes. The male Sungrebe, uniquely, carries the babies in flight in pouches of skin beneath its wings, even when the nestlings are still blind and helpless.

Close relatives of the heliornithids is the huge family of the RALLI-DAE—no fewer than 133 species that are arranged, for convenience, in three series. In the first are the rails—waterbirds that live the world over and have narrow bodies like angelfish, which help them to squeeze through dense vegetation; they have loud voices and are much more seen than heard. The second group includes the gallinules and crakes, including the widespread swamphens, as big as farmyard cockerels, with their stout red beak. The third includes the many kinds of coots, which prefer more open water and have rounded bodies.

All alone in the family EURYPYGIDAE is the Sun Bittern from South America. It looks and hunts like a heron and is called Sun Bittern because when it flies or displays, it reveals bright golden-buff patches on its rounded wings.

Sister group to the Eurypygidae is the family RHYNOCHETIDAE—again with just one species: the Kagu from New Caledonia. It resembles a small, stocky, snow-white heron with long red legs—and has a magnificent wig of plumes on its head, like white dreadlocks. It lives in dense vegeta-

Coots are wonderfully aggressive—and have some wonderfully aggressive relatives.

tion and is more heard than seen, the mated pairs dueting in the morning with calls reminiscent of both roosters and dogs.

The two species of seriemas in the family CARIAMIDAE are long-legged and long-necked stalkers of frogs, large insects and small birds, rodents, and the occasional snake. They are South America's answer to Africa's Secretary-bird (which is related to the hawks).

But the best of the seriemas was in the past. Apparently related to the modern kinds was *Phorusrhacos*, which lived in Patagonia around 20 million years ago (the first of its bones known to science came from mid-Miocene rocks in the Santa Cruz Formation of Argentina). *Phorusrhacos* stood at around 2.5 meters (8 feet) and weighed 130 kilograms (286 pounds). In general shape it was like a cassowary, with long legs and very short wings; but it had an enormous head, around 60 centimeters (2 feet) long, including a huge hooked, eagle-like beak. In short, *Phorusrhacos* seems to have been one of the most formidable avian predators of all time—indeed, one of the most formidable of all land predators. Again, we see what birds *might* have been like if they had not had to compete with mammals. *Phorusrhacos* is commonly given its own family, the PHORUSRHACIDAE, but it could well be placed in the Cariamidae, and it certainly belongs among the gruiforms. I am often struck by the general hooliganism of coots. But when we see that *Phorusrhacos* is lurking in their family tree, like some mad uncle, their aggression seems less surprising.

We might also in passing mention *Diatryma*, from North America, France, and Germany. It looked like *Phorusrhacos*, with a huge head and commensurately formidable beak, although it was much sturdier and slightly shorter (the biggest at around 1.8 meters or about 6 feet). However, although *Diatryma* was once classed as a seriema, it now seems that it probably was not—yet another case of convergence. *Diatryma* may well have been a galliform—a giant fowl.

Obvious gruiforms but with no close relatives are the bustards in the family OTIDIDAE. Bustards surely originated in Africa, where twenty-one of the twenty-five remaining species now breed; but they have spread, albeit patchily, through semi-desert and grassland to open acacia woodland throughout Eurasia and down into Australia. The biggest kinds, like the Great Bustard, are slow and sturdy walkers, yet they can fly with power and speed and indeed are among the world's heaviest flying birds—bigger even than a Wild Turkey.

The MESITORNITHIDAE are the smallest gruiforms, thrush-size, and live on the forest floor in Madagascar. They can fly, but like tinamous they generally prefer not to. Little is known about them, but there are recent reports of pairs of females sharing one male and laying in the same nest.

Finally, the quail-like buttonquails, of the family TURNICIDAE, are generally included among the gruiforms. But Cracraft suggests they belong with the charadriforms, so I will discuss them later.

HERONS, EGRETS, AND BITTERNS; STORKS, IBISES, AND SPOONBILLS; THE HAMMERHEAD, THE SHOEBILL, AND FLAMINGOS: ORDER CICONIIFORMES

The Ciconiiformes, as traditionally defined, includes six families that seem very different and probably are. The order surely needs re-sorting. But the families themselves seem valid enough.

Family ARDEIDAE contains sixty-two species of heron and bittern, either standing at the water's edge on long legs, waiting to spear passing fish, or, as some species do, stalking their prey with varying degrees of stealth. Some have become more land-based and arboreal, with a shorter neck, legs, toes, and bill. On their broad wings, beating slow and deep, herons are fine, long-distance fliers. Such is their diversity, the ardeids are divided into five subfamilies.

Subfamily Ardeinae includes the forty-two species of typical herons and egrets, such as the Grey Heron of the Old World, the Great Blue Heron of North America, and the Cocoi Heron of South America.

Subfamily Botaurinae contains the thirteen species of bitterns—including the smallest of all the ardeids, the Least Bittern, at 28 centimeters (11 inches). Bitterns are like small, compact, shy, solitary, cryptic versions of herons, ambushers of fair-sized fish as well as of small fish, frogs, and insects, and given when alarmed to standing stock-still among the reeds, beak pointing straight up, looking for all the world like reeds themselves. Sometimes they even sway, as if with the wind, to complete the illusion.

The three remaining subfamilies are less well known. The Tigrisomatinae are another solitary and secretive lot that live in tropical swamps in the New World and are named for their striped plumage. The Agami Heron, with its very long neck and beak, has its own subfamily, the

Agamiinae, which, so their DNA suggests, separated from the rest of the Ardeidae a very long time ago. So, too, did the Boat-billed Heron, which also has its own subfamily (the Cochleariinae). Its beak is indeed shaped like a boat, and although it generally feeds like other herons, standing still and lunging, it also scoops up prey in beakfuls of water and mud.

A spooky encounter: in many an Indian village you may come face-to-face with an adjutant, a stork on the prowl for scraps.

The second great ciconiiform family, the CICONIIDAE, includes the nineteen species of storks, spread around the world on every continent except Antarctica. Again, the Ciconiidae is an ancient family, known from 55 million years ago and presumably much older—old enough perhaps to have known the dinosaurs. The present-day ciconiids are divided into two subfamilies: the Cinoniini, which are the "true" storks; and the Myxteriini, which are the wood storks.

There are thirteen species in the Ciconiini. The greatest numbers and the greatest variety are from the tropical wetlands of Asia and Africa, although the White Stork (which in truth is white and black) and the Black Stork (which is black and white) venture deep into Europe—until fairly recent times as far as Sweden. But both the White and the Black migrate south to Africa for the winter, and some stay in Africa year-round. Storks are great migrators, riding the thermals.

The Myxteriini includes the six species of wood storks. They are slimmer, and include the openbill storks of Africa and Asia, whose mandibles bow away from each other so the middle of the bill is never quite shut, adapted for catching large aquatic snails. Openbills in Africa ride on hippos. The Painted Storks rising in great flocks from Indian wetlands are one of the world's greatest wildlife spectacles. The New World has just one species—the Wood Stork.

Family THRESKIORNITHIDAE again contains two subfamilies. The Theskiornithinae are the ibises, twenty-seven species of them, that typically probe in the mud for invertebrates, with long, curved beaks. The White Ibis from North America and the Scarlet Ibis from South America meet in northern Venezuela and interbreed, to form yet another hybrid zone. But most ibises belong to the Old World. Most widespread is the Glossy Ibis, which has also found its way to North America (more in Chapter 5). Several are seriously endangered, including the Bald Ibis, or Waldrapp, now reduced to one colony in Turkey. But Waldrapps are being intensively conserved, not least by captive breeding, and attempts are afoot to re-establish them in North Africa. Several ibises have been flightless, but all of the flightless kinds are now extinct. Seventeenth-century sailors reported a "white Dodo" on the island of Reunion, 200 kilometers (125 miles) from Mauritius in the Indian Ocean, but French ornithologists now suggest that it was probably a flightless ibis.[2]

The second of the two subfamilies is the Plataleinae, with six species of the spoonbill. In size and shape they broadly resemble ibises, but they have very different beaks, ending in a spatula—a cunning feeding device.

Family SCOPIDAE has just one species: the Hammerhead, or Hammerkop, from sub-Saharan Africa and Madagascar. It looks like a small, brown, short-legged stork with a long thick crest at the back of its head— so that with the beak in front, the head does indeed resemble a hammer. Famously, the Hammerhead builds a huge nest like a haystack in low trees, which it adds to year after year. It flies well and is good at sizing up the land, and so exploits new watery feeding places as soon as they become available—dams, canals, irrigation ditches, whatever. The Hammerhead is among the minority of birds that are doing well.

Out on a phylogenetic limb, apparently with no very close relatives, is the Shoebill, *Balaeniceps rex*, also known as the Whale-headed Stork, sole incumbent of the family BALAENICIPITIDAE. It is big—up to 1.7 meters (5.5 feet) tall—and is named for its massive, clog-like beak, 20 centimeters (8 inches) long and very deep, sharp-edged and with a hooked tip; the beak is ideal for catching lungfish (more again in Chapter 5). The Shoebill, like the storks, can soar very high on the thermals, neck tucked in like a heron.

You cannot mistake flamingos, five species of them in the family PHOENICOPTERIDAE: the James's, or Puna; the Andean; the Lesser; the Chilean; and the Greater (which is divided into two subspecies, Greater and Caribbean). Mostly they feed around coastal wetlands, but some live inland and two—James's and the Andean—live around high Andean lakes. You can't mistake their long necks and long pink legs with knobbly knees and their bent beaks, which they hold upside down for filter-feeding. They scoop up algae-rich water from the surface, then squeeze it out through the lamellae that fringe the beak with the thick, muscular, spiny tongue. They may also trample the mud with their webbed feet to stir up the esculent creatures within. The "algae" that makes up most of their diet indeed includes true algae—simple plants, such as the *Spirulina* that forms such spectacular blooms in East African lakes, rich in protein and in highly desirable polyunsaturated fatty acids. But it also includes equally nutritious cyanobacteria, formerly known as "blue-green algae." The carotene in their diet—the same as occurs in green leaves and more obviously in mangos and carrots—lends the pink or bright-red hue to their feathers. In old-fashioned zoos, when flamingos were fed unsuitable fare, they tended to turn a disappointing white. Incidentally, young flamingos are born with a conventional beak—straight and pink—and acquire the kink and the black coloration after they fledge. Fossils dating from the Mid-Miocene, 10 million years ago, show that flamingos once ranged through all of Europe, North America, and Australia, as well the places where they still live.

I have discussed flamingos among the Ciconiiformes because that is where they are conventionally placed. But as shown in Figure 4, modern molecular studies link them to grebes—an idea that is gaining strength, although it seems most counterintuitive. In their time they have also been linked to all kinds of other birds, even ducks. Flamingos may well need a quite new order of their own.

PELICANS, GANNETS AND BOOBIES, TROPICBIRDS, CORMORANTS, FRIGATEBIRDS, AND DARTERS (aka ANHINGAS OR SNAKEBIRDS): ORDER PELECANIFORMES

The order Pelecaniformes as conventionally understood contains six families of waterbirds. Each family is a convincing clade, but DNA studies now suggest that the different families may not have a great deal to do with each other—and some seem closer to various members of the conventional ciconiiforms. But for convenience I will treat the Pelecaniformes here as if it were a true taxon. So here are the families.

There are just seven species of pelican in the family PELECANIDAE—or eight, if the Peruvian is divided off from the Brown. All are placed in the single genus *Pelecanus*, and they live virtually worldwide. Their name comes from the Greek *pelekon* meaning "axe"—a reference to the huge beak, although in truth it acts more like a scoop, able to hold a mass of water almost equal to the bird's own body weight. The Great White at 15 kilograms (33 pounds) is the heaviest of all seabirds. Pelicans may look clumsy, but in truth they are superb fliers, especially at soaring, gliding in like pteranodons against the evening sun, diving into the sea after fish almost like gannets, happy to journey more than 150 kilometers (100 miles) in a day after food. Some species migrate—and may cross deserts or mountains—but none, although they are seabirds, will cross long stretches of sea. Despite their big webbed feet, some nest in trees. They live in colonies, with a complex social life. Superb, fantastical creatures.

Family SULIDAE are the gannets and boobies—nine species, in three genera. Gannets in particular are wonderful, dashing birds with beaks that the great gannet-watcher Bryan Nelson compares both to garden shears and to bolt cutters: able to seize and subdue the strongest and slipperiest of fish, for which they plunge from a great height (more in Chapter 5). The sulids breed from Arctic to sub-Antarctic, and the boobies have specially strong presence in the tropics. Some, such as Abbott's Booby, like to breed away from the madding crowd, but most breed in teeming colonies. Those on the rocks and islets around Scotland are among the world's great spectacles. Breeding colonies of Peruvian Boobies may be a million strong, the birds packed in at three or four per square meter. They, plus the Guanay Cormorant and the Peruvian Pelican, are the prime source of guano.

The three species of tropicbirds in the family PHAETHONTIDAE are like gleaming white angels: two meters (6 feet) across their scimitar wings, with two long central tail feathers like streamers, riding the tradewinds over vast areas in the warmer latitudes of the Pacific, Atlantic, and Indian Oceans. They are strictly aerial marine—feeble on land but never bothering with the land except to breed. They do not merely glide like albatrosses but maintain a constant, rapid beat. Their courtship flights are fabulous. Like gannets, they plunge into the sea, sometimes from a height of 50 meters (160 feet), sometimes spiraling, to seize often quite large fish in their stout, sharp, serrated beak, which is yellow or bright red according to species. Fabulous birds.

Family PHALACROCORACIDAE are the thirty-nine species of cormorants and shags, which between them exploit every kind of waterway, from tropical storm drain to the wild Antarctic Ocean—but they avoid open ocean. They hunt underwater, and for this they go to great lengths to increase their weight. The Galápagos Cormorant, heaviest of the family, is flightless—the nearest thing among modern seabirds to *Hesperornis*. Twenty-nine of the thirty-nine species are entirely marine, or almost so; four are freshwater specialists; and the other six happily move from one to the other, often far inland. Some roost and nest in trees—a common sight in Britain.

The five species of frigatebirds within the family FREGATIDAE patrol the tropical oceans right around the globe. They are basically fishers, scooping prey from the surface, but are also notorious pirates, well content to harass and pillage seafood and nesting material gathered by other seabirds.

The darters, all in the genus *Anhinga*, form the family ANHINGIDAE. Some say there are four species, one in North and South America, and

Anhingas, alias "snakebirds," hunt underwater like cats. Their heavy bodies keep them almost submerged at all times.

three in Africa, India and Southeast Asia, and Australasia. Others say, from various sources of evidence, that the three Old World species should be lumped into one. They are underwater stealth hunters of fresh water, tropical to warm temperate, and look a bit like cormorants except that they are even heavier and can swim even lower in the water, with just their necks showing (hence the alternative name of "snakebirds"). Their feet are set well back for underwater swimming like a cormorant's or a grebe's, but their legs are long and they climb in trees with remarkable agility. Despite their weight they soar on thermal updrafts like crosses in the sky, with neck extended and long tail and wings outstretched.

PLOVERS AND LAPWINGS, SANDPIPERS, SNIPES, CURLEWS, DOWITCHERS, PHALAROPES, AVOCETS AND STILTS, JACANAS, PAINTED SNIPES, OYSTERCATCHERS, THE CRAB PLOVER, STONE CURLEWS, PRATINCOLES AND COURSERS, SEED SNIPES, THE PLAINS WANDERER, SHEATHBILLS, GULLS, TERNS, SKUAS AND JAEGERS, SKIMMERS, AND AUKS: ORDER CHARADRIIFORMES

The Charadriiformes is even more enormous and even more diverse than the Gruiformes, with no fewer than eighteen families. But, again, they seem to be a true clade. DNA studies suggest just one adjustment: the order should probably include the buttonquails, which are commonly included in the gruiforms. But then, the buttonquails have an odd mixture of features (somewhat quail-like, somewhat rail-like, somewhat plover-like) and were always difficult to place.

Family CHARADRIIDAE includes forty-one species of plovers within the subfamily Charadriinae, and twenty-four species of lapwing in the subfamily Vanellinae (although some of the Vanellinae are sometimes called "plovers" just to confuse). They are all called "waders," although most of them search for food along the water's edge and some plovers and a lot of the lapwings are grassland species with no special affinity for water at all. Plovers have some interesting breeding habits, including polyandry—one female with several males. Many plovers and particularly lapwings are known for feigning injury, to lure predators away from their nests. The trick may misfire, however—the wily predator concluding from the display that there must be eggs about. Their eggs are a prized delicacy. There are plenty of plovers and lapwings in the Northern Hemisphere, but more in the Southern, and the family as a whole almost certainly arose in the ancient southern supercontinent of Gondwana, perhaps in Antarctica, which now is the only continent where they no longer venture.

The family SCOLOPACIDAE has been around for at least 40 million years—plenty of time to produce the modern variety: eighty-six species of snipes, sandpipers, curlews, turnstones, dowitchers, godwits, redshanks, woodcocks, and more. In winter, sandpipers commonly account for more than half of the 2 million waders that throng the estuaries of Atlantic Europe, and North America has plenty, too. They all run quickly and swim well, but when the tide comes in they retreat first into the high salt marshes and then sometimes onto arable farmland. Their breeding is wonderfully various: some females lay up to three clutches in a year, each incubated by a different male; while in others, most famously the ruff, the males are polygynous and attract mates in collective mating displays known as "leks." Most of the family are prodigious migrators.

The three species of phalarope in the family PHALAROPIDAE have two principal claims to fame. First, they practice what is contentiously known as "gender reversal": the females are bigger, more colorful, do the courting, and leave child care to the males; more in Chapter 7. Second, their plumage traps so much air that they float like corks and in strong winds are blown from the surface of the sea and hither and thither. So, outside the breeding season, they may turn up as vagrants almost any-where; they literally blow in with the breeze (and are sometimes known as "galebirds"). Yet this doesn't stop them from embarking on heroic migrations. After breeding, Wilson's Phalarope in North America gather by the hundred thousand in saline lakes such as Utah's Great Salt Lake to molt and fatten before setting off to South America, while 2 million Northern Phalaropes may gather in Newfoundland's Bay of Fundy before migrating south. (Incidentally, the British call the Grey Phalarope the Grey Phalarope, because in winter it is indeed gray, but the Americans call the same bird the Red Phalarope because in summer breeding plumage, it is indeed red.)

Seven species of avocets and stilts, plus the Ibisbill, form the family RECURVIROSTRIDAE (although the Ibisbill is often placed in its own family, the IBIDORHYNCHIDAE). *Recurvirostridae* means "bent-back beak," which is what the avocets have—quaintly turning upward at the end. They are supremely elegant, with the longest legs of any shorebird, hunting in deep water for small creatures, as described in Chapter 5. The Black-winged Stilt is found in all six habitable continents—and in New Zealand is hybridizing with the Black Stilt, which is Critically Endangered. Others are much more localized: the Andean Avocet feeds only in lakes and marshes above 3,600 meters (about 12,000 feet). Some species are prodigious migrators.

There are eight species of jacanas in the family JACANIDAE. They are also known as "lily-trotters," for throughout the tropics they do indeed stride on long legs and extraordinarily long toes atop floating vegetation. They can be brightly colored, but—like brightly patterned coral fish—they can also be hard to see. As they land they flick their wings and then seem to disappear, distracting the eye the way cuttlefish do as they emit a cloud of sepia and then shoot off (or lie still). Young jacanas when alarmed can swim beneath the water with only the tips of their beaks showing—a trick that among birds seems unique. Some jacanas lead model, monogamous family lives, but in others the females are polyandrous and, like phalaropes, spectacularly reverse the conventional gender roles (more in Chapter 7).

There are only two species of painted snipes, in the family ROSTRAL-IDAE, both in tropical wetlands: the Greater in the Old World and the South American in South America. It isn't clear who the painted snipes are related to: they have been linked to true snipes, but also to jacanas and sheathbills. Their marvelous coloring is complicated: again, highly conspicuous against a plain background, but disruptive and all but invisible in their wild surroundings, especially as they tend to be active at evening time and they freeze when disturbed, like stone curlews.

You don't have to be a birder to spot an oystercatcher—generally crow-size, in flocks except when breeding, versatile mollusk-feeders with long, stout, typically bright red straight beaks—eleven species in the family HAEMATOPODIDAE. They are mostly coastal, but they are happy in brackish and fresh water, too, from all around the world both tropical and temperate. More are black and white, but four species are all-black (including the Alaskan species), while one species, the Variable Oystercatcher, includes both black-and-white and all-black forms.

The Crab Plover—just one species in the family DROMADIDAE—crunches crabs with a great black beak like heavy-duty pruning shears, along dunes, mudflats, coral reefs, and estuaries around the coasts of Madagascar, East Africa, Arabia, western India, and Sri Lanka. Since they focus on fiddler crabs, which come out at night, Crab Plovers are nocturnal, too. Crab Plovers are gregarious, stick together, and have sometimes mobbed human beings.

Stone curlews, otherwise known as "thick-knees"—because their long legs are indeed interrupted by knobbly knees—form the family BURHINIDAE: nine species in all continents except North America. They look a bit like curlews, though with fairly short, straight beaks; but they have big heads and huge round eyes under striking eyebrows, for hunting on the ground in twilight and by night—for they are partially nocturnal; and they tend to stand stock still, staring, in a most disconcerting manner, as if frozen to the spot. They are usually at home in dry, open country (of the kind that is becoming more scarce), but a couple of species like the banks of rivers and lakes and a couple more like coasts.

The eighteen species of pratincoles and coursers in the family GLAREOLIDAE between them live all over the Old World, both in deserts and at the water's edge. They are brilliant fliers with long pointed wings and forked tails, and they feed on insects on the wing like swifts or marsh terns. But unlike terns and especially unlike swifts they run fast, albeit on short legs (apart from the Long-legged Pratincole) and also take insects from the ground. The coursers, by contrast, have long legs and short tails—dedicated runners that take to the air only when forced to, and inclined, like stone curlews, to stop and stare at intervals. The Egyptian

Plover is also in this group—said by Herodotus to take morsels from between the teeth of crocodiles.

The THINOCORIDAE is yet another small group—just four species of seed snipes from the highlands of western South America. In many ways they are most unlike the other charadriforms. The two members of the genus *Thinocorus*, are like larks—the males even have a similar display flight. The two *Attagis* species are like small partridges—and, most unusually for birds in the Charadriiformes—they feed on the succulent parts of plants. *Attagis* species have a shutter or operculum over the nostrils to keep the dust out.

The Plains Wanderer from Australia has an entire family to itself—the PEDIONOMIDAE. The Plains Wanderer looks a bit like a rail, only more round-bodied, and indeed has sometimes been placed in the Gruiformes, but DNA studies suggest it is a charadriiform, perhaps allied to the seed snipes. As with the phalaropes, the females initiate courtship and the males look after the nests and young, although the females may help. They again tend to stand still when alarmed and have largely fallen foul of introduced foxes.

The two species of sheathbill—the top half of the beak is enwrapped in a horny "sheath"—form the family CHIONIDIDAE. They are extreme southerners, on islands in the Southern Ocean and on the Antarctic and South American mainlands. They look cuddly enough—something like a chicken and something like a pigeon—but in truth are holy terrors.

The gulls—around fifty-one species, although their taxonomy is fluid—form the family LARIDAE. Among them they form an entire "suite" from the Little Gull with its slender bill to the enormous, tough Great Black-backed with its heavy, slightly hooked all-purpose beak. The biggest of the ten genera is *Larus*, whose members generally have white heads, like the Herring and the Great and Lesser Black-backed Gulls. The "dark-hooded" or "masked" gulls in general are more slightly built, and in the summer in the breeding season they have a chocolate-brown or sooty-black head, like the Black-headed Gull of Europe, and the Laughing, Franklin's, and Bonaparte's Gulls of North America. Between them, gulls are spread all around the globe, but with far more in the north and not so many in the tropics, perhaps because the food supply is less certain. Though gulls are often called "seagulls," some are positive landlubbers. Kittiwakes spend their lives at sea diving for fish and nesting on the most precarious of sea cliffs, but even they may sometimes nest on window ledges. Herring Gulls routinely nest on buildings and are prominent members of seaside towns. The Great Black-headed Gull and the Relict Gull breed on islands on inland lakes and seas in the Asian steppes, many hundreds of kilometers from the nearest ocean. Franklin's Gull and others migrate over the Great Plains of North America to and from the highlands of Mexico, nesting and roosting in vast flocks in the open for an all-round view—sometimes on airfields, to the angst of all concerned. The most recent occurred out of LaGuardia Airport in New York City; a plane hit Canada Geese and had to ditch in the Hudson River.

Forty-four species of terns—the sea terns (or "sea swallows") and the various marsh terns—make up the family STERNIDAE, like slim gulls with long, pointed, swiftlike wings and forked tails. Like hawks and kingfishers, terns show us a complete spectrum of feeding strategies (see Chapter 5). Terns live all over the world; they don't go so near to the poles as some gulls do, but they are fonder than gulls are of the tropics and subtropics. The Roseate and Caspian Terns breed on all continents and thus are among the most cosmopolitan birds of all. Others are local. Some are prodigious migrators—notably the Arctic Tern (see Chapter 8). Most terns are monogamous and pair for life. Often they nest in mixed colonies of different species—other terns and nonterns. Some, inevitably it seems, are endangered.

The STERCORARIIDAE—seven species of the big, brutish skuas and the somewhat smaller versions known in the United States as jaegers—are nature's hooligans: serious predators and pirates. They are like big gulls, from which they are presumably descended, with a heavy beak that ends in a hook and webbed feet with long sharp claws. The Great Skua, heavyweight of the group, known in the Shetland Isles as the Bonxie, lives all around the North Pole but cannot move too far south because it is too well insulated. At the other end of the world, skuas venture closer to the South Pole than any other vertebrate. The name "Stercorariidae" derives from the Latin name for excrement (as in *stercoraceous*), and indeed the skuas and jaegers are stinkers. Yet they are mostly monogamous and mate for life (although about one in ten marriages break up each year when the pairs fail to breed). But the Brown Skuas around New Zealand commonly practice biandry—one female with two males.

The three species of skimmer that make up the RHYCHOPIDAE live on coasts and on big rivers throughout the tropics except in Australia. With their long wings—two and a half times the length of the bird—and their brilliant beaks and legs they are exceedingly handsome. They have a unique style of fishing with their extraordinary beaks.

The twenty-three species of seabirds within the family ALCIDAE are colloquially referred to as the "auks." They include the birds that are called auks (the most famous of which is extinct—the flightless Great Auk), and the auks include the smaller auklets and Dovekie. Also among the alcids are the puffins (three species), the guillemots (along with the murres and murrelets), and the Razorbill. Alcids are the northern equivalent of the penguins—North Atlantic, North Pacific, Arctic—except that most of them have retained the power of flight. Again the alcids form a "suite" of predators with the different species taking different kinds of prey, and their beaks are accordingly specialized. They nest in very various places, from narrow ledges, as in the Thick-billed Murre; to high coniferous trees, miles inland, as in the Marbled Murrelet, which is threatened by logging in the North Pacific; to long burrows with breeding chambers at the end, which they excavate themselves, like the puffins. Alcids may well have originated in the Pacific, where all of the genera still live except that of the Razorbill, though there was an extinct razorbill in California. All in all they are among the most abundant of seabirds and often the dominant players in northern waters; the populations of Common and Thick-billed Murres, Least and Crested Auklets, and Dovekies all exceed 10 million.

If I were following Joel Cracraft's tree slavishly (from *Assembling the Tree of Life*), I would discuss the flamingos at this point, together with the grebes. But I have already discussed them ultra-conservatively among the order of the storks. So on to the grebes.

THE GREBES: *ORDER PODICIPEDIFORMES*

The grebes—twenty-two species, all in the family PODICIPEDIFORMI-DAE—arose at least 70 million years ago. They surely originated in South America, where some still live in rainforest clearings and high in the remotest Andean lakes, but many now live in the north, where the bigger ones in particular, like the Great Crested of Eurasia and the Western and Clark's Grebes of North America, bring a serious touch of exoticism to rivers, lakes, marshes, estuaries, and sometimes the coast. I love the small ones, too, like the Little Grebe and the Black-necked. Grebes hunt underwater. They propel themselves with feet set well back on the body (grebes tend to be fairly hopeless on land) on highly flexible ankles, the toes not webbed like a duck or a gull but lobed, with flanges on either side of the toes. When frightened, they may even hide underwater. Most species can fly rapidly but prefer not to unless they are migrating—and they may migrate long distances. Sometimes on their migrations they mistake roads for rivers, land, and then are stranded. Two entire species and one subspecies are flightless. The Great Crested and the Clark's and Western of North America are best known for their extraordinary, extravagant, wonderfully choreographed and synchronized courtship dances, while the smaller species are famed for their vocalizations.

NOW FOR AN ORDER that seems to sit oddly with all these waterbirds. But modern DNA studies and some other similarities suggest that it belongs among them nonetheless.

NEW WORLD VULTURES, OLD WORLD VULTURES, EAGLES AND HAWKS, FALCONS, THE OSPREY, AND THE SECRETARY-BIRD: ORDER FALCONIFORMES

The Falconiformes are the raptors, where *raptor* means "seizing"—for they grab their prey with their talons rather than their beaks, as a gull or a swift would do. The owls are not included among the falconiforms. But even without the owls, the five families that are included are very diverse—and may not be closely related to each other at all.

The New World vultures and condors of the CATHARTIDAE now live exclusively in the Americas, all the way from Canada to Tierra del Fuego. But they may not have originated in the New World. The oldest known fossils of New World vultures—35 to 40 million years old (Late Eocene or Early Oligocene)—come not from the Americas, but from France; while the oldest New World fossils, from California and Kansas, are only 2 to 5 million years old (Pliocene). By the same token, Old World vultures clearly lived in the New World until about 10,000 years ago. In general, where animals live now, whether vultures or rhinos or elephants or jaguars, is just how things have turned out.

The Cathartidae today includes seven species. Five are called vultures. The Turkey Vulture and the Black Vulture are found in the United States.

The other three—the Greater and Lesser Yellow-headed Vultures, and the almost-white King Vulture with its multicolored bare-skinned head— live in the Neotropics (tropical Central and South America). The remaining two cathartids are called condors: the California and the Andean. The condors are the biggest flying birds with a wingspan of around 3 meters (10 feet) and a weight of up to 10.5 kilograms (23 pounds); some albatrosses have longer wings but are far lighter, while Mute Swans and bustards may match them for weight but not for wingspan. Yet there were giant relatives of the condors known as the teratorns (albeit usually placed in their own family, the TERATORNITHIDAE)—and the largest, *Argentavis magnificens*, from the Miocene of Argentina (12 to 5 million years ago), had an astonishing wingspan of up to 8.3 meters (27 feet) and must have weighed up to 100 kilograms (220 pounds), as much as a heavyweight boxer. Because its wings were so vast the body mass per unit area of wing (the wing loading), at 11.5 kilograms (25 pounds) per square meter, was well within the range required for flight. It was by far the biggest bird that ever flew (or at least the biggest that we know about).

The Secretary-bird, widespread in sub-Saharan Africa, now has the family SAGITTARIIDAE all to itself—though again, extinct species are known from France, dating from 20 to 30 million years ago. (If you had a time machine, I reckon France in the Oligocene would be a good place to visit.) In body and head the Secretary-bird resembles a hawk but it stands on long legs and can be more than a meter (3 feet) tall. It is called a Secretary-bird because of the long black-tipped plumes at the back of its head, like an old-fashioned clerk with his quill pen behind his ear. As in many raptors, the skin of the face is largely bare and brightly colored; and as befits a bird that lives in semi-desert, it has long eyelashes to keep out the dust. My wife opined, when we met one face-to-face in a local aviary, that it was a terrible old tart. But it's a formidable tart, which catches small animals on the ground by giving them a sharp kick, and grabs them when they are suitably subdued. Since it commonly catches snakes, this is a wise precaution. Most if not all raptors rely to some extent on impact to subdue their prey, but the only other predator I know that routinely kills by kicking or punching is a kind of prawn that crashes its way through the carapace of crabs. The Secretary-bird is a great flier, too, as raptors tend to be, riding the thermals with its legs thrust straight behind it, like a stork— except that, unstork-like, the central feathers of the tail are even longer than the legs. Like some eagles, it can stage some spectacular aerobatics—swinging upward, like a pendulum, then diving steep, not quite in the class of a peregrine but in similar vein. At night the Secretary-bird typically roosts atop a low tree at the heart of its extensive territory, which it guards jealously by kicking potential intruders.

The Osprey (aka "Fish Hawk") also has a whole family to itself—the PANDIONIDAE; but, unlike the Secretary-bird, the Osprey lives worldwide—in every continent except Antarctica. Ospreys are tremendous fishers; in the courting season, males show off their prowess to females with aerial displays that include the brandishing of fish. Most of the Ospreys in the Northern Hemisphere migrate south for the winter—the European breeders mostly to Africa, and the North Americans to Central and South America (though in each population, a few stay behind). Ospreys suffered hugely in the early days of organochlorine and organo-phosphorus pesticides, and are shot because they take game fish. On the whole, though, Ospreys seem to be increasing and spreading. They are magnificent birds.

The FALCONIDAE includes sixty-three species of falcons—and although falcons are sometimes confused with hawks, they are very different. The Falconidae lack the distinctive brow ridges of the hawks and eagles, and the wings of true falcons in the genus *Falco* are generally longer than those of hawks and pointed. Falcons are very much birds of open country, whereas hawks are often more at home in forest. There are plenty of falconids in the Northern Hemisphere, but there are far more in the Southern—so that is where they probably originated, as so many bird groups seem to have done.

The Falconidae is divided into two subfamilies.

Subfamily Falconinae includes the true falcons of the genus *Falco*, plus three genera of pygmy falcons and the falconets—who may well be ancestral to the true falcons. Quite a few species of *Falco*—including the Gyrfalcon, the Peregrine, the Eurasian Kestrel, the Hobby, and the Merlin—occur in both Eurasia and North America. But most of the Falconinae live in Africa, including ten species of kestrels.

The subfamily Polyborinae are the caracaras and forest-falcons, plus the Laughing Falcon. The caracaras are big, conspicuous, and often common in their native Latin America where, like crows or New World vultures, they tend to be generalist feeders with a penchant for garbage dumps and carrion. The forest falcons are long legged and more hawklike than most falconids and hunt in the canopy for small birds and lizards—and also follow columns of ants to pick up the insects and other small animals that they disturb. The Laughing Falcon feeds mainly on arboreal snakes. A snake that lives in a tree must feel it is pretty safe—but not necessarily, it seems. Phylogenetically and physically, the Laughing Falcon seems to provide a link between the caracaras and the forest falcons.

The biggest family of the Falconiformes with 234 species—a big family by any standards—is the ACCIPITRIDAE. The family is often divided into eight groups, although they are not true taxa. Indeed, phylogenetically speaking they probably overlap. But for practical purposes the groupings are useful, so here they are!

First, there are fifty-eight species of "true hawks," including sparrowhawks and goshawks.

Then come the Buteonine hawks—so called because they include the genus *Buteo*. *Buteo* includes the big soaring birds that the British call "buzzards," although Americans—at least in movies—tend to call any New World vultures buzzards. America's famous Red-tailed Hawk is a *Buteo*—extra-famous these days because one pair has taken to nesting on a fashionable apartment building to the east of New York's Central Park, seriously upstaging the celebrity residents. Some eagles also belong in this buteonine group, including the giant Harpy Eagle of South American forests, which feeds on monkeys and sloths, and the equally prodigious Great Philippine Eagle. At the other extreme, the small, white South American hawks of the genus *Leucopternis* are buteonines too, and eat insects and small reptiles.

Third are the thirty-three species of "true" or Aquiline eagles, which include the fabulous Golden Eagle, both of Eurasia and North America, the Martial Eagle, and the Wedge-tailed Eagle, which patrols some of the most inhospitable country in Australia, typically breeding only at five- or six-year intervals—when the weather happens to be particularly propitious. Some are known as "booted eagles" because they have feathers right down to their feet. The mighty Haast's Eagle of New Zealand also belonged among the booted eagles.

Fourth, we have the kites and honey-buzzards: twenty-nine species in fifteen genera. Britain has two natives in this group: the Red Kite and the Honey Buzzard. The Red Kite became extremely rare in the early twentieth century, and for decades was likely to be found only in Wales; but it was reintroduced to the northwest of London in the early 1990s and has already become a common sight along the main road (the M40) into Oxford.

Then come the Old World vultures—fifteen species of them. All have bald or partially bald faces and some—like the Griffon Vulture of North Africa, southern Europe, and Asia—also have bald necks. Others, though, like the Palm-nut Vulture of Africa and the handsome Bearded Vulture or Lammergeier of southern Europe, Africa, India, and Tibet, have a well-feathered head and neck. Those with featherless heads and necks can thrust their beaks into bloodied corpses without getting too sticky; but the Lammergeier feeds on bones and the Palm-nut Vulture feeds on palm nuts, so they don't need such an adaptation. You can always tell vultures from eagles by their feet. Eagles have powerful feet—sometimes huge, like size 12 boots, as in the Golden Eagle, for striking and seizing large prey; and sometimes short-toed, as in the kind that catch snakes. But vultures' feet are feeble. They are just for standing on.

Sixth in the catalogue are the ten species of fish eagles, which include such wonders as the Bald Eagle, symbol of the United States; the White-tailed Eagle, with an even bigger wingspan than the Golden Eagle, and newly introduced to Scottish islands and now to the mainland; and the mighty Steller's Sea-Eagle, biggest of all the living accipitrids.

Six species of snake-eagles and serpent-eagles form the seventh subgroup, including the Short-toed Snake-Eagle and the Crested Serpent-Eagle. They do indeed kill snakes, mostly in forest. For this they have short toes and heavily scaled legs, and big, owl-like heads. The Bateleur is in this group too, though it patrols the African savannah for small prey in general and for carrion. Probably belonging among the snake-eagles are the African and Madagascar harrier-hawks, which have peculiar double-jointed legs that enable them to winkle small creatures from holes in trees and rocks.

Finally, there are the sixteen species of harriers and harrier-hawks (we might call them "true" harrier-hawks), including the Marsh Harrier, Montague's Harrier, and so on. In Britain, Marsh Harriers cause much excitement among other birds and birdwatchers alike, as they patrol the wetlands in search of prey. In New Zealand a virtually identical harrier has taken to eating roadkills, in great gangs, and is commonplace. Birds are wonderfully adaptable.

ALBATROSSES, SHEARWATERS, PETRELS, STORM-PETRELS, AND DIVING-PETRELS: ORDER PROCELLARIFORMES

Procellariforms include four highly various families of specialist seabirds. Some look gull-like and some look auk-like, but all have the characteristic facial feature that proclaims their affinity, one to another: nostrils contained within tubes along the top of the beak—usually fused into one tube, though in albatrosses separated on either side of the bill. With their "tube noses," a procellariform is a procellariform is a procellariform.

There are many procellariforms in the Northern Hemisphere—petrels and shearwaters—but the order yet again seems basically Southern. Albatrosses and diving petrels live within roughly the same latitudes as penguins; like penguins, they favor cold currents even in low latitudes—for cold currents well up from the depths, so carry more nutrients, and so harbor more plankton and the squid and fish that feed on plankton. But procellariforms may not be quite so inveterately Southern as they seem. Back in the Pleistocene, 1.8 million to 10,000 years ago, albatrosses lived and bred in the North Atlantic. Even now, the Laysan Albatross breeds in the North Pacific; it is the second most common seabird on the Hawaiian Islands. The Galápagos Albatross breeds on the Equator—the Galápagos Islands; but then, the Galápagos Islands are bathed by the cold Humboldt Current from the south (which also accounts for the presence of the Galápagos Penguin). All the procellariforms are slow breeders. Most, even the smaller species, do not breed until they are three or four years old, while albatrosses are liable to be five or six before they breed or even as old as fifteen. When they do breed, they lay only a single egg. Thus, if their populations are reduced—and there are many threats in the modern world—they are slow to bounce back.

There are twenty-one species of albatross in the family DIOMEDEIDAE—although in the days before DNA studies, when ornithologists had to rely on appearance, it seemed there were only fourteen. The name "albatross" is from the Portuguese *alcatraz*, meaning "large seabird"; and *alcatraz* in turn derives from the Arabic *al-cadous*, meaning "pelican." Albatrosses stay at sea for most of their lives, gliding without effort and fishing as they go. Typically they mate for life, and if they do "divorce," which a couple may do if more than once they fail to breed, they typically remain single for three or four years before mating again—which takes a lot out of their reproductive life even though they live to about thirty. The total breeding cycle of the Wandering Albatross takes more than a year so that they, and at least nine other species of albatross, are able to breed only on alternate years. Slow breeding and the hazards of modern life mean that seventeen albatross species are now ranked as Vulnerable or worse.

The PROCELLARIDAE—seventy-nine species of shearwaters, petrels, prions, and the Northern and Southern Fulmars—extends from pole to pole: the most widespread of all bird families. Confusingly, the largest genus (including Manx Shearwater, Sooty Shearwater, Audubon's Shearwater, and others) is called *Puffinus*, though it has nothing to do with puffins (which are auks). Short-tailed and Sooty Shearwaters breed in Australia and New Zealand and avoid the southern winter by migrating to the north (into either the Atlantic or the Pacific). Cory's and Manx Shearwaters breed in the North Atlantic and sit out the northern winter in the South Atlantic. The Snow Petrel nests about 440 kilometers (280 miles) inland in Antarctica. The huge and formidable Northern and Southern Giant Petrel—both from the deep south—are fierce and ugly predators: they feed on the stranded carcasses of seals and whales, but they are also arch-pillagers of eggs and chicks and are big enough, with their 2-meter (6-foot) wingspan, to kill a King Penguin. They are sometimes known as "stinkers."

Some procellarids are doing well—the spread of Northern Fulmars in Britain since about 1950 is one of the classic success stories of British ornithology—but thirty-six of the seventy-nine species are now ranked as Vulnerable or worse.

The twenty-one species of storm-petrels, family HYDROBATIDAE, are the smallest of all seabirds: up to a mere 25 centimeters (10 inches) long. In bad weather they shelter in the lee of ships, which perhaps is how they came by their name. There are two clear groups, one in each hemisphere. The Northerners have pointed wings and shortish legs, and feed from the surface like terns. The southerners have rounded wings and longer legs, often dangling, and hover just above the surface, pattering the surface with their feet apparently to disturb or attract potential prey. So they seem to walk on water. It is commonly believed that the term *petrel* is derived from "St. Peter's bird"—for Peter also walked on water. But I am reliably informed that this is not true. It derives from the "pitter-patter" sound of their feet. Storm-petrels nest in holes in colonies where they lay a single egg that is relatively huge—only the kiwi's is bigger, relative to the mother's body weight. They prefer to visit their chicks at night, perhaps to avoid skuas and Herring Gulls who would otherwise gobble them up. You can hear them inside their burrows, chuckling and purring. America's National Audubon Society has cashed in on this: they have lured Leach's Storm-Petrels to new islands, with sound recordings of flourishing colonies.

The four species of diving-petrels of the family PELECANOIDIDAE that live around Australia, New Zealand, and the west coast of South America are truly brilliant birds. They are small—almost as small as a storm-petrel—and "fly" underwater after fish, like penguins, down to about 60 meters (200 feet); but also, quite unlike penguins, they fly through the air as well, shooting from the surface like sea-to-air missiles, with wings whirring like bumblebees—and then for good measure they may plunge into the face of advancing waves and out the other side. Again like storm-petrels, the Common and the Georgian Diving-Petrels fly into their colonies at night and then, though it's after lights-out, they chatter away. The Georgian squeaks and the Common says "kuaka," which the Maoris adopted as the name for the species. Peruvian Diving-Petrels used to excavate their burrows in guano, but most of the world's guano has been harvested and now they are Endangered.

FAIRLY CLOSELY RELATED to the procellariforms as a whole (though exactly how closely is still not clear) are the divers, alias loons.

THE LOONS OR DIVERS: ORDER GAVIIFORMES

In the order Gaviiformes are just five species of *Gavia* in the family GAVI-IDAE, which the British call "divers" and the Americans call "loons." They are beautiful, evocative waterbirds, slim and sleek with sharp-pointed beaks, all from the Northern Hemisphere, in both the Old and New Worlds, in freshwater lakes as well as the northern seas. They plunge from the surface to hunt underwater by sight, mainly for fish, in dives of around 45 seconds, though sometimes much longer, propelling themselves with their rear-set feet, like an outboard motor. They may use their wings for steering, too, but they can still fly. Their family life seems most agreeable. The two sexes call to each other in mournful synchrony—"duetting." They produce two eggs that hatch within twenty-four hours of each other, and both chicks leave the nest a day or so after the younger chick hatches.

For the next two weeks they may hitch a ride on their parents' backs, like baby grebes, or shelter under the warmth of a wing. They are pleasant birds all around.

APPARENTLY RELATED both to the procellariforms and to the divers, although again the relationship is not quite clear, are the penguins. They look bizarre—only some of the auks remotely resemble them—and were once thought to be quite distinct from other birds. But their DNA (as well as features of their anatomy) tells a different story.

THE PENGUINS: ORDER SPHENISCIFORMES

Penguins, as the New Zealand penguin watcher Lloyd Spencer Davis observes in his admirable *The Plight of the Penguin*, are birds trying to be fish. They are not as fishlike as whales, which breed at sea, and neither do they match the whales for size. The biggest penguin ever, the extinct *Platydyptes* from the Oligocene of New Zealand, was less than the height of a man—not even as big as a modest seal. Since water removes the burden of gravity, it seems strange that waterbirds in general have remained so small. Perhaps, again, the reason is that they lay eggs. Whales and the long-gone but very fish-like ichthyosaurs give or gave birth at sea to live young, but eggs—at least of the avian kind—would surely perish if laid at sea. So all birds have to come on land to breed. Even so, one might suppose, diving birds could still be far larger than they are. Elephant seals are remarkably clumsy on land, but still they contrive to be huge. So perhaps penguins failed to achieve large size because the competition has been too great—the leopard seals and sea lions got there first. But it remains a bit of a puzzle.

Penguins, all seventeen species of them in the single family SPHENIS-CIDAE, are birds of the south. Most species are based between 45 and 60 degrees of latitude—the southern equivalent of, say, southern Canada, or of Britain up to the Shetlands. But the greatest numbers (albeit from just a few species) are around Antarctica and its islands—down to 70 degrees south. Only the Galápagos Penguin ventures north of the Equator, but only by a whisker, since the Galápagos are roughly on the Equator. So penguins came to the attention of Europe only in the fifteenth and sixteenth centuries with the first global explorers: Vasco da Gama was first to report the Jackass Penguin of South Africa, and Ferdinand Magellan spotted the

Magellanic Penguin on southern South America, but many remained un-known until the eighteenth century, when sailors in bigger, better, and swifter ships made it to the deeper south.

Penguins prefer cold water. They are kept warm by three layers of dense feathers, trapping air, with a layer of fat beneath. The blood vessels in their feet and legs are highly efficient heat exchangers, as in a polar bear: hot blood going outward through the arteries transfers its heat to cold blood coming in via the veins, so penguins lose little heat through their feet even though they swim in freezing seas and may stand for months on polar ice, as the Emperor Penguin does when incubating. Even the penguins that live in temperate or even tropical latitudes—Australia, New Zealand, South Africa—contrive to find cold currents. For warm-blooded animals in the tropics, overheating is always a problem: the pen-guins that live farthest from the poles have big flippers that help them to lose heat, in the same way that African elephants have big flapping ears. They also tend to rest in burrows by day, like desert owls and gophers.

All penguins are submarine hunters and accordingly are wonderful swimmers. Emperor, Jackass, and Adelie Penguins have been timed at 5 to

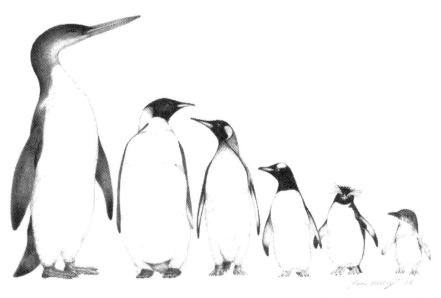

The largest penguin ever was Platydyptes, on the left. The largest living penguin, next in line, is the Emperor, and then the King. The Gentoo and the Rockhopper Penguin, the latter with its crests, are middle-sized, and smallest of all is the Little Blue.

10 kilometers per hour (3 to 6 miles per hour)—which may not sound enormous but is well up to Olympic swimming speed and relative to body length is up to dolphin speed. Often they "porpoise," leaping out of the water as they go, perhaps to save energy in short bursts (less drag in air than in water) and perhaps to make life harder for the leopard seals, sea lions, and killer whales that are happy to pick them off. Like auks they "fly" underwater with wings adapted as paddles and steer with their tails. As in grebes and divers, and other such swimmers, the penguins' legs are far back on their bodies—but unlike all other birds they have turned this to good account and evolved to walk upright, like humans. They may excite mirth, but some can walk and toboggan on their bellies over rocks and ice for many a mile. Emperor Penguins walk scores of kilometers across Antarctic sea ice to their breeding grounds. I have watched Yellow-eyed Penguins in New Zealand sidling up the beach as they must every night, with Southern fur seals to the right of them and sea lions to the left of them, properly wary of both, before dashing with all possible speed into the dunes to roost. In New Zealand, too, the Little Blue Penguins come ashore to roost in the forest—and on Maud Island, they find themselves roosting alongside the extraordinary ground parrots, the Kakapos. We think of penguins as birds of the cold and parrots as birds of the tropics, and so in general they are, but in New Zealand and Australia the two extreme forms live side by side.

THAT BRINGS US to the end of the first great subdivision of the Neoaves. The next great subdivision is occupied by a single order: that of the owls.

THE OWLS: ORDER STRIGIFORMES

Order Strigiformes contains two families of owls: family STRIGIDAE is just called "owls," while the TYTONIDAE is the barn owls and bay owls. They live throughout the world except Antarctica, and the Barn Owl is among a very short short list of birds that breed in every continent (others include the Osprey and the Roseate Tern). The number of recognized species is now 189, but because they are mostly nocturnal and all are cryptically colored, it is hard to sort them out, and new kinds are being discovered all the time, especially in the tropics. So the final cast list is anyone's guess. Even so, twenty-three of those that are known are threatened with extinction, with seven Critically Endangered. North America's Spotted Owl became a *cause célèbre* in the early 1990s—George H. W. Bush warning his fellow countrymen that if environmental measures were put in place "we'll be up to our necks in owls." More and more I am struck by the astonishing ignorance of powerful people in high places.

The largest owls, like the eighteen species of eagle owls and the related Snowy Owl, are a hundred times bigger than the smallest, the sparrow-size Least Pygmy Owl of South American forests, and the Elf Owl of the southwestern United States. But they are all unmistakably owls, with big wide heads that they swivel almost 360 degrees; huge forward-pointing eyes, excellent for gathering light; compact beaks; plump bodies; powerful, raptorial feet; and brilliantly designed wing feathers with loose barbs that enable them to fly, sometimes at great speed, in eerie silence. More of what they eat and how they catch it in Chapter 5.

THE NEXT GREAT branch of the Neoaves includes two orders that, like the ciconiiforms and pelecaniforms, need serious sorting. Conventionally, we have the Apodiformes, which includes the swifts, treeswifts, and hummingbirds; and the Caprimulgiformes, which currently are taken to include the nightjars, poorwills, nighthawks, frogmouths, owlet-nightjars, potoos, and the Oilbird. But they should probably be fused into one order—or else divided into several orders but along quite new lines. But just to be consistent, I will discuss the two conventional orders in turn as if they were bona fide taxa, and point out the problems as we go.

SWIFTS, TREESWIFTS, AND HUMMINGBIRDS: ORDER APODIFORMES

Order Apodiformes includes three families of remarkable fliers.

The first, the APODIDAE, includes the ninety-two species of swifts, which live the world over. Their superficial resemblance to swallows and martins is one of the most stunning examples of convergence in all of nature. Swifts and swallows are quite unrelated (swallows are passerines), but both have scimitar wings that give tremendous lift—although the swifts' are even longer. Both have forked tails for maneuverability—though the swifts' on the whole are less forked. You can tell swifts, too, because they are noisy. They "scream": they recognize each other by their voices. The young ones form screaming parties, as if to annoy the nesting adults, like teenagers revving their motorcycles. Swallows and martins have short legs good for perching, but the swift's legs are so short it can hardly perch at all; the *apod* part of the family name means "no feet." Indeed, it is often said that they spend their entire lives on the wing—but I am informed by expert ornithological chums that this is true only of the European Swift, and perhaps one or two others. Most species cling to buildings at night to roost. Swifts and swallows both are supreme "hawkers"—flying after insects on the wing without pausing in between, as bee-eaters generally do. The beak of the swift looks very short, but it opens into a tremendous gape—which indeed is characteristic both of the apodiforms and the caprimulgiforms. The swift does not simply trawl the air, open-mouthed, like a filter-feeding basking shark after plankton. It picks out its quarry, like a cheetah homing in on a particular gazelle.

Swifts, too, have extended the art of nest-building: they use their own saliva to make little cups that they stick to the roofs of caves and the eaves of houses (and some of the smaller swift relatives stick them to leaves). The nests of small swifts—swiflets—in Asia are the stuff of bird's nest soup. More of swifts in later chapters.

In the family HEMIPROCNIDAE are the four species of treeswifts that live in India and out to the Solomon Islands. Their name means "half swallow." They perch and have long tails like swallows, but their wings are more swiftlike and they stick their remarkable, tiny, paper-thin swiftish nests to thin branches: too thin to support a predatory snake. Three of the four treeswifts catch insects like a flycatcher—in short sallies from a perch. But the Whiskered Treeswift is a hawker like a swift or a swallow, except that it chases its prey through the forest canopy, which swifts would not.

Family TROCHILIDAE is the hummingbirds, all from the New World, from Alaska to Tierra del Fuego (they are not exclusively tropical) and out into the West Indies. Ninety percent of the 328 species are "typical" hummingbirds in the subfamily Trochilinae. The remaining 10 percent are the hermit hummingbirds, in the Phaethornithinae. All are small; the smallest, the Reddish Hermit of Guyana and Brazil and the Bee Hummingbird of Cuba, are the smallest warm-blooded vertebrates of all at less than 2 grams—you would get about fourteen to the ounce. Yet they are supreme, long-distance migrators. They are also the world's most specialist avian nectar feeders.

NIGHTJARS, POORWILLS, NIGHTHAWKS, FROGMOUTHS,
OWLET-NIGHTJARS, POTOOS, AND THE OILBIRD:
ORDER CAPRIMULGIFORMES

Five families make up the traditional though somewhat unconvincing order of the Caprimulgiformes.

Family CAPRIMULGIDAE includes eighty-nine species. The nightjars from the subfamily in the Caprimulginae spread over most of the world; and the nighthawks form the Chordeilinae, which is usually thought to be entirely New World, although some Old World nightjars in fact seem more like nighthawks. The nightjars and the nighthawks both eat insects—especially big ones—and have bristles around the mouth to protect their faces against flailing legs. In general they favor the tropics, which is where most insects live, though some migrate far into Europe or even up into Canada, or to the deep south of Argentina, to breed. They are, of course, largely nocturnal or certainly crepuscular. More in various contexts later.

In the family PODARGIDAE are the twelve species of frogmouths, in two distinct genera that look alike although their DNA shows they split apart a very long time ago. One genus is Asian, and the other is from Australia and New Guinea. They look like squat nightjars and hunt in much the same way, but are even more nocturnal. But frogmouths are best known for their camouflage: their plumage is mottled like bark and they stand stock still on a branch, beak pointing upward, for all the world like a broken stump.

Even by caprimulgiform standards the owlet-nightjars of the family AEGOTHELIDAE are extraordinarily elusive. Only one of the nine species seems known at all—the widespread Australian kind. The rest, from dense forest in New Guinea and some outlying islands, are rarely if ever seen, though most are thought to be Critically Endangered. Until 1998 the New Caledonian Owlet-Nightjar was known only from a single specimen that was caught in 1880. Owlet-nightjars look a bit like owls and a bit like nightjars, but with rounded wings, long thin legs and toes, and a long tail. Like nightjars they catch insects in their wide gaping beaks, though always in short sallies from perches—never hawking: catching prey on the wing. Localized and recondite though they now are, however, it seems that the owlet-nightjars may have flourished in the past. There are fossils from Eocene France, dating from 38 million years ago and one—

perhaps flightless—lived in New Zealand until a few thousand years ago. For reasons unknown and possibly for no reason at all, the eggs of owlet-nightjars have very thick shells.

The family NYCTIBIIDAE is yet another family of wide-gaped consumers of large insects—the seven species of potoos, from tropical America and the Caribbean. Again they are creatures of the night, and the larger potoos catch the odd bat. Like other night-lovers, potoos call loudly and distinctly. Some say "potoo," which is how they got their name, but others boom or bark or rasp or whistle. Like frogmouths, potoos stand stock-still when frightened like stumps of wood, although with their sleek feathers they also look remarkably toadlike. They close their eyes when they do this—though a small notch in the upper lid enables them to keep watch. It's amazing what refinements natural selection can lead to.

Finally, the extraordinary Oilbird—just one species in the family STEATNORTHINIDAE, from north and west South America and Trinidad. Like other caprimulgiforms they have a strong bill with bristles around the mouth, long wings and long tail, up to half a meter (20 inches) long. But by day they live in huge colonies deep in caves and treat intruders to snarling, screaming alarm calls that can all but deafen, besides frightening all but the most resolute out of their wits. Local people regard them as devils—they are known in Trinidad as *diablotín*. In their caves they also use sonar, like dolphins—you can hear the clicks. At dusk they emerge in a great screaming mass to hunt, but not for insects: for fruit. Indeed, among all birds, they are the only nocturnal fruit eaters—like fruit bats. Despite the dark they find the fruit mainly by sight, but perhaps also by scent. They even build their saucer-shaped nests out of fruit stones—and they roost in them at night: their nests are not just for breeding. Oilbird chicks, stuffed on the oily fruits of palms, soon grow fatter than their parents; more in Chapter 5.

The latest DNA evidence suggests that the swifts and hummingbirds are sister groups, and that the owlet-nightjars are the sister group of those two together. Potoos and nightjars seem to be sister groups too, with the Oilbird sister to those two—so these three form a second group. The frogmouths then form a group all by themselves. If the DNA evidence is right, then the present apportionment between the two conventional orders makes no sense.

PIGEONS AND DOVES: *ORDER COLUMBIFORMES*

Most places in the world have two or three local species of pigeon; after all, there are 309 species, all in the family COLUMBIDAE. Most pigeons are highly adaptable (though a few are specialists) and are fabulous fliers, and some, like Europe's Turtle Dove and America's Mourning Dove, are long-distance migrants. Sometimes the Columbidae are divided into subfamilies, but these may or may not be true taxa. The "typical" pigeons eat mainly seeds and include the Rock Dove, ancestors of the domestic pigeon; the wood-pigeons; and the smallest of all the pigeons, the sparrow-size ground-doves. The bright-colored fruit pigeons live throughout the Old World tropics—excellent seed dispersers. Then there are three species of crowned-pigeons, which live only in New Guinea—the biggest of the living pigeons by far, at more than 2 kilograms (4.5 pounds). All alone is the peculiar Tooth-billed Pigeon of Samoa. In most species the sexes are alike but some are highly dimorphic—like the Orange Dove of Fiji, where the males are bright orange and the females are dark green (and both have yellow heads).

CLOSELY RELATED TO the pigeons are the sandgrouse. The two might even be combined in one order (though there seems no pressing reason to do so).

SANDGROUSE: *ORDER PTEROCLIDIFORMES*

The sixteen species of sandgrouse in the single family PTEROCLODIDAE are supremely adapted to semi-desert, searching through most of the daylight hours for small, protein-rich seeds and other such stuff in the drylands of southern Europe, Africa, India, and China. They are beautifully protected against the daytime sun and the cold of winter and of nighttime by a dense undercoat of dark brown feathers that are arranged, not in discrete tracks like the feathers of most birds, but continuously, more like the fur of mammals (though still feathery). In some ways they are grouselike, with feathers around the nostrils to keep out the dust and feathered feet like snowshoes—and indeed have at times been classed among the gamebirds. Some, including Burchell's Sandgrouse of the Kalahari, can float on water like ducks and fly from it direct—so sandgrouse at times have also been linked to shorebirds. Many dust-bathe—the Pin-tailed Sandgrouse of southern Europe lie on their backs, with their feet in the air. Every now and again the population of Pallas's Sandgrouse in the steppes of Central Asia grows too large for the surroundings and then the birds "erupt," lemming-like, into Beijing or Europe. In the late 1880s, some even bred in Britain. They are quaint birds all around.

THE NEXT GREAT branch of the Neoaves contains the order of the parrots.

PARROTS, LOVEBIRDS, MACAWS, COCKATOOS: ORDER PSITTACIFORMES

Just one family again in the Psittaciformes—the PSITTACIDAE: 356 known species of parrots, lovebirds, macaws, cockatoos, and so on and so on. Again, though, several leading authorities now place the cockatoos in their own family—the CACATUIIDAE. Mostly the parrots are birds of the tropics—in the Americas, Australia, and New Guinea. Despite what Hollywood is wont to imply (no jungle without a parrot), there are relatively few in Asia or Africa. But many are happy to weather the cold. The Austral Parakeet lives in Tierra del Fuego and others live high on the cold mountains of the Andes, Ethiopia, and the Himalayas, while the Kea of New Zealand's Southern Alps positively enjoys the snow. The most northerly of recent times (apart from "escapees") was the Carolina Parakeet of North America, common in the early nineteenth century but shot to extinction by the early twentieth as an agricultural pest. The oldest fossil that is definitely a parrot was found in 55-million-year-old Eocene clay in Essex, northeast of London. It is named *Pulcrapollia*, meaning "pretty Polly." But then, the Eocene in general was tropical—roughly as the present world is predicted to be in a few decades' time. An even older fossil from Wyoming, which might possibly be a parrot, was a contemporary of the later dinosaurs, dating from the Cretaceous.

The huge array of species is grouped in seven subfamilies—as follows.

Subfamily Psittacinae—with 265 of the 356 species—includes the psittacids that are commonly known as parrots, lovebirds, macaws, amazons, parakeets, rosellas, the Budgerigar, and the extraordinary hanging-parrots of Asia, which roost upside down like bats (and look like leaves from a distance, which is good camouflage). Subfamily Lorinae—fifty-four species of them—is the lories and lorikeets. The Nestorinae has just four closely related species, all in one genus: the Kea, the two species of kaka from New Zealand, and the Blue Bonnet, or Bluebonnet, of Australia. Subfamily Strigopinae contains just one species—New Zealand's extraordinary, flightless, giant ground parrot, the Kakapo, which is the heaviest of all the parrots at up to 3 kilograms (7 pounds), but in the past, New Zealand had other species of ground parrots. In subfamily Cacuitinae are the cockatoos, known for their "erectile" crests and their intelligence—and their emphatic brightness, white or black or salmon pink; twenty species of them. Subfamily Nyphicinae includes just one species, the Cockatiel. Subfamily Micropsittinae contains the fig-parrots and pygmy-parrots, eleven species in all, the smallest weighing only 10 grams (a third of an ounce), spread through the tropics worldwide and into the southern United States. So the biggest parrot (the Kakapo) is 300 times bigger than the smallest.

Much more of parrots in later chapters.

CUCKOOS, THE HOATZIN, AND THE TURACOS: ORDER CUCULIFORMES

Three families are conventionally bundled into the Cuculiformes, although recent DNA studies suggest that they are not closely related to each other, or to anyone else. Each, probably, should have its own order. But for the time being, here they all are.

The 140 cuckoos in the CUCULIDAE are mostly tropical: the people of the north know only the hardy outliers (and they tend to migrate south for the winter, like the European Cuckoo, which takes off for Africa in the

Characteristically, Red-and-green Macaws fly arrow-straight, line ahead.

autumn). Like all the cuculiforms the cuckoos are zygodactylous (two toes pointing forward and two backward, as in parrots), which makes them very versatile. Some run up reeds like pipits while others run along the ground like fowl. So the Cuculidae are very diverse, and hence are divided into six subfamilies, which some taxonomists feel should form six separate families.

Three of the subfamilies are from the Old World. The biggest of them with fifty-four species—in fact the biggest of all the subfamilies—is that of the Old World cuckoos. The second largest contains the twenty-eight (some say thirty) species of coucals from Africa, southeast Asia, and Australasia. Coucals seem to be close to a grand common ancestor of the Cucilidae: it seems that the cuckoos evolved from coucal-like ancestors. The third subfamily includes twenty-six species of couas and malkohas—the former from Madagascar, the latter from Southeast Asia.

Three of the subfamilies are from the New World. The first includes the eighteen species of New World cuckoos (which in various ways are clearly distinct from the Old World cuckoos). The second includes the three species of ani, black-feathered but glossy and with parrot-like beaks, plus the Guira Cuckoo. The third New World subfamily contains ten species of ground cuckoos, which include the two species of roadrunner—the larger and more northerly of which, the Greater Roadrunner of the United States, is famous enough to warrant its own Warner Brothers cartoon.

All fifty-four of the Old World cuckoos practice "brood parasitism": they commonly lay their eggs in the nests of other species, whom they leave to foster the chicks, including of course the European Cuckoo, which has given us the word *cuckold*. But at least some of the time, some of the Old World cuckoos also raise their own young, in conventional style. Of the remaining eighty-six species in the family Cuculidae only three are brood parasites—and all of them are ground-cuckoos from the New World. It seems very likely that the Old World cuckoos and the New World ground-cuckoos evolved their brood parasitism separately—and perhaps the Old World cuckoos evolved their brood parasitism independently more than once. Some of the other cuculids also have interesting sexual and social lives even though they are not brood parasites.

The extraordinary Hoatzin—roughly pronounced "HWAT-son"—of northern South America has the family OPISTHOCOMIDAE all to itself. Hoatzins look like chickens with a loose crest of feathers on the head, which gives them an appearance of permanent surprise, but they live in trees along the banks of the Amazon and the Orinoco. Remarkably, Hoatzin chicks retain two claws on the wrists (the main bend in the wing) with which to clamber through the branches like slightly inept monkeys. Most other birds have these claws only as embryos. When alarmed, the baby birds drop into the river below, then swim ashore and clamber back up again. They lose their childish wing claws soon after they learn to fly, at about two months. Uniquely among birds, the Hoatzin is a "foregut digester": fermenting vegetation in an expanded chamber of the stomach like a cow.

The twenty-three species of turacos that form the family MU- SOPHAGIDAE have bright and beautiful feathers that are uniquely pigmented—with pigment compounded from ingested copper. One of the pigments, found in fourteen of the species, is a rich green and indeed is the only green pigment among all the birds: other green birds (of which, of course, there are many) achieve greenness by refracting the light through their feathers. The other copper-based pigment is crimson, and it appears in most species as flashes on the wings and on ornaments of the head. It takes the young turacos about a year to acquire the adult coloring—probably because they need that long to get the copper. But then turacos live across Central Africa, in or near one of the world's richest copper belts. The nestlings of turacos have tiny claws on their wing joints, Hoatzin-like. More later.

THE FINAL FLOURISH

The whole tree of the Neoaves is completed by a group of five related orders, of which two are small and local (both African) but nonetheless of great interest, and three are huge, various, and of immense ecological import all over the world.

The first of the seriously big and important orders is the Piciformes: the woodpeckers and all their related families. The second is the Coraciiformes, the kingfishers and their relatives. The Coraciiformes seem to be linked to the two small groups, the Coliiformes (which are the mousebirds) and the Trogoniformes (the trogons). Finally we come to the vast and ubiquitous order of the Passeriformes, alias passerines, which includes 60 percent of all living birds.

Between the orders of the woodpeckers and the kingfishers we see many extraordinary parallels and convergences. It's a reasonable guess that both orders originated from a common ancestor who roughly resembled a primitive kingfisher and caught insects on the wing or on the ground from short sallies from perches. The modern Kookaburra, which is a kingfisher, shows the way the first ever kingfisher might have hunted. Most modern kingfishers, however, have graduated from this primitive beginning to become fishers—like, of course, the flashing blue European Kingfisher, which is what most of us mean by "kingfisher." (North America has lovely kingfishers, too, but they are not quite so iridescently blue.) The modern woodpeckers went in a quite different direction: they became supreme peckers of wood (or sometimes of the ground or of termite nests—wherever tasty insects hide themselves). So it is that modern kingfishers, in the coraciiform family ALCEDINIDAE, and modern woodpeckers, in the piciform family PICIDAE, do not get in each other's way. They may and often do share the same habitat, with no competition. Accordingly, both families live all over the world, often side by side.

But within both orders, we find several families of smallish birds that all roughly look like kingfishers and that, in essence, retain the primitive ways. They continue, in the main, to catch insects in short flights from perches. Among the piciforms, the families of the puffbirds and nunbirds, and of the jacamars, continue to do this. Among the coraciiforms, the motmots, todies, bee-eaters, rollers, and the Cuckoo Roller have stuck to the original ways, albeit with some considerable refinements. When birds from different groups live similar lives, they can get in each other's way. It's no surprise to find that none of the seven piciform and coraciiform families of small insect chasers occur worldwide. The puffbirds and nunbirds and the jacamars from the piciforms are exclusively New World— basically from the American tropics. The motmots and todies from the coraciiforms are also New World; while the bee-eaters, rollers, and the Cuckoo Roller are exclusively Old World. It's as if the two sets of families from the related orders had agreed to carve up the world between them.

The most spectacular families from the two orders make the point beautifully. The toucans (and the closely related barbets) are piciforms, and the hornbills are coraciiforms. The parallels between the two are uncanny. Both have enormous, largely ornamental beaks. In both, the smaller species tend to be insect eaters, and the biggest (with a few notable exceptions) are fruit eaters. Again they have carved up the world between them.

The toucans are exclusively New World; the hornbills are exclusively Old World. No jungle seems big enough for the both of them.

So here are the final five orders:

TOUCANS, HONEYGUIDES, BARBETS, PUFFBIRDS, JACAMARS, AND WOODPECKERS: *ORDER PICIFORMES*

There are six families in the Piciformes.

Family RAMPHASTIDAE includes the thirty-four species of toucans—exclusive to the American tropics (the "neotropics"). The huge but light-weight beak is ideal for picking fruit: the bird can sit on a branch that can take its weight and reach out to twigs that cannot. The beaks are color-coded—each species seems to recognize its own kind. Some beaks are all the colors of the rainbow. Yet several species have identical beaks and plumage and can be told apart only by their calls, one species yelling and another barking. Toucans are playful and cooperative, like parrots. A favorite game is a form of quasi-cooperative volleyball, as they toss fruit to one another, catch it in their beaks, and pass it on.

The barbets—all eighty-two species of them, from Africa, south Asia, and both Americas—seem closely related to the toucans, but nonetheless are generally given their own family, the CAPITONIDAE. The name "bar-bet" has the same root as *beard*, as in *barber* and *Berber*, for most species have bristles around the gape of the mouth and sometimes covering the nostrils, which in some Asian species extend beyond the bill. They have branched out from their origins and feed largely on fruit, though some eat flowers and nectar as well and most and probably all eat some insects.

The honeyguides—fifteen species in sub-Saharan Africa and two in Asia—form the family INDICATORIDAE. They are called honeyguides because, extraordinarily, they solicit the help of passersby—people or other animals—to help them break into bees' nests. They are also notorious brood parasites.

The family BUCCONIDAE includes the thirty-four species of puff-birds, the somewhat larger nunbirds, and the Monklet, all from tropical America. Puffbirds are sit-and-wait hunters, mostly catching insects in a short dash. Some species save energy by sitting remarkably still—and re-main sitting even if people approach them and handle them. But then some of them smell bad, so perhaps their potential predators give them a wide birth. The Monklet, too, is extremely lethargic most of the time, but

The beak of the toucan—this one is the Toco Toucan—is partly functional (long-distance fruit picking) and partly for show (which is functional, too, because it attracts mates).

then goes after relatively large prey—the same basic strategy as a lion (though while lions chase zebras, Monklets chase large insects). Nunbirds are far more active, sometimes hovering like flycatchers.

The eighteen species of jacamars in the family GALBULIDAE from the New World tropics and subtropics seem closely related to puffbirds, but in starkest contrast are jewel-like with metallic colors and wonderfully lively—always looking up and around when perched, and given in flight to wild aerobatics.

Family PICIDAE includes the 218 species of woodpecker. They live everywhere that there are trees apart from Australia and New Zealand, and shun the highest latitudes; one lives in rocky landscapes where there are no trees. There are three subfamilies. Subfamily Picinae includes 185 species of "true" woodpeckers—keystone species that help to set the tone of their entire environments and turn up throughout the rest of this book. Piculets, thirty-one species in the subfamily Picumninae, are much like miniature but somewhat feeble woodpeckers—lacking the stiff tail, for instance. The two species of wrynecks in the subfamily Jynginae—both in the genus *Jynx*—are called "wrynecks" because of the way they twist and sway the neck, and hiss if they are threatened in the nest. This is all very snakelike and apparently does scare some of the predators some of the time.

KINGFISHERS, MOTMOTS, BEE-EATERS, ROLLERS, THE CUCKOO
ROLLER, THE HOOPOES, WOODHOOPOES, AND HORNBILLS:
ORDER CORACIIFORMES

Until fairly recent times—until the Pliocene, which began around 5 million years ago—the Coraciiformes was probably the most diverse and widespread of all bird orders: the world's principal small-to-smallish birds. Then for whatever reason, or combination of reasons, they began to lose out to the passerines, the perching birds. The coraciiforms are still a huge force in the world, with nine families totaling more than 200 species in a wide variety of habitats, many of which are key players within their own ecosystems. Yet the ones we see now, glorious and influential as they can be, are surely just a remnant of former glories. Coraciiformes in general have a squat body, short neck, big head, and strong beak, with the second and third toes partially fused into one. Some divide them into nine families, and some into ten.

Best known and most widespread are the eighty-six species of kingfishers in the family ALCEDINIDAE. Mostly they live in the tropics—most still live around New Guinea, where the family most likely arose—but now they spread far to the north, to Finland in the Old World and Alaska in the New, and far down into South America. By no means are all fishers: the family as a whole shows a wonderful range of feeding techniques.

The ten, highly diverse species of motmots in the family MOMOTIDAE now live only in Central and South America, but they must have been widespread once: there is a fossil from Switzerland. They are very handsome, like small kingfishers with a bright green or turquoise back, some with long central tail feathers that lose the barbs at the end so they hang like wires and swing from side to side. As we have seen, they mostly catch insects on the wing. Like kingfishers they nest in burrows dug in banks or in the ground.

Todies—five species in the family TODIDAE, all from the Caribbean—are quaint, very tame, and approachable birds that look like small motmots. All again are brilliant green above, and all have a crimson throat patch that puffs out when they call. But on the breast they are color-coded by species: whitish, pale gray, brownish, yellow, green, pink. They have short tails (unlike motmots) and again catch insects on the wing in their long, straight, flat beak, plus the odd lizard. Much of the day they sit around, like the piciform puffbirds. Both sexes make a whirring noise with their wings as they fly, perhaps for courtship. Compare the drumming of the male snipe.

So to the bee-eaters: twenty-four species in the family MEROPIDAE, all from the Old World—Eurasia, Africa, Madagascar, New Guinea, with just one species in Australia: the Rainbow Bee-eater or Rainbowbird, which is seen as a herald of spring. They are dashing birds, bright and sometimes dazzling like small kingfishers, though all have a black mask over the eyes as if to emphasize their raffishness. Like the kingfishers they are mainly tropical and probably arose in Southeast Asia, then spread to Africa and then Europe. Mainly they eat honeybees—and their range, through the Old World, corresponds with that of the bees. More in Chapter 5.

The world's seventeen species of roller live all over the warmer parts of the Old World: Africa, Eurasia, New Guinea, and eastern Australia. They are sometimes grouped into one family, the CORACIIDAE, although some prefer to hive off the ground-rollers, which live in Madagascar, into their own family, the BRACHYPTERACIIDAE. The Cuckoo Roller, also known as the Courol, is also exclusive to Madagascar and the nearby Comoro islands, and has its own family, the LEPTOSOMATIDAE. The three families (or two) seem to form a natural grouping, which for people with a taste for such things might be called a superfamily. They all resemble small, colorful crows; they are called rollers because of the way the "true" rollers (genus *Coracias*) and the broad-billed rollers (genus *Eurystomas*) tumble most marvelously in their courtship flights. They spend a lot of time in the air, but they drop to the ground to feed on small animals.

There is just one species of Hoopoe in the family UPUPIDAE, albeit spread over Eurasia, Madagascar, and Africa (though the African one is sometimes placed in a separate species). Their characteristic crest gives their heads the profile of a pick—as shown in murals in Ancient Egypt and in Minoan Crete. Again they eat invertebrates, but unlike most coraciiforms they probe for them with their curlew-like beaks—more like Green Woodpeckers. Often they sun themselves on the ground with wings and tails spread, long bill pointing to the sky, slightly open. They call like doves from treetops or rooftops—the name "hoopoe," and indeed Upupidae, is onomatopoeic. They fly like butterflies on short rounded wings, but like many a bird that seems to fly feebly (and indeed like some butterflies), they migrate great distances—for example, from Europe down into Africa. They can be agile in flight when they need to be, but still they are picked off, in significant numbers, by Sooty Falcons. Hoopoes nest in cavities wherever they can find them, including old woodpecker holes and drainpipes, in termite mounds, and often in the walls of houses even in busy villages. They keep the entrance narrow to repel intruders. The chicks actively repel intruders, too; they hiss, they spray excreta (compare the spitting of fulmars), and they secrete a foul liquid like rotting meat from the preen gland—an exercise in chemical weaponry.

The family PHOENICULIDAE with its eight species of woodhoopoes and scimitar-bills is among the very few bird families endemic, or exclusive, to Africa. Apart from their beaks—straight in the woodhoopoes proper, curved in the scimitar-bills—the two groups look very alike. But they behave differently: most woodhoopoes tend to gather in small extended family flocks, noisy and cackling; the scimitar-bills, by contrast, are generally solitary. Their DNA reveals that in truth the two subgroups separated from their common ancestor about 10 million years ago, way back in the Miocene—the great age of coraciiforms. Both groups hunt for insects in the bark of trees, sometimes supporting themselves with both tail and wings as they clamber through the small twigs. Not many birds use their wings for climbing. Why not?

The biggest coraciiforms are among the fifty-four species of hornbills in the family BUCEROTIDAE. Again, they are exclusively Old World. Just under half of the species live in Africa while the other half are Asian, from the Arabian Peninsula east to New Guinea. They seem closely related to Hoopoes and woodhoopoes—I like to think of the Hoopoe as a miniature hornbill. It's likely that their common ancestor was African.

Hornbills are named for the casque: the projection, often huge, that runs lengthwise along the top of the beak. In many the casque is a simple ridge, but in some it is cylindrical, or turned up at the end, or folded, or inflated, and sometimes is bigger than the beak itself. In the Great Hornbill of Asia in the genus *Buceros*, the casque has two prongs at the front, while the casque of the Rhinoceros Hornbill, also *Buceros*, turns up at the end. In males the casque is bigger and in old males bigger still—it grows with age like the tusks of elephants. In most species the casque is like honeycomb inside, strong but light. In the Black-casqued Hornbill of Africa and its relatives in the genus *Ceratogymna*, the huge casque opens directly in the mouth and so apparently serves as an echo chamber, enhancing the call (compare the various elaborations of the trachea which act as amplifiers in some other birds). In the Helmeted Hornbill of Asia the casque is solid, and known as "hornbill ivory," though it is made of keratin, like feathers, hair, and rhino horn—not dentine, like the ivory of an elephant or a walrus. It accounts for about 10 percent of the bird's body weight and is prized and elaborately carved by the local tribespeople, which is not good news for the bird. For their own purposes, male Helmeted Hornbills use their massive beak gear in midair battles for territory, like flying billy goats. More mundanely, they may also use the casque to dig small animals from rotting wood. But they also feed on figs. Figs, all 750 or so species of them, grow throughout the tropics—and the world's fauna would be seriously impoverished without them.

The two ground-hornbills are among the world's most formidable avian hunters, cornering their prey like wolves. They differ from other hornbills in other ways, too, (they even have one more vertebra in the neck) and are often placed in their own subfamily, the Bucorvinae. They come from Africa but seem more like *Buceros*, an Asian genus: both carry the same species of feather-lice that is an unsavory clue to family ties, but is often significant. In both, too, the preen gland is densely covered in feathers, which enables the bird to spread its oils more efficiently. The ground-hornbills use these oils to color the bill—red, orange, or yellow. This is the only example I can think of, in all of nature, of an animal using cosmetics. Darwin suggested that everything humans do has its precedent in other creatures. Who would have thought, though, to look for precedents for such a quirk among ground-hornbills?

THE TROGONS, INCLUDING THE QUETZALS:
ORDER TROGONIFORMES

Just one family—the TROGONIDAE—with thirty-seven species completes the order of the Trogoniforms. Some look a bit like parrots, beautifully bright and long-tailed, and some are more like finches, though they obviously are related to neither. They live throughout the tropics and subtropics, though more New World than Old. They are creatures of the trees, typically hovering briefly before plucking fruit, which they often swallow whole (regurgitating big seeds later), or whatever small animals they may find on leaves or branches. Best-known of the trogons and among the most beautiful of all birds is the Resplendent Quetzal of Mexico and Central America. The male is metallic green above and crimson below, with a long curving tail that in the breeding season is twice as long as its body, and hangs languidly when it perches and ripples as it flies. Traditional peoples prized quetzal feathers more highly than gold, but they harvested plumes from live birds, which they then released. Among the Mayas, to kill a quetzal was a capital offense. But the Spanish soon put a stop to such nonsense, and for the next 400 years plunder was the only rule, so now the Resplendent Quetzal lives only in the remotest mountains—with the last few threatened by the spread of coffee plantations. The four other quetzals of South America are just as bright but lack the tail.

THE MOUSEBIRDS: ORDER COLIIFORMES

In the order Coliiformes is just one family, the COLIIDAE—the six strange and charming species of mousebirds. They look like mice—dumpy, fluffy, brown or gray—and behave like mice, running under or through thick bushes in woods or scrub and sometimes in gardens, typically in permanent social groups of three to twenty. They like to hang under twigs as parrots often do, rather than to sit on top, and again like parrots, they use their feet to transfer food to the short, stubby bill. Mousebirds are extremely social. They often preen each other ("allo-preening"), and for this they hang belly to belly, either side of a twig, most quaintly. They are cooperative breeders: the young of previous broods, sons and daughters, stay around to help raise the new generation. The eggs, two to four of them, are among the smallest relative to body weight of all birds. Sometimes both parents incubate together—togetherness is definitely their thing. Sometimes helpers incubate. The Speckled and the Red-faced Mousebirds, also from Namibia, like to nest near nests of aggressive wasps. This a favorite trick of many kinds of bird. The wasps keep the predators away.

Mousebirds are very common and successful where they occur—here and there in Africa, south of the Sahara. But since they generally prefer to stay at home, they have not spread elsewhere.

THE PERCHING BIRDS: *ORDER PASSERIFORMES*

No one knows when the passerines first appeared—how could they?—and estimates vary enormously. Alan Feduccia emphasizes that the earliest known passerine fossils date from 40 to 50 million years ago. Others guess on the basis of zoogeography—where the ancient and modern types occur—and from the range of variation within their DNA that they must have originated in the Cretaceous, perhaps as much as 80 million years ago, so the first ones might have perched on the backs of dinosaurs. Either way, it seems they made a fairly quiet start, but then multiplied rapidly in the Pliocene to become the most various order by far: 60 percent of all living birds are passerines. This, you might well conclude, is because they are innately superior—extremely agile, often intelligent (they include the crows), and fabulous communicators (they include the songbirds).

But although the passerines do have many assets, they have probably become so various mainly because they are good at being small, and there is more room in the world for viable populations of small creatures than big ones. The biggest passerine is the Greenland race of the Raven, but it rarely exceeds 2 kilograms (4.4 pounds). Their rise these past few million years seems to correlate with global cooling and drying out, and the spread of grassland at the expense of forest. Many passerines—buntings, sparrows, finches, weavers, goldfinches—have evolved to become specialist consumers of small seeds. Their chief rivals in the small-bird stakes, coraciiforms and piciforms, are mainly insectivorous or eaters of fruit—and either way tend to favor the diminishing woodlands.

What makes a passerine a passerine? In 1982, Robert J. Raikow of the University of Pittsburgh described no fewer than eighteen distinctive characters that passerines share, but he concluded that only five of them were unique to passerines and could therefore be used to define them.[3] One feature that unites them all is the arrangement of bones in the palate (a special variation of the neognathan palate). Another is their peculiar sperm—passerine spermatozoa have long coiled heads, and they tend to cluster.

But of all their unique features, one is outstanding: their feet. Passerines have a very long and strong hind toe (the hallux), which they can wrap around a twig (while a cormorant, say, can merely balance). Parrots and rollers and barbets can do this, too, of course, but they are zygodacty-lous—two toes forward and two back. Also, passerines have a peculiar ar-

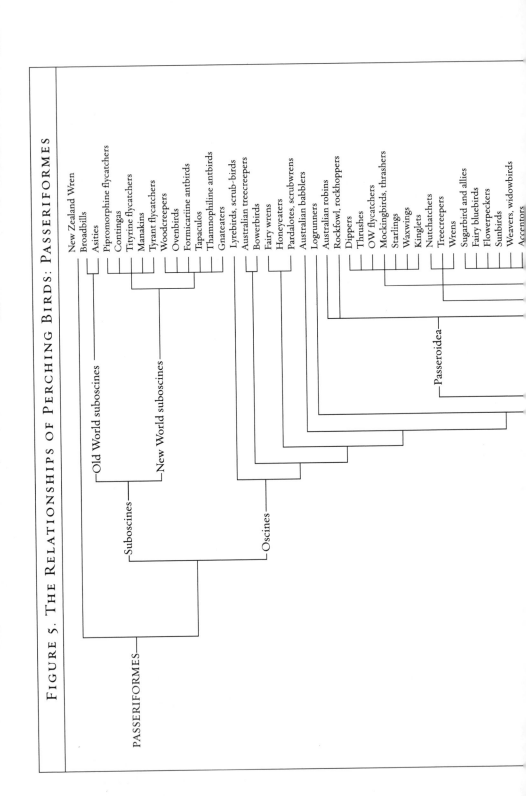

FIGURE 5. THE RELATIONSHIPS OF PERCHING BIRDS: PASSERIFORMES

Buntings, sparrows
NW orioles, blackbirds (icterids)
NW Warblers
Chaffinches, bramblings
Wagtails, pippets
Tits, chickadees
Larks
Swallows
Cisticolid warblers
Bulbuls
Sylviid warblers
White eyes
Babblers
Mud-nest builders
Melampittas
Monarch flycathers
Crows, jays
Shrikes
Rhipidurine flycatchers
Drongos
Wattle-eyes, batises
Ioras
Helmetshrikes
Vanga shrikes
Bush shrikes
Currawongs, butcherbirds
Wood-swallows
OW orioles
Whistlers, shrike-thrushes
Cuckooshrikes
Tit berrypicker
Shrike-tits
Erpornis
Vireos
Pitohuis
Crested Bellbird
Whipbirds
Sittellas
Berrypicker, longbills
NZ wattlebirds
Cnemophillines

Corvoidea

About 60 percent of all living birds belong to the order Passeriformes, colloquially known as passerines. Their relationships are very difficult to sort out, but molecular studies added to centuries of traditional ornithology are making things clearer. This is the present state of understanding, as presented by Cracraft and Donoghue (2004).

In general shape, the phylogenetic tree of passerines follows the pioneer studies of Sibley and Ahlquist (1990), but Cracraft suggests many refinements. All is explained in the text, but here are a few salients. The New Zealand Wren is a very small bird but has the distinction of being the sister to all the rest (or so it seems). Most of the passerines belong to the group known as the oscines—the "true" songbirds. The oscines in turn include a number of primitive groups, such as the bowerbirds and lyrebirds, but most of them belong either to the Passeroidea, which include the finches, larks, and so on; or to the Corvoidea, which include the crows, shrikes, and so on. There is a great deal of convergent evolution within the passerines—not least between various corvoids and various passeroids.

rangement of muscles and tendons in their legs that causes their feet to grip the branch securely whenever they land. They do not have to cling on. A sparrow may die while perching and stay perching. It is the same kind of trick by which an albatross locks its wings in the gliding position and so can cover an ocean with almost no effort. On the other hand, the arrangement that enables passerines to perch so well also reduces the fine control of their toes (although it is not clear what fine control most birds need in their toes). This arrangement explains why the passerines are known specifically as "perching birds" even though they are not the only birds that perch.

As Robert Raikow's account makes clear, the passerines do emerge as a true clade. Their unique, common ancestor, the first ever passerine, almost certainly arose at the Australasian end of the ancient southern supercontinent of Gondwana. Some fossils suggest this. So, too, does the distribution of living passerines. A high proportion of them live only in Australasia, including many entire families: lyrebirds and scrub-birds, Australian treecreepers, bowerbirds, fairywrens, honeyeaters, scrubwrens, Australian babblers, and logrunners. Some Australasian families contain only one or a very few species. This suggests that those families diverged early from the rest, but have since fallen on hard times and lost most of the variety they might once have included. Finally, many of the most primitive living passerines are from Australasia: birds that apparently have not deviated far from the ancestral state and have stayed near their ancestral home this past 80 million years or so.

The 6,000 or so species are conventionally grouped into about eighty families. The present classification is immensely valuable and must be on the right lines, though there are many loose ends. Some of the conventionally recognized families will surely be broken up, as indeed is happening before our eyes. Others will be amalgamated and regrouped. But the eighty-two families that Perrins (2003) recognizes, and are described here, will certainly do.

Sibley and Ahlquist, in their seminal volume of 1990, *Phylogeny and Classification of Birds*, divided the passerines into two groups: the Suboscines and the Oscines. They differ in the structure of the syrinx (the voice box). The oscines are the ones known as the "songbirds" and indeed include the world's most noted songsters, like thrushes and the Nightingale and wrens and so on. But the suboscines sing, too. Sibley and Ahlquist then divided the suboscines into two: the New World suboscines and the Old World suboscines. They also divided the oscines into two: those that are vaguely sparrowlike, which they called the Passerida, and those that are more crowlike, which they called the Corvida.

The scheme I am following here is by Joel Cracraft and his colleagues, first published in 2004 and explained in the caption to Figure 5. It follows Sibley and Ahlquist (1990) in general form, but there are some significant alterations.

First, in Cracraft's scheme, the New Zealand wrens emerge as the sister group of all the rest. The fact that they live in New Zealand, close to their Gondwanan origin, is surely significant. Sibley and Ahlquist simply placed the New Zealand wrens at the primitive end of the Old World suboscines.

Cracraft has also modified Sibley and Ahlquist's view of the oscines. Most of the oscines are still placed either in the Corvida or the Passerida. But there is a mixed bag of eight families—all from Australasia!—that don't seem to belong to either. Sister group of all the oscines are the lyrebirds and scrub-birds of Australia (which Cracraft places in the same family, although Perrins separates them). Then come the Australian treecreepers and the bowerbirds. Then the fairywrens, honeyeaters, and the pardaloes-cum-scrubwrens. Then the Australian babblers, and then the logrunners. All of these are oscines, but they are primitive outliers within the oscine suborder nonetheless.

The following account is divided up in line with Cracraft's scheme.

SISTER TO ALL THE PASSERINES: THE NEW ZEALAND WRENS

The New Zealand wrens, family XENICIDAE, should be treasured not simply as fine little birds but also because of their unique evolutionary status. About 80 million years ago, so it seems, the world's first population of passerines split into two. One group evolved to become crows, shrikes, finches, swallows, and all the rest—6,000 species of all shapes and ways of life. All that remains of the other group is the New Zealand wrens. The two that are still left to us are the small, yellow Rock Wren and the small, yellow Rifleman, while a third, the Bush Wren, has not been seen for certain since 1966 and is probably extinct. Certainly extinct is the most famous New Zealand wren of all, the Stephen Island Wren, which, uniquely in the history of birds, was made known to science by a domestic cat and then finished off entirely a few months later by the same cat.

THE OLD WORLD SUBOSCINES

There are only three families of Old World suboscines and these are sometimes combined into two, as follows.

The thirty-two species of pitta in the family PITTIDAE live in the tropical forests of Asia and Australia, with two in Africa. "Pitta" is a local Indian name for "small bird"—not unlike the western birder's LBJ ("little brown job"). But some pittas are beautiful, like jewels.

Brightly colored, too, are the four species of asity, indigenous to Madagascar; and the stocky broadbills, from China, India, and sub-Saharan Africa. The asities are traditionally placed in the family PHILEPITTIDAE, and the broadbills are traditionally in the EURYLAIMIDAE, but both tend now to be combined into one family—albeit one with a very odd and scattered distribution. Asities have long curved beaks with long tongues for sipping nectar, like sunbirds and honeycreepers (which are both passerines) and hummingbirds (which of course are not). Broadbills eat insects and other arthropods or fruit, like coraciiforms.

THE NEW WORLD SUBOSCINES

The New World suboscines are a far bigger group, with nine families.

The TYRANNIDAE is a huge family of enormous ecological impact containing nearly 440 species of tyrant flycatchers, extending from the north of Canada to the south of Patagonia, but especially in Amazonian forest. Some tyrant flycatchers sometimes lay their eggs in other birds' nests—though usually they dump on other birds of the same species, so it's tit for tat. But some tyrant flycatchers also turn birds of other species out of their nests and take them over, even if the nest's true owner is already brooding. Like many birds, they like to build their own nests near to wasps' nests, for protection.

The cotingas—ninety-four species in the family COTINGIDAE—are mainly South American, often very brightly colored but again hugely diverse. Best known are the polygynous and spectacularly scarlet cocks-of-the-rock, which tout for mates in the quasi-cooperative groups of males known as leks, of which more in Chapter 6.

Related to the contingas are the fifty species of manakins in the family PIPRIDAE, also from the American tropics, mainly Brazil. Their elaborate courtship, including the spectacular Catherine wheel mating dance conducted by the dominant and his sidekick, is described in Chapter 6.

The FURNARIIDAE, the 217 species of ovenbirds, range from Central America all the way down into the Falklands. They are known for their wondrous nests, some of which are just like clay ovens.

Closely related to the ovenbirds are the fifty-two species of woodcreepers in the family DENDROCOLAPTIDAE—yet another group of Latin Americans, though most diverse in the Amazon, where they probe for insects with their long, sickle beaks like diminutive curlews or ibises.

Two quite distinct and not particularly closely related families from tropical America are both known as antbirds: the THAMNOPHILIDAE, with 204 known species; and the FORMICARIIDAE, sometimes known as "ground antbirds," with sixty-two species. Some members of both families make their living by following columns of army ants.

The gnateaters, eight species, also catching insects in tropical South America, were at one time classed among the antbirds but now have their own family, the CONOPOPHAGIDAE. In those parts, too, catching insects and spiders and eating the occasional plant, are the fifty-five species of tapaculos in the family RHINOCRYPTIDAE (which seems to mean "hidden nose").

THE PRIMITIVE OSCINES

Now comes a group of eight families that Sibley and Ahlquist placed in the suborder Passerida, but which Cracraft suggests should be classed as primitive oscines that cannot be classed as either passerids or corvids. All of these primitive outlying families are Australasian—from either Australia itself or New Guinea or New Zealand. Here is evidence indeed that passerines as a whole began in Gondwana and very possibly in Australasia.

As you can see in Figure 5, the lyrebirds and scrub-birds seem to be the sister group of all other oscines.

Just two species of lyrebird, Albert's and the Superb, form the family MENURIDAE. They look vaguely like gamebirds, with whom they were once classed, with pheasant-like tails, which in times of excitement do indeed assume the form of a lyre. But passerines they very definitely are. They are known for the flamboyance of their courtship and in particular for their spectacular powers of mimicry.

The two scrub-birds, ATRICHORNITHIDAE, are vaguely thrushlike but stiff-tailed birds that both live deep in the bush and—strangely—live 3,000 kilometers (1,800 miles) apart from each other. Presumably there once were scrub-birds between the two present outposts that have since disappeared. One of them, the Noisy Scrub-bird, became known to Europeans only in the mid- to late nineteenth century, but then seemed to go extinct—only to turn up again in 1961. Intensive conservation has now restored its ailing fortunes to something like safety levels.

Next in line are the seven Australian treecreepers in the family CLI-MACTERIDAE. Again they have endearing breeding habits. Youngsters—usually sons—stay around to help their parents with the next brood, to form cooperatives of up to eight individuals. Quaintly, some Australian treecreepers seek to deter predators (or so it seems) by smearing the entrance to the nest hole with snakeskin or the wings of insects.

The twenty species of bowerbirds in the family PTILONORHYNCHI-DAE are known for the marvelously elaborate love palaces that the males of some species build purely for the purposes of seduction.

The fairywrens, grasswrens, and emuwren, twenty-seven species in all, form the family MALURIDAE. The fairywrens are beautiful, in many shades of blue, including the tail of the Splendid Fairywren, which it spreads into a turquoise fan. Like wolves, most fairywrens live in groups of adults and youngsters, but only one of the several females lays eggs, while the youngsters help to look after later broods. Unlike wolves, the senior male fairywren does not have exclusive rights to the boss female: her eggs may have two or several fathers, some of whom may not be in the social group. She picks out the fathers in inspections carried out before dawn. When the young are fledged after a mere ten days or so, they cannot fly, and when a predator approaches, the whole group runs away like mice—squeaking for good measure—and trailing their tails.

More Australians—but also in New Guinea, Indonesia, New Zealand, and Hawaii—are the MELIPHAGIDAE: 177 species of honeyeaters, plus the five species of Australian chats, who once were placed in a separate family, the EPTHIANURIDAE. Varied and adaptable as they are, they show all the marks of a group that was once even more successful and is now reduced to relics—for fourteen out of the forty genera of honeyeaters contain only one species each, and another ten have only two species each. They eat nectar as well as small insects. In the four species of honeyeaters known as "miners," up to thirty individuals may help a single breeding pair to raise their brood.

The ACANTHIZIDAE, also known as the PARDALOTIDAE (also known as pardaloids), are pleasant little birds, not spectacular, but almost half of them—seven from a total of sixteen species—are Vulnerable or worse, and one of them was last seen in 1936. Like many a small creature they have been swept aside without anybody noticing, apart from a few naturalists.

The ORTHONYCHIDAE have spread from Australia and New Guinea into Southeast Asia; the sixteen species of logrunners and their relatives, elusive birds of the undergrowth. So, too, the POMATOSTOMIDAE: five species of Australo-Papuan babblers, which resemble the babblers of Asia but are of a quite different family—another example of convergence.

THOROUGHLY MODERN OSCINES

The most derived of all the oscines are the sparrow-like Passerida and the crow-like Corvida. They both seem to have arisen from the same lineage of primitive oscines, and so are sister groups.

THE SPARROW-LIKE OSCINES: THE PASSERIDA

The Passerida seems to divide into three main lineages. The first includes the Australo-Papuan robins (Australasia again). In the second are the rock-jumpers and rockfowl—both African. The third lineage includes all the rest.

THE FIRST LINEAGE OF PASSERIDA

The Australo-Papuan robins, all forty-four species of basically "perch-and-pounce" insect eaters, form the family of the EOPSALTRIDAE, also known as the PETROICIDAE. Of course, they have nothing to do with the European Robins or the various American robins, all of which are in the thrush family, Turdidae, though some of them do have bright red breasts. By 1929, the Chatham Island or Black Robin was reduced to four males and one female on its native Chatham Island. All were transferred to nearby Mangere Island and now there are around 250, but it is a pity that such heroic measures are sometimes necessary.

THE SECOND LINEAGE OF PASSERIDA

Two species of rock-jumpers (genus *Chaetops*), from southwest to southern Africa, plus two species of rockfowl (genus *Picathartes*), from west to central Africa, are conventionally grouped within the single family PICATHARTIDAE; but in truth the two genera are very different and they separated one from another about 30 million years ago, so they should surely be given separate families. Rock-jumpers look like warblers and in the past have sometimes been classed alongside them—or with babblers or thrushes. Rockfowl can be twice as big, with brightly colored bald heads, and are among the many birds that follow army ants to catch whatever is disturbed. But although the two genera seem so different, DNA studies suggest they are indeed related—and not closely related to anything else.

THE THIRD LINEAGE OF PASSERIDA

The third lineage of the Passerida contains no fewer than thirty-eight families.

Family PASSERIDAE includes thirty-six species of sparrows and snowfinches, originally from Eurasia and Africa but long since introduced into the Americas, Australasia, and many islands.

Archetypally passerine are the TURDIDAE, the thrushes and robins (American and European), chats, wheatears, Fieldfare, Nightingale, and European Blackbird: 304 species in all, through all the world. But the Turdidae has few copper-bottom features that unequivocally declare their relationship, or distinguish them from, say, babblers, warblers, or flycatchers—so probably the family is not a true clade.

Close relatives of the thrushes, or so it's thought, are the thirty-six species of mockingbirds, catbirds, and thrashers, in the New World family MIMIDAE. Among them are some some of the bird world's finest mimics. The Northern Mockingbird is known for its imitations of frogs, pianos, car alarms, and human voices, a mocker indeed.

The dippers, five species of them from the Americas, Europe, and North Africa, have their own family: the CINCLIDAE. They are the only passerines that forage actively for food underwater. They swim underwater, as many other birds do, of course, but they are also the only birds that I know that actually cling to the bottom and walk along like old-fashioned deep-sea divers. They are called dippers, apparently, not because they take dips but because they stand on stones in the middle of the stream and bob their bodies up and down, while flashing their white eyelids.

The Old World flycatchers, 115 species of them in the family MUSCICAPIDAE, catch insects on the wing in woodland and forest, sallying forth from perches. When the weather is cold, flycatchers must hover in the canopy, which requires a lot more energy. Most flycatchers seem to be monogamous, but the Pied and Collared Flycatchers have a fine line in bigamy.

The wonderful, intelligent, lively, and mischievous starlings, mynahs, and oxpeckers, 114 species of them, form the family STURNIDAE: basically Old Old World—all over it, including a small presence in Australia—but now introduced to southern Australia, New Zealand, and many tropical islands, and well established in North America. Much more of them in later chapters.

The waxwings and silky-flycatchers—seven species in the BOMBYCILLIDAE—come from Eurasia and North and Central America. They feed on insects, flowers, and fruits—and like many creatures that feed on fruits, which tend to occur patchily but are abundant when they do, the waxwings are generally gregarious, foraging in flocks.

The REGULIDAE contains only six species, which include the Firecrest, the Goldcrest, and the kinglets. They come both from the Old World and from the New, and are among the smallest of all birds: sharp beak invertebrate eaters of mainly coniferous forests, pretty but not flashy and difficult to see, although they tend to have bright-colored heads, including the golden crown (meaning the top of the head) of the Goldcrest.

In the family SITTIDAE are the twenty-five species of nuthatches (genus *Sitta*) and of wallcreepers (*Tichodroma*), from Europe, North Africa, Asia, and North America. Mainly they hunt for insects on the bark of trees—and can run headfirst down the tree as well as up, as very few birds can do (and not many mammals, either, apart from squirrels and the Madagascan civet known as the fossa, both of which can swivel their hind feet backward). Nuthatches also like hazelnuts, hence the name.

The Holarctic treecreepers, six species of them from North America, Africa, Europe, and Asia, form the family CERTHIIDAE. They also hunt insects on the trunks of trees but run only upward, often ascending in a spiral, looking for all the world like mice; then, when they are as high as it is reasonable to go, they glide to the base of another tree and start again. Somewhat similar though rarely actually creeping are the three species of Philippine rhabdornises, which are endemic to the Philippines. They are now placed in their own family, the RHABDORNITHIDAE.

Britain's tiny Jenny Wren, with its cocked tail, is almost our national bird (one of the few to make it onto the backs of coins, in the olden days), but it should in truth be called the Winter Wren and is the only Old World species out of the eighty-three wrens from the family TROGLODYTIDAE. The rest all live in America, especially Central America, which presumably is where the family first arose; the Winter Wren presumably reached Eurasia via Alaska into Siberia and so on down as far as North Africa. Also wrenlike to look at are the gnatwrens and gnatcatchers, fourteen species from the American tropics, including parts of the Caribbean, who form the family POLIOTILINIDAE.

The NECTARINIIDAE are the sunbirds, spiderhunters, and sugar-birds—132 species throughout the Old World tropics from Africa to northeastern Australia, and up into the Himalayas. The sunbirds are among the Old World's answer to the New World's hummingbirds: small and brightly colored drinkers of nectar that also pollinate the plants they feed from.

The ten species of leafbirds in the family IRENIDAE are vaguely starlinglike in appearance and live in Asia from Pakistan through to the Philippines. They are very beautiful, in greens and turquoise and yellows.

The fifty-five species of the family DICAEIDAE are from South Asia, New Guinea, and Australia and include the flowerpeckers, among them the longbills and berrypeckers from New Guinea. Indeed they eat berries and nectar, but also insects and spiders. They are prominent among the small birds of Australia—not least the lovely little Mistletoebird, which builds a hanging nest like a purse.

The PLOCEIDAE are the weavers, including the bishops and widow-birds: 118 species of them in all from sub-Saharan Africa, Madagascar, with one in Arabia, and more in Asia. They are named for their extraordinary, intricately woven hanging nests, all of different design depending on species. Some are brood parasites, laying their eggs in other weavers' nests. Red-billed Queleas drift in huge swirling flocks across Africa, in search of the insects and seeds that accompany the rains, and are major pests. Much more of weavers in later chapters.

Family PRUNELLIDAE—thirteen species in the genus *Prunella*—are the accentors, including the Dunnock, which formerly, and wrongly, was known as the "hedge sparrow." Accentors are mostly mountain birds from Europe, northern Asia, and Africa north of the Sahara—apart from the largely lowland Dunnock. I have a yen to see the Robin Accentor in particular, at home among the rhododendrons of the Himalayas. Accentors, including the tweedy Dunnocks, have a wonderful sex and family life, with what seems to be serious tension between the sexes.

The tanagers and tanager finches—413 species in the family THRAUPIDAE—occupy the New World from Canada to almost the farthest south, sometimes appearing in wonderful multicolored flocks in tropical America, eating fruit and flowers and scattering seeds. Their DNA suggests that they are a true clade, and yet they are immensely various—and able to evolve very quickly. "Darwin's finches" are tanagers—fourteen species of them, exclusive to the Cocos Islands and Galápagos. The Warbler Finch has a narrow bill and gleans vegetation for insects and spiders. Some "ground-finches" pluck ticks from giant tortoises and iguanas. The Sharp-beaked Finch plucks new feathers from seabirds and drinks the blood. The Woodpecker Finch is one of many birds that use tools—it uses twigs or cactus spines to probe insects from decaying wood, doing with its twig what a woodpecker does with its unaided beak (and there are no woodpeckers on Galápagos, so the niche is going begging). The Vegetarian Finch eats only leaves, fruit, and buds—the Galápagos' answer to the Bullfinch. The ancestor of all these Galápagos "tanager finches" may have been Caribbean. All this variation has arisen in the past 5 million years; the Galápagos weren't there before then (and this is what the DNA evidence suggests, too).

The CARDINALIDAE are the cardinal grosbeaks: forty-three species through both Americas. Many are stunning. The Indigo Bunting (not a true bunting at all) is a beautiful deep blue and lives in the eastern United States, while the Lazuli Bunting, similar but with an orange breast, is from the western United States. The two species meet in the middle and hybridize to provide yet another hybrid zone. Some cardinalids feed on the free seed supplied by suburban bird-lovers, and so are increasing their range, like the Northern Cardinals of the eastern United States, whose males are a deep orangey scarlet and have a pointed crest like the hat of a Tibetan lama—jewels against the winter snow. The Dickcissel migrates south from the United States to eat the seeds of the Argentinian pampas, but tends to stop in Venezuela to eat crops and so alas is a major pest.

The family ICTERIDAE is again exclusive to the New World, all the way from Alaska to Cape Horn, though mostly in tropical Central America. There are only 102 species, but they are extraordinarily various: meadowlarks, bobolinks, American orioles, the brilliantly colored troupials, caciques, oropendolas, baywings, cowbirds, American blackbirds, and grackles. Many resemble Old World birds—some are like Old World orioles, some like larks, while the grackles, some of the oropendolas, and the cowbirds are remarkably like crows—but all are examples of convergence. Between them the icterids seem to exploit all the niches that are exploited by all the other passerines put together. Much more in later chapters.

In the family PARULIDAE are the 116 species of wood-warblers, also known as New World warblers, in both Americas from Canada to Argentina. They are all small and pointy-beaked, and mostly eat insects, although some take fruit and nectar, too. Some, like the Slate-throated Redstart, are bright and easy to spot; but an awful lot of the North American species, like the Bay-breasted Warbler, the Blackpoll Warbler, the Pine Warbler, and the Palm Warbler, all in the genus *Dendroica*, are greenish or yellowish with dark streaks and hard to tell apart. The Parulidae seem to be closely related to the buntings (Emberizidae), the tanagers (Thraupidae), and the icterids (Icteridae). But they are not at all close to the Old World warblers, in the family Sylviidae, although they do feed and generally behave like them.

In the family EMBERIZIDAE are 291 more species of small birds with conical beaks that eat seeds and insects. The family is named after the bunting genus *Emberiza*. America's buntings include the New World sparrows, towhees, juncos, and longspurs—all sparrowlike in general mien, though some have striking plumage, including a smart cinnamon and black in some of the towhees. Towhees are also known for the way they scratch the ground in search of food, staying in the same spot, like a Michael Jackson moonwalk. Old World buntings include the Corn Bunting, the Yellowhammer, and the Cirl Bunting. Some birds that do not belong to the Emberizidae are called buntings, including the Indigo Bunting.

The family CARDUELIDAE (the true finches) and the family DREPANIDIDAE (Hawaiian honeycreepers) both seem to have descended from the FRINGILLIDAE—now represented only by two species of chaffinch and the Brambling.[4] This trio in turn seem closely related to the buntings of the family EMBERIZIDAE. (Incidentally, too, the Red-Legged Honeycreeper, *Cyanerpes cyaneus*, is a tanager.)

The fringillids nowadays are strictly Old World (unless introduced elsewhere)—the chaffinches confined to Europe, where the common species is among the commonest woodland birds, while the Brambling also spreads into Asia. In the manner of small, primitive birds, the fringillids mainly eat insects. The carduelid finches—about 136 species of them—live worldwide, although they have missed Madagascar and they live in Australia and New Zealand only because they were introduced. They have majored in seed eating, which in general requires more special-ist beaks, as in hawfinches and bullfinches, of which more later, and the crossbills mentioned in Chapter 1. In the deep past, a group of siskins were blown to the New World, and now, in Central and South America, they form a "suite" of twelve very similar species that may well have de-rived from a single flock. The drepanidids—the Hawaiian honeycreep-ers—apparently all descended from a chaffinch-like ancestor who arrived on Hawaii about 5 million years ago. Now there are twenty-two species, astonishingly various in beaks and plumage, song, and way of life. Some are carduelid-like seed eaters, some of which specialize in particular trees. Some have thin, straight beaks like warblers for picking insects from bark. The longer, stronger beak of the Po'o-uli is for gleaning land snails and other invertebrates from the bark of trees. In the Hook-billed 'o'u, the bill is basically finchlike but longer and stronger for eating fruit. The Laysan and Nihoan "Finches" eat insects, a variety of vegetation, and the eggs of seabirds. In many, the tongue is tubular with a tip like a brush for sucking nectar—hence "honeycreeper"—but is also good for insects.

To the MOTACILLIDAE belong the pipits, neat and slender, that live in every habitable continent; the wagtails (which do indeed wag their long tails), which are mostly Old World; and the longclaws, which look pipit-like but have heavier beaks and are exclusively African—60 to 70 species in all. Wagtails and longclaws in the wild follow zebras and other big animals for the insects they kick up, and on farms follow cattle and tractors. Pipits are known for their wondrous fluttering ascending flights and parachute drops, accompanied by tremendous song, intended both to attract mates and to repel rivals. Males have been known to rise to more than 100 meters (330 feet) and sing for up to three hours. Some wagtails and pipits move only locally between summer and winter, but some are great migrators, forming aggregations of up to 70,000 birds on the stopovers en route to the tropics to while away the northern winters.

The "true" tits, fifty-three species of them—including America's chickadees—form the family PARIDAE. Again they live both in the New World (North America into Mexico) and the Old (Eurasia and Africa south of the Sahara), and everywhere are much-loved garden birds that feed from feeders and nest in birdhouses. In summer they mostly feed on insects, though in temperate climates in winter they turn to seeds and fruit—and are wont to cache both insects and seeds under bark, for later use. Hume's Ground-Jay, which lives above the treeline on and around the Tibetan plateau, looks nothing like a tit. It has a medium-length beak that turns downwards. But details of its anatomy, and its DNA, proclaim that it belongs with the Paridae.

The thirteen species in the family REMIZIDAE are the penduline tits—pointy-beaked insect eaters from North and Central America, Africa, and Eurasia. They are named for the nests that some species build, which hang from branches and are used, by some African people, as purses.

The seven species of long-tailed tits in the family AEGITHALIDAE complete the core trio of small birds known as tits. Again they live in both the New World (North and Central America) and Old (Eurasia). You will often see them in flocks of up to a dozen, dipping in flight across the clearing. Smallest are the Bushtits of western North America, which weigh less than 6 grams (seventy to the pound) and they need to eat 80 percent of their body weight in insects a day just to keep warm. Many die in cold weather. Long-tailed tits take many days to build fine, purse-like nests of moss and feathers—typically more than 2,000 feathers per nest—which they bind with spiders' webs and decorate with lichen. Truly they are creatures of fairy tales.

The eighty or so species of larks form the family ALAUDIDAE, from all over the Old World, including Australia, and especially in the grasslands of Africa; while the Skylark and the Horned Lark live in the New World, too. Larks are very diverse. The Hoopoe Lark resembles a Hoopoe. The Thick-billed cracks hard seeds like a Hawfinch. Some, including the Bifasciated Lark, which has a long curved beak, dig for insects. Some larks fly away when threatened, some crouch and rely on camouflage, some run away on foot, and some just walk. It's all a matter of style. Skylarks sing on the wing as many larks do, but some sing from the ground, like the Calandra Larks, which were kept in the Mediterranean like canaries to sing to their captors.

Family HIRUNDINIDAE is the swallows and martins—all eighty-nine species of them living all over the world apart from some remote islands, and not too near the poles. They look swiftlike with their scimitar wings and their forked tails, which make them highly maneuverable; and many species, like swifts, build cup-like nests stuck to rocks or walls, though made of mud rather than saliva—a stunning example of convergence. But the river-martins, with longer legs, bigger feet, and thicker bills, look more like regular perching birds and may be the closest to the ancestral types. The family probably arose in Africa—which still has twenty-nine endemic species, although Central and South America have twenty-one endemics. A few breed in more than one continent: the Sand Martin breeds in Europe, Asia, North America, and North Africa. Swallows and martins are great migrators. Most species are basically monogamous, though some are polygynous and the Cliff Swallows are positively bohemian.

There are 140 species of bulbuls in the family PYCNONOTIDAE, prominent in Africa and all across Asia. They are sparrow-size to thrush-size, pretty though not generally spectacular, with hair-like feathers at the back of the neck that in some species form a crest and long fluffy feathers on the back. They are catholic feeders on berries, pollen, insects, and spiders, and sometimes the eggs of other birds. As in many families, family life is various, from monogamous pairs, to leks and cooperative breeding groups.

There are 389 Old World warblers in the family SYLVIIDAE, from all over the Old World, including Australia (the Spinifex-bird) and New Zealand (the Fernbird), and some from the Americas, with isolated species on many a tropical island. Many are brightly plumaged, but on the whole they are more heard than seen because they are small and generally like to stay hidden, emerging only to dash after insects. The tailorbirds of Asia use vegetable fibers or spiders' webs to sew large leaves into cone-shaped nests, puncturing the edges with their beaks and then threading the twine. Many warblers are great migrators. Alas, more than one in ten species are threatened. But the Sylviidae seems to be splitting up. The largest genus, with forty-five species, is *Cisticola*, streaky birds well known in African grasslands. More and more the *Cisticola* species are being referred to simply as "cisticolas"; and so it is that the erstwhile Fan-tailed Warbler is now called the Zitting Cisticola, named for its "zitting" song (which you will be pleased to know I have heard and seen performed in southern Spain). Many these days place the cisticolas in their own family, the CISTICOLIDAE.

Also African and Asian, but again in Australia and Oceania, are the White-eyes, who include the speirops and the Mountain Black-eye: eighty-eight species of them in the family ZOSTEROPIDAE. They are small, greenish, and unspectacularly pretty with white rings around their eyes, again taking berries and insects with their sharp little beaks—sometimes probing, sometimes hawking—and also sipping nectar with their brush-tipped tongues.

The babblers are a big and prominent group—285 species of them in the family TIMALIIDAE. They are mainly Old World—Africa, Madagascar, Asia—but one has found its way to western North America. The laughingthrushes, fifty-four species of them, belong in the TIMALIIDAE, as does Britain's so-called Bearded Tit. Babblers are famously social and interdependent, even by avian standards—of which more in Chapter 7.

There is only one species of Palmchat, *Dulus dominicus*, in its own family, the DULIDAE. It lives only on Hispaniola in the West Indies—but where it does live, it is extremely common, looking vaguely thrushlike and feeding on berries, flowers, and insects and making large, jerry-built communal nests. But DNA studies suggest that the Palmchat might best be placed in the family of the waxwings and silky-flycatchers, the BOMBYCILLIDAE.

Also with its own family—HYPOCOLIIDAE—but also probably belonging among the waxwings is the Grey Hypocolius from the Middle East, which migrates to northwest India and the Arabian peninsula. It really is smoothly gray with black cheeks, a handsome bird, again thrush-size, again eating berries and insects, and often nesting in palms.

The ESTRILDIDAE is the family of the waxbills, 140 species of them;

nowadays it also includes the sixteen species of whydahs, which used to form the separate family VIDUIDAE but are now demoted to a subfamily (Viduinae) within the ESTRILDIDAE. Waxbills live in sub-Saharan Africa, Madagascar, Arabia, Asia, Australasia, and out into some Pacific islands. They are small, and their stout little beaks really do seem modeled from wax. They are often very bright and extremely diverse—each species apparently determined to stand out from the others, like minor league baseball teams. Some are polymorphic—different colors within the same species—like the Gouldian Finch of Australia. Waxbills are much-prized cagebirds, and some have escaped and become acclimatized to new areas around the world. They use their domed nests for roosting as well as for raising young. The males often sleep on the top like old men chilling out on the verandah.

The whydahs live only in sub-Saharan Africa. They, too, are often beautiful to look at and, though also small, have much longer tails. But they are serious brood parasites, like cuckoos.

THE CROW-LIKE OSCINES: THE CORVIDA

The Corvida are named after the family CORVIDAE—118 species of crows, including jays, magpies, and choughs. They are tough, intelligent, versatile, and highly influential throughout Eurasia (apart from the high Arctic) and both Americas (apart from the south of South America). Between them they live everywhere: Alpine Choughs live in mountains while ravens, the largest of all passerines, feature in the Old Testament, killing lambs in the semi-desert. Many crows, including jackdaws and Florida Scrub-Jays, have intricate social lives. Some are among the cleverest of all birds—the New Caledonian Crow is a match for the chimp. Since crows feature throughout the rest of this book I won't go on about them here.

Two linked and recondite families are the GRALLINIDAE, with two species of Magpie-larks alias "mudlarks," plus the CORCORACIDAE, the White-winged Chough and the Apostlebird, collectively known as Australian "mudnesters." They do indeed come from Australia, and thereabouts, and make beautiful nests of mud, neat as soup bowls, firmly clamped and cemented onto horizontal branches. White-winged Choughs cannot raise their young successfully without young helpers and are known to kidnap fledglings from other nests to act as au pairs or indeed as slaves if they have none of their own. (Europe's "choughs" are in the family Corvidae.)

The ninety-eight species of monarch flycatchers in the family MONARCHIDAE, from Africa, tropical Asia, and Australasia, are renowned for their bright blue colors and their enormous nervous energy: tails constantly fanned, wings constantly aquiver, chasing insects through the air with tremendous zeal.

Thirty species of shrikes form the family LANIIDAE, from Africa, Europe, and North America, hooked-beaked and preying mainly on insects but also on any creature, frog, lizard, rodent, or bird up to a size as large as themselves. Famously—of all the birds that do this, shrikes are the most famous: they impale their prey on thorns for later attention or, in these mechanical times, on barbed wire. Do they do this simply for convenience or do they like their food wind-dried, like jerky? It is not a silly question. Dried food is more concentrated, and if it has been attacked by microbes, it can be richer in vitamins. Many shrikes in Europe and North America are suffering badly from industrialized farming, like the rest of us.

The forty-two species in the RHIPIDURIDAE are the fantail flycatchers, from India, southern China and Southeast Asia, New Guinea, Australia, New Zealand, and some Pacific islands. They positively impose themselves on people—flying close and hovering, or perching just a few meters away with their tails coquettishly fanned; they are charming wild companions on woodland walks in New Zealand, as I discovered a few years back. But, like the European Robins keeping company with the gardener, their love is of the cupboard variety: it's not us they like but the insects we attract and disturb. More in Chapter 5.

The drongos in the family DICRURIDAE, twenty-four species of them, are dark and glossy starling-like or crow-like eaters of insects and small vertebrates from Asia, Australia, sub-Saharan Africa, and Madagascar—their name is Madagascan. The Black Drongo is famously fearless, berating even eagles that approach its nest. Some drongos are excellent mimics. The Fork-tailed imitates the alarm calls of other birds to drive them from their food—which it then pinches. In its turn, however, the Fork-tailed Drongo is a prime host to the eggs of the African Cuckoo, while the Black Drongo is parasitized by another cuckoo, the Common Koel.

The VANGIDAE, forty-one species of vangas, wattle-eyes, and batises, exclusively from Africa and Madagascar, have evolved into many forms, in the manner of the Hawaiian honeycreepers or the Galápagos finches. Some have short, broad beaks for catching insects on the wing; some have large, shrike-like beaks for hunting down large insects and small vertebrates; and the Sickle-billed Vanga pokes its long curved beak into crevices after insects just like an African woodhoopoe—yet another fine example of convergence. The vangids have much in common with the helmetshrikes in the family PRIONOPIDAE, and some would combine the two. The helmetshrikes (eight species of them) live in sub-Saharan Africa. Mainly they eat insects, but the White Helmetshrike has a particular penchant for geckos.

The forty-six species of bushshrikes in the family MALACONOTIDAE, also including gonoleks, boubous, tchagras, puffbacks, and brubrus, are also from sub-Saharan Africa, with one out on its own in the Arabian peninsula. Many are very beautiful. Some have brilliant songs, including the duetting gonoleks. All have a hooked and notched beak like the shrikes and feed on insects and other small animals.

Some link the four species of iora in the family AEGITHINIDAE to the leafbirds, in the IRENIDAE. They look very similar, after all, and do much the same things in much the same region of the world. But the leafbirds are classed as passerids and DNA studies suggest that the ioras are corvids.

The CRACTIDAE includes the fourteen species of butcherbirds and related types such as the well-known Australasian Magpie, the large currawongs, and the Bornean Bristlehead, all endemic to Australia, New Guinea, and Borneo. Mainly they feed on insects but the butcherbirds also catch frogs and small birds and in any case, like shrikes, they often impale their catch on thorns for later consumption. The Grey Butcherbird does a fine line in mimicry, not just of other birds but also of dogs and horses. It is up there with the mynahs, mockingbirds, and lyrebirds (and indeed parrots). The Australasian Magpie lives in open country and may nest near human habitations—although it does sometimes attack people, which for would-be commensals is not in general a smart move (commensals being creatures that contrive to live in the company of other creatures, and rely on them to a greater or lesser extent).

Closely related to the cractids, it seems, are the eleven species of woodswallows in the family ARTAMIDAE, from tropical Asia and into New Guinea and Australia. Like Barn Swallows, from a quite different family, woodswallows can stay aloft for hours, scooping up insects in their broad beaks, and with their brush-like tongues could presumably feed on nectar, too, though it isn't clear that they actually do this. Maybe there's a clue here to the origin of swifts and hummingbirds—closely related families, one the supreme hawker of insects and the other the perfect nectar-sipper. The link, perhaps, lies in the extraordinary powers of flight that both require.

The twenty-eight species of Old World orioles—absolutely not to be confused with the New World orioles—form the family ORIOLIDAE. There are two genera, one of which is generally known as the figbirds. The name "oriole" derives from the Latin for "gold," and indeed orioles tend to be yellow (or golden) with the males especially bright, and splashed with yellow and black. They live through most of the warmer bits of the Old World, but because they favor the canopy of woods or forests, they are hard to spot—though good birders pick up on their flutey songs from miles away.

The PACHYCEPHALIDAE—the name means "thick head," for they all have large, rounded heads—includes sixty-two species of birds such as the whistlers, the Bellbird, the shrike-thrushes, sitellas, and so on: again, they are all from Australia and Southeast Asia and include some excellent songsters. On the whole they are very like tits—another fine example of convergence.

The CAMPEPHAGIDAE are the cuckoo-shrikes, alias "caterpillar birds": eighty-five species of them, including minivets, woodshrikes, and several more. They are closely related to orioles—the two families are often combined—and, like orioles, live throughout the warmer parts of the Old World. They are sparrow to pigeon size and tend to have sharp, robust beaks slightly hooked at the end, like shrikes. Mostly (like shrikes) they eat spiders and insects (mostly caterpillars), with the odd lizard or frog, and some like fruit.

The fifty-two species of vireos—"true" vireos, greenlets, shrike-vireos, and peppershrikes—in the family VIREONIDAE of the United States look a bit like heavyweight wood-warblers, though they seem to be closer to crows. Some are favorite targets of the Brown-headed Cowbird, which lays its eggs in their nests. Bell's Vireo often responds by burying the cowbird's eggs within the nest and smothering them. Some vireos throw the cowbird eggs out. But the main threat to them—as ever—is loss of habitat.

The New Zealand wattlebirds, three species in the family CALLAEI-

DAE from the dense forest of the North Island, are best known for the one that is probably extinct—the Huia, last seen in 1907. The Huia is famed for its sexual dimorphism—not in body size or plumage but in the size and shape of the beak. The male's was fairly short, slightly down-curved, and heavy, while the female's was much longer and slender, and more down curved. The male used his beak to chisel insects from rotten wood while the female delicately probed. Perhaps the two sexes cooperated—one chiseling, the other probing, both sharing the spoils. But the Huia suffered from the beauty of its feathers, which appealed first to the Maoris, and from its sexual dimorphism, which appealed to the keepers and curators of European museums. Fatal.

Also loosely related to the crows are the forty-two species of birds-of-paradise in the family PARADISAEIDAE, mainly from New Guinea, which is where the family undoubtedly arose, but also from Australia and Indonesia. Some are monogamous and inconspicuous, but most are polygynous, with extraordinarily flamboyant plumage and extraordinary courtship routines to go with them.

That completes the cast list of modern birds.

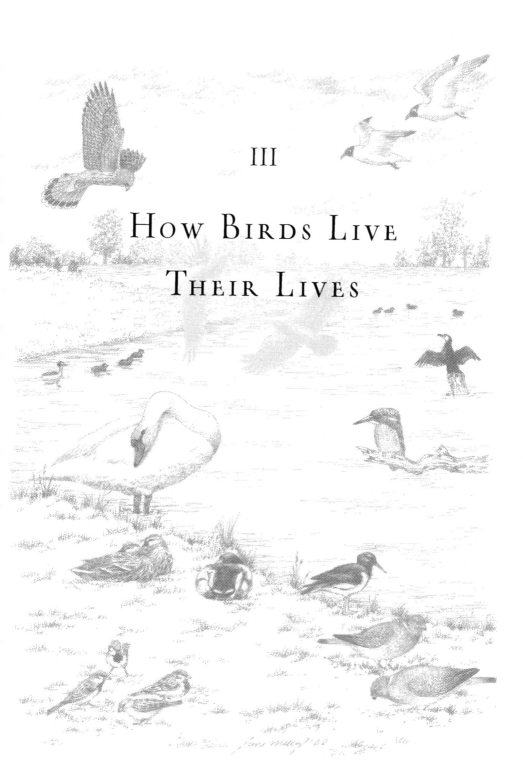

III

How Birds Live
Their Lives

Tough, muscular lungfish, deep in their African swamps, seem safe from predators—but not from the formidable Shoebill.

5

THE EATING MACHINE

ONE EVENING OUR LOCAL PUB HAD RUN OUT OF PEANUTS and was offering Japanese cocktail biscuits instead. They were brightly colored and quite disgusting—so I threw them down for the peacocks, which roam around the garden (I am told by a friend who keeps peacocks that despite appearances they are as tough as old boots and can eat anything). The peacocks ate all the yellow biscuits first, then the red ones, and then the blue ones—and thus they illustrated all the fundamental principles of food and feeding.

Animals have to eat the right things: the things that will nourish them and to which they are especially adapted (for food is not nourishing unless you are adapted to it). Peacocks are fowl, and fowl are basically seed eaters with a penchant for grain; and biscuits are a reasonable approximation. Owls would surely have disdained such provender. Between them, though, birds eat just about everything that nature provides that is remotely edible—and this is more remarkable than it seems, for the provender of nature is extraordinarily variable and yet all creatures, in the end, have the same requirements for protein, energy, essential fats, and micronutrients whether they live on dead zebras or winkles or the seeds of thistles. But then, human beings are remarkable, too, in this respect. Some of us—like traditional Inuit—live almost entirely on flesh, while others—like many traditional Asians—are almost entirely vegan.

Animals also need the right equipment to eat what they eat; and fowl have pointed beaks ideal for pecking, and good close vision, and muscular gizzards just above the stomach, which they pack with gravel to crush hard seeds and (in the case of peacocks that tout for cocktail biscuits) with a

battery of enzymes that enable them to cope even with what the food industry likes to call "permitted colorants."

Then, also, animals need the right technique if they are to get enough
to eat. Walking about, searching and pecking is the right technique for a
bird that eats seeds from the ground. It would not be appropriate for a
peacock to hover like a kestrel even if it could, or quarter the ground like
a barn owl, or swoop upon the unsuspecting biscuits like an Osprey, spectacular though this would be. Searching and pecking is what peacocks do
and what is most appropriate.

Finally, crucially, but subtly, all animals must seek to adopt an "optimum foraging strategy." They must not expend more time or energy in
searching for food than the food is worth. Neither, in their necessary
search for food, must they run unnecessary risks. Brilliant technique
would be no good at all if linked to poor strategy.

So why did the peacocks choose the biscuits in the order they did—
first yellow, then red, then blue? Like all birds (and unlike most mammals),
they have perfectly good color vision and may simply have preferred
yellow. Yellow is the most grain-colored, while the color blue is fairly
rare in nature apart from the sea and the sky and is rarely associated with
good food—certainly not the bright royal blue of those horrible biscuits.
So perhaps the birds were right to be suspicious of blue. I suspect, though,
that a different principle was at work: that these peacocks were demonstrating the particular foraging strategy known as the "specific peck response."

Animals that feed on small items face two particular logistic problems:
that no individual item is worth lingering over, and that in any one collection of small items only some are liable to be edible, and so they are forced
to waste time in making choices. Thus fowl feeding on grain scattered on
the ground have to distinguish as quickly as possible between wheat (or
cocktail biscuits) that is edible and gravel that is not. Birds solve both
these problems by developing a "search image." They rapidly get it into
their heads what the desirable items look like and then concentrate on
those items alone, which means they can then feed quickly. When birds
are faced with a mass of different items, some of which they know are edible and others that may not be, they exhibit the "specific peck response":
they simply home in on whichever of the edible items happens to be the
most common, and ignore all the rest.

The specific peck response has been demonstrated many times in the laboratory and in the field. If you give pigeons a lot of barley with just a few peas, they go for the barley and ignore the peas. If you give them a lot of peas and not much barley, they focus on the peas. In the wild, many perching birds may breakfast, lunch, and dine almost exclusively on grasshoppers for weeks at a time while grasshoppers are in season, even though there may be a lot of other things around as well. It just isn't worth working out whether a particular beetle is tasty or not when there are grasshoppers around that are known to be good. In classic experiments, Professor (then Dr.) Bryan Clarke, then at Oxford, showed that thrushes faced with pellets of artificial food identical except for color would always home in on the commonest kind. Thrushes in the wild commonly eat snails, and Clarke went on to show that some snails, especially those of the genus *Cepaea*, benefit from having many different ground-colors and patterns of stripes, because the thrushes always homed in on the ones that were most common at any one time. Clarke's work thus showed more generally how natural selection sometimes favors creatures specifically because they are in some ways different from their fellows—a phenomenon known as "frequency-dependent selection." This in turn leads to a "balanced polymorphism" in the prey—a more or less fixed proportion of different colors and patterns or modes of behavior within any one species or "suite" of species. The proportions of each kind are fixed (more or less) simply because natural selection, in this instance, favors minorities—for if the minorities grow into majorities, then they are clobbered until they become rare again, whereupon they become relatively safe. There are many examples of frequency-dependent selection and of balanced polymorphism in nature, in many different contexts. The general point is, as the French say, *vive la différence*.

With such thoughts in mind, I hypothesized that the pub peacock ate the biscuits in the order yellow, red, and blue not because the blue ones were particularly disgusting even by the standards of Japanese cocktail snacks but because, in the sample offered, they were the rarest (presumably because blue dye is the most expensive).

I could have tested this, which means that this hypothesis is within the purlieus of true science: for the hallmark of truly scientific hypotheses, according to the philosopher Sir Karl Popper, is that they should be testable and, in principle, disprovable. I could have tested my idea by buying several packets of cocktail biscuits and giving the birds a great many blue ones and very few yellow ones. The specific peck hypothesis says that they would go for the blue ones first and ignore the yellow ones until the blue ones were nearly all gone (when the yellow ones become the more common). In contrast, the color preference hypothesis predicts that the peacocks would go for the yellow ones even though they were rare, like children picking out the red ones from a bag of jelly beans. I regret I did not do this experiment. As the great twentieth-century biologist Sir Peter Medawar observed, "Science is the art of the soluble," meaning among other things that scientists can test only those ideas that they can afford to finance, and to buy enough biscuits I would have needed a grant. Besides, it was morally unacceptable to encourage their manufacturers. So the specific peck response goes untested in these particular peacocks, but it is testable in principle and has been well supported in many other contexts and species.

Once you are alert to the idea of optimum foraging strategy, you can see it in action everywhere. Note, for example, how a blackbird, taking berries in late summer from a cotoneaster bush, flits at what seem like random intervals from twig to twig or from bush to bush. Is the bird simply a scatterbrain? Does it suffer from attention deficit? That is the uncharitable view. More probably, its fidgetiness is a deliberate tactic, built in to its day-to-day foraging behavior. As it feeds it is quite likely that some groundling predator—a weasel, a suburban cat—or perhaps a Sparrowhawk is lining it up. So don't hang about. Take a few berries, then hop off and take a few from somewhere else, and force the gardener's pussycat to begin his mischief all over again. Similarly, birds that commonly feed on bigger fruit—hornbills taking figs, for instance—typically fly off with it to eat it somewhere else. It would be all too easy for predators to lie in wait where the fruit is. Fruit bats active by night apply a similar strategy as they strive to avoid the attention of tropical owls. Many tropical trees bear their flowers and then their fruits directly from the trunks to make it easier for the bats to feed on them (a habit known as "cauliflory," or "flowering from the stem"): the cocoa tree does this. The relationship between the fruit-eating birds or bats and the fruit-bearing trees is "mutualistic"— a mutually beneficial symbiosis. Because they carry the fruit away, and commonly reject the seeds (or fail to digest them), fruit bats and fruit-eating birds are among the principal seed dispersers. Cassowaries largely subsist on the large tree fruits of their native New Guinea and North Australian forests; and as the cassowaries become rarer, the trees whose seeds they scatter must suffer, too.

It can also pay a bird to protect its food supply by establishing a territory, sometimes for all of the year, sometimes only in the breeding season. Again, it's all a matter of cost-effectiveness: whether it is better to leave the food resource unprotected, which is obviously risky; or protect the boundaries, which reduces the risk but takes a lot of time and effort. The European Robin is outstanding among familiar birds that clearly do think territories are worthwhile. Robins look sweet with their bright red breasts—brighter in winter than in late summer so they positively glow against the snow, the females just as bright as the males (contrary to common belief). So they feature on Christmas cards—commonly depicted, quaint as little children on a sleigh ride, huddled in rows on frozen washing lines. Yet the last person a European Robin would ever be seen in the company of is another robin, except for his or her mate, in due season. In the nonbreeding season, both sexes maintain their own feeding territories and defend them fiercely, and in the breeding season, pairs defend a territory together (with the male doing most of the scrapping, if any intruder is foolish enough to ignore warnings). The limits of each territory are not obvious to the human eye; if there is a robin in one suburban garden and another next door, we can be fairly sure that the boundary that separates them is not the garden fence but follows some invisible line that only the robins know.

Come the autumn some (but not all) of the European Robins in any one population migrate south, over distances that seem somewhat arbitrary. Some of the migrants cross the sea (notably between England and France), and when they do they tend, as migrant birds commonly do, to hitch lifts on passing boats. Most hitchhikers fly off after a few hours of rest, but in *The Life of the Robin*, a classic text, the Oxford ornithologist David Lack describes a couple of males who in 1833 hitched a ride on a coastal ship out of Yorkshire and stayed aboard for several days. They took up residence at either end of the boat and slugged it out in perpetual song and shows of aggression. The dividing line between their two notional territories was roughly at the level of the bridge.

Yet many birds find it far more advantageous to cooperate, to forage or hunt in groups—sometimes with other birds or even with other animals of different species. Although birds of prey—"raptors"—are commonly solitary and sometimes fiercely territorial, some of them are highly social, in all kinds of contexts.

None is more social than North America's Harris's Hawk, *Parabuteo unicinctus.* Harris's Hawk virtually never hunts alone and only rarely in pairs, but usually in teams of four to six. In one field study in New Mexico, the birds generally assembled in trees in the mornings in family groups—Ma, Pa, and offspring up to three years old. Then they broke into subgroups of two to three birds, which would hop from tree to tree, all moving in the same general direction and all closely watched by all the others. When one group spotted a jackrabbit—typical prey—all the hawks converged to catch it by various maneuvers used in various combinations. In the "surprise pounce," several hawks would strike a rabbit that had broken cover. If a rabbit found cover, then the hunters employed the "flush and ambush": one or more hawks chased it out while others waited to strike. If the prey fled across open ground, the birds employed the "relay attack": individual hawks swooping at intervals and driving the hapless prey into the path of others, alternating the role of lead pursuer. Analysis showed that the birds hunted far more efficiently in groups of five to six than if they hunted merely in pairs.

An adult black-tailed jackrabbit may weight twice as much as an adult female Harris's Hawk and three times as much as a male Harris's Hawk, and rabbits in general can be powerful combatants (*Watership Down* got it right). But after such treatment a rabbit is too exhausted and confused to resist. But although many birds cooperate in hunting, it seems rare for birds of prey to bring down prey that is larger than themselves even when hunting in teams. Wolves and dogs do cooperate to bring down deer, including moose, and lions work in gangs to fell African buffalo, which are truly formidable; but most predatory birds seem to cooperate, when they do, simply to increase the efficiency with which they catch small creatures. Indeed, apart from Harris's Hawks, the only other avian hunters I can think of that cooperate to bring down animals appreciably bigger than themselves are the ground-hornbills, of which more later.

Many more birds are "commensal" feeders: they depend largely or entirely on the activities of other species in order to feed—and those "other species" are sometimes human. House Sparrows and feral pigeons are an obvious example. Both nest in crevices of the kind that cities provide in abundance; both are catholic feeders, happy with scraps; and both are behaviorally versatile and more than able to put up with human company.

More specifically, many big animals, including us, disturb insects and other invertebrates as we go about our lives—and many different birds of many kinds cash in. Kestrels commonly hunt by motorways—the vibrations of the traffic disturb the worms and other small creatures on which they feed. Cattle Egrets follow cattle for the insects they kick up. Gulls follow tractors and trawlers. European Robins, stroppy though they are with their own kind, often seem uncommonly attached to gardeners, perching cutely on the wheelbarrow's edge waiting for worms while the gardener toils. They can make engaging pets—David Lack's book carries a picture of Lord Grey, founder of the Edward Grey Centre at Oxford University, world-renowned among ornithologists, with his pet robin perched on his hat. But why are robins so attached to us? Because they think we are pigs. In the wild, robins are woodland birds and they follow wild boars as they dig with their snouts for roots and truffles.

One of my outstanding memories of Beijing was in the Forbidden City, where thousands of people (including me) milled about in the courtyard; above us was a thick though largely invisible layer of small insects, attracted to the abundant sweat and possible donations of blood; and among them—clearly visible—whizzed a positive swarm of dragonflies; and above them again was a dense shoal of swifts catching whatever the dragonflies missed (and possibly taking the odd dragonfly). I say "shoal" of swifts because the different layers of creatures with us at the bottom were arranged like plankton, and the swifts, uppermost, in that particular capacity were commensals.

The antbirds include two large and related families from Central and South America—they are called "antbirds" because some of them at least make much of their living by following columns of army ants. Few phenomena in nature disturb insects and other creatures quite so decisively as army ants do. Birds of many other species may join in the fun, including some puffbirds and nunbirds, which may also follow troops of monkeys.

So it goes on, for birds of many kinds are marvelous opportunists, and opportunism is an important component of strategy. Oxpeckers, relatives of starlings, take blood-gorged ticks like miniature bonbons from the backs of African cattle and antelope. The Grey (or Red) Phalarope picks sealice from the backs of whales. Many seabirds do their best fishing when predatory fish or whales or porpoises go on a rampage beneath the waves, sending prey fish scurrying to the surface. On land, but on the same principle, Goshawks in Africa follow ratels, alias honey-badgers, which in truth are powerful predators (powerful enough to see off leopards), and being swifter and more agile, pinch a lot of the creatures that the ratels send running before the ratels can get to them.

It is good strategy, too, not to waste energy. Some birds reduce their expenditure simply by sitting as still as possible—none more so than nunbirds and puffbirds. Hummingbirds, which have such a demand for energy, ease up at night by allowing their temperature to fall. (As do some mammals, such as sloths. Since sloths are so inactive all the time, you may wonder why they bother. But sloths have a very poor diet—unpleasant leaves of tropical trees—and so they have difficulty taking in enough energy.) America's Common Poorwill (from the nightjar family, Caprimulgidae) allows its body temperature to fall and become torpid. Indeed, folklore had it that the Common Poorwill hibernates through the winter—which ornithologists thought most unlikely because that's not the kind of thing that birds do. But in 1947, a Common Poorwill was found immobile in a cleft in a rock in southern California, and now they are known to stay like this for months with a body temperature down to 18°C (64.4°F), for all the world like hibernating bats.

In fact, a bird's entire life may be seen as an exercise in foraging strategy. The strategy for a young bird is simply to feed and grow, courtesy of its tireless parents. The peculiar Oilbirds, relatives of nightjars, feed their chicks exclusively on the oily fruits of palms and some other trees—such rich fare that within seventy days the chicks are about 50 percent heavier than the parents. (I am reminded of some of the children I saw in Beijing—munched out on burgers and Coke and far outweighing their wiry parents, raised in the days of Mao Tse-tung, when a bowl of rice was a good meal.) Then over the next thirty days or so the young Oilbirds lose the extra weight, develop full plumage, and grow their long wings, ready to fly. Emperor Penguin chicks, too, selflessly fed by their fathers, finish up heavier than the adults—but then they must fast while they grow their adult feathers before taking to the sea for themselves. When birds finally reach adulthood they often find it hard to get enough to eat in a day, without more effort and risk than the food is worth. When the going does seem easy, and the bird is getting fat, you can be sure that it is preparing for some future exigency—finding mates, building nests, laying eggs, rearing young, migrating, surviving the next bout of drought or cold, or molting and refledging. They have to do everything that life demands in the right way and at the right time. Birds do not carry diaries, but they run their whole lives by the clock and the calendar nonetheless, because they need to; that is life's necessary strategy. "Free as a bird," we say. But if a wild bird does not do the right things at the right time, it will die or fail to reproduce. In truth, we who envy the birds their freedom are freer than they are.

WHAT IS GOOD TO EAT?

Some foods are more rewarding than others. Some provide much more energy, weight for weight. Carbohydrates—meaning sugars and the molecules that are made from them, notably starch—provide about 400 kilocalories (kcals) per 100 grams. But only a few birds ever get to eat pure sugar. Mostly they consume sugar while it is still in the plant, accompanied by water and "fiber," which mainly consists of cellulose—the stuff of which cell walls are built. The water and fiber "dilute" the energy content. Fat is far richer—about 900 kcals per 100 grams—but again animals rarely get to eat it in pure form. The fat in plants (usually oils) comes packed in fibrous seeds. Only a few birds do get to eat pure animal fat. Vultures get to peck at the hard fat—suet—around the kidneys of zebras or cattle, while sheathbills may gorge themselves on the fat of dead whales, not least around whaling stations.

In plants, seeds are the most concentrated source of energy and protein; they, after all, are intended to provide baby plants with a food store. A typical seed, like barley, provides about 360 kcals per 100 grams with 8 percent protein. A typical tuber, like potatoes, provides around 90 kcals per 100 grams with 2 percent protein. But cabbage, not a bad example of a green leaf, provides only 20 kcals per 100 grams and is less than 2 percent protein. Lettuce is down to 12 kcals per 100 grams, with 1 percent protein. Many leaves are feeble provender, indeed.

Meat and fish in general provide roughly as much energy weight for weight as seeds, and far more protein than most. Beef, for example, provides about 300 kcals and 16 grams of protein per 100 grams. Sardines, the kind of fish that many birds like, provide about 220 kcals per 100 grams, with about 24 percent protein.

Now physics comes into play. Birds in general need to eat a lot compared to most animals because they expend such a lot of energy keeping warm and flying—a hugely energetic pursuit. Big birds eat more than little birds, of course, but *weight for weight*, little birds need more. All warm bodies lose heat through the body surface and the rate of loss is proportional to the difference between the inside temperature and the outside.

Since birds like to stay at 40 to 42°C (104 to 108°F)—somewhat warmer than human beings—and since the outside may be 40 degrees or more cooler than this, they are liable to lose heat quickly. Even in the tropics, especially in deserts and mountains, the nights can be very cold.

In addition, small bodies, whether birds or cups of coffee, cool quicker than big ones because they have a far higher area of body surface relative to their total volume. So a mouse may metabolize hundreds of times more rapidly than an elephant—and indeed, if an elephant metabolized as rapidly as a mouse it would reduce itself to charcoal, since it would be unable to lose the surplus heat through its (relatively) small surface area. Similarly, a hummingbird metabolizes many times faster than an Ostrich. Indeed, some very small mammals and birds need to eat at least their own body weight each day just to stay alive, which mainly means just to stay warm. Since there is obviously an upper limit on how much an animal can eat in twenty-four hours, or in the limited hours in which feeding is possible, small animals in general need their food to be as concentrated as possible, providing as much energy as possible per unit weight. Bigger animals, needing less energy per unit of body weight, can make do with food that is less concentrated. In general, small birds tend to eat animal fare—notably insects—and small birds that eat mainly plant fare, like true finches of the family Carduelidae, must concentrate on seeds. Woodpigeons may make a fine mess of the cabbages in the backyard, but no bird smaller than a grouse (and no mammal much smaller than a rabbit) can afford to live on leaves alone.

All in all, there are pros and cons in every form of diet and every way of feeding; and although some birds such as Carrion Crows and Black-headed Gulls are big enough and clever enough to feed on just about anything that presents itself (just as human beings are able to do), most are more or less obliged to specialize. Once birds—or any creature—start to specialize, natural selection often tends to make them more specialized, or better and better at plowing one particular furrow. Yet in many surprising ways, birds and other animals of all kinds that may seem highly specialized sometimes explore quite new ways of life—and once they break out of the mold, their descendants are free once more to evolve in quite new ways.

Sometimes we see this happening before our eyes. So it was that in Britain a few decades ago Blue Tits learned how to pry the tops from bottles of milk that were left on doorsteps and feast on the cream within.

News of how to do this apparently spread from bird to bird, and soon a lot of them were doing it. Glossy Ibis that live in southern Spain traditionally caught fish and other small aquatic creatures during the summer and migrated to Africa for the winter. But many of them have taken to living largely on rice, which is widely grown in shallow paddy fields in southern Spain, and to eating the American crayfish, which has been introduced into waterways throughout Europe. With a little saffron, garlic, and olive oil they would have themselves a paella. In addition, as global warming makes itself felt, many of Spain's Glossy Ibis have given up migrating and stay in Spain throughout the year. Once an animal adopts a new way of life, it exposes itself to a whole new set of selective pressures—and so changes in behavior can lead to evolutionary change. So while most mockingbirds spend a lot of time at ground level, under the bushes, looking for invertebrates, the Galápagos Mockingbird likes carrion and small crabs. The Hood Mockingbird, also from the Galápagos, eats the eggs of many seabirds and even of the Galápagos's iguanas, both land and marine. It has also been known to eat the chick of a Masked Booby. Who knows where this may lead? If we came back in a few million years, would we see Hood Mockingbirds marauding like skuas, the terrors of the skies?

Within most orders of birds—and indeed, within many families—we can see how some shift of habit in some ancestor has allowed quite new lifestyles to evolve, with the anatomy to go with them. The order of the Anseriformes—swans, geese, and ducks—has explored the complete spectrum of possible diets. Geese mainly graze on land—and so does just one group of ducks: the wigeons. Most other ducks are dabblers. Some dabblers take whatever they can find from the surface, as shovelers do—filtering nutritious surface creatures through fringes on the edge of the tongue. Some shove their heads under and pull out weeds, like Mallards and pintails. Swans graze but mainly pull weeds from the bottom, and can reach far deeper than ducks. Pochards and Tufted Ducks dive for their food, after weeds, mollusks, and small fish. The eider ducks that live in northern waters dive to the bottom to depths up to 20 meters (66 feet) after shellfish, echinoderms, and crabs. Mergansers have narrow, unduck-like, saw-edged beaks and dive after fish. The anseriforms are related to the fowl, in the order Galliformes, which in the main are seed eaters. Their common ancestor was presumably some kind of generalist feeder that diverged in all kinds of directions.

We see comparable shifts from that one diet to another within the kingfishers, terns, gulls, shorebirds of many kinds, raptors of many kinds, nightjars, treeswifts, finches—indeed, almost everywhere you look. Typically we see more specialist and technically difficult ways of feeding evolving from simpler and more generalist ways. But sometimes (as in terns), the more derived (highly evolved) types have reverted to become more generalist again. All these principles are at work both among birds that have specialized in eating other animals, of one kind and another, and among those that have gone down the vegetarian route. Let's begin with the meat eaters.

HOW TO EAT OTHER ANIMALS

In many ways flesh is the best food—both nutritious and fiber-free. It requires relatively little processing once it is down the throat, so carnivores in general can make do with a short, straight gut with few elaborations, and only a small cecum or none at all (a cecum being a diversion of the hind-gut where difficult material can be digested at leisure). A small gut means less weight—especially appropriate for predators that need to be agile.

But animal flesh is rarely easy to come by. Animals can hide or run away. They can fight back, too, with jaws, claws, beaks, and stings, and various forms of noxiousness. No animal, predator or otherwise, can afford to expend more energy and time than the food is worth—so in general it seems best to hunt for prey that is as large as possible. But no creature can afford to run too many risks—and predators are always likely to be killed by the very creatures they seek to kill. All, then, must strike a balance: reward versus time and energy versus risk. It's not easy.

In general, the easiest and safest way for a predator to be a flesh eater is to catch insects and other invertebrates on the ground. This, probably, was what *Archaeopteryx* did, the most primitive bird of all—even though it could fly. From this mundane beginning, birds evolved scores of new ways of feeding.

Some birds stayed with insects, but developed ever more elaborate and technically adept ways of catching them. Others shifted to larger prey, on land and sea (but many then reverted to smaller prey again). Some began to specialize in mollusks, which have nutritious flesh like animals yet do not generally run away (apart from squid) and so seem ideal—

except, of course, that the ones that do not run are usually encased in thick calcareous shells or else, like slugs, are unappetizingly slimy. Other birds became vegetarian, which physiologically is harder.

Many that stayed with invertebrates found ways of probing deeper into the mud or plants where they lurk. Starlings pick leatherjackets—the worm-like larvae of crane flies—from beneath the surface of the soil by thrusting in their beaks and then opening a slot in the soil by "gaping": by means of a special arrangement of muscles around the skull they can open their beaks with at least as much force as they can close them and so create a convenient aperture. Their eyes are positioned so that they can see through the gape of the open beak. America's meadowlarks probe the soil in the same way. Jamaican Blackbirds and the Solitary Cacique perform similar keyhole surgery on deadwood, in their search for grubs. But Jamaican Blackbirds and meadowlarks are quite unrelated to starlings; they belong to the wondrously various New World family of passerines, the Icteridae. Starling gape and icterid gape are another fine example of convergence.

But of all the birds that dig for grubs in bark and timber, there is none to compare with the woodpeckers. Woodpeckers are built for chiseling into trees. First, they cling very firmly and climb with great agility with their parrot-like, zygodactylous feet. The stiff-feathered tail, pressed hard against the trunk, completes the triangle to make a firm base. Their skulls and necks are fitted with shock absorbers so that they can peck with impunity, without scrambling their brains as a nonspecialist would do if it tried to peck its way into solid timber. Their beaks are powerful chisels. Finally, their tongues are enormously long and tipped with barbs to pry insects out of bark and to dig into their tunnels beneath. But evolution never stands still, and some woodpeckers have adapted that great skill to other purposes. Some, like Europe's Green Woodpecker, feed on the ground, while others dig down into the ground after ants and termites. Some eat other animal fare as well—including nestlings—and some also eat fruit and berries. America's famous Acorn Woodpecker largely eats acorns, which it stores ("caches") in little custom-built holes in the trunks of trees. Acorn Woodpeckers live communally and share these larder trees, any of which may contain up to 50,000 little pits, each with its acorn to be eaten in bad times—a remarkable sight.

In countries where woodpeckers live, they tend to commandeer the wood-pecking, insect-probing niche. In countries without woodpeckers, other species may contrive to fill this role. Among them are the rosellas and cockatoos of Australia—and, as we have seen, the Woodpecker Finch of the Galápagos.

Conceptually more difficult than probing for insects and their larvae is to catch them on the wing—and the simplest method is to make short flights from perches. Shrikes do this; so do bee-eaters; and so do the Old World flycatchers, of woodland and forest. Flycatchers are very efficient: Spotted Flycatchers can catch an insect every 18 seconds (as measured by stopwatch). When the weather is cold, insects are to be found mainly in the canopy and there the flycatcher must hover—and so they find themselves using a lot more energy than they otherwise would just at the time when they need it most. Old World warblers—all 369 species of them—generally hunt in much the same way. But they are small and elusive, more often heard than seen.

"Hawking"—catching insects (and sometimes bats) on the wing—seems far harder, yet many birds from many different orders have managed it: nightjars and swifts, which are related to hummingbirds; martins and swallows, which are passerines. All of them prefer to hunt in the open sky, but the Whiskered Treeswift refines the technique still further and hawks for insects in the congested canopy. This requires reflexes and agility that are truly breathtaking.

But some insects are big and strong like beetles, or dangerous like bees and wasps. Some birds that specialize in such prey have long beaks, to keep them at beak's length, as Hoopoes and bee-eaters do. Those with shorter beaks commonly have bristles around the base of the beak to provide a face guard, and keep the thrashing legs and jaws away from the face and eyes—which is what the "beard" of the barbet is for. Bee-eaters bash and wipe their prey on a twig before eating it, to discharge the sting's poison and preferably remove the sting all together—although the Carmine Bee-eater of Africa manages to remove the sting of bees in flight, which is a party-trick, indeed (and, for good measure, it will take the occasional fish). Carmine Bee-eaters also specialize in catching insects flushed by bushfires.

Each bee-eater gets through about 200 bees a day, and since they operate in flocks, and since a typical hive will hold around 30,000 bees, a flock of bee-eaters can make considerable inroads. Yet I remember once in Spain a row of beehives at the foot of a low sandstone scarpin in which dozens of bee-eaters had built their nest burrows. So why didn't the bee-keeper move his hives or block the nest holes? Perhaps because bee-eaters also eat wasps and hornets, which attack bees. So perhaps, despite appearances, the bee-eaters pay their way.

The potoos, too, are relatives of the nightjars and also wide-gaped consumers of moths, beetles, and crickets, which they mainly take on the wing. But the larger potoos also take the odd bat. Bats on the whole are given a hard time by a wide variety of birds, which perhaps is why bats are more or less exclusively nocturnal unless forced out by day—but potoos, along with owls, have followed them into the night. On the other hand, zoologists at Aberdeen University in Scotland, long renowned for their research on bats, have recently shown that noctules, big bats from Europe, may catch small birds on the wing, so bats do at least get some revenge.

Some invertebrates are not fast or dangerous—but if they are neither, they are bound to have some other protective trick or they would not be here. Earthworms, tasty and nutritious, live underground. But in the forests of New Zealand, kiwis probe for earthworms with their long beaks, also taking the odd insect and spider that comes their way, plus a few seeds and fleshy fruits. They pick up their food fastidiously with the tip of the beak like a Chinese gourmet and then, most unlike a Chinese gourmet, toss it in jerks to the throat for swallowing. Kiwis are among a minority (though a surprisingly large minority) of birds that are most active at night (although small birds in particular prefer to fly at night when on migration). They hunt by scent and for this their nostrils are uniquely placed—near the tip of the beak, closest to their intended prey. Yet kiwis are not the only birds that sniff out their prey. The caracaras and some of the New World vultures also have a fine sense of smell. Thus in my favorite local aviary I have watched Turkey Vultures sniffing under bushes for carrion, like badgers or armadillos.

But the principal probers after worms and small mollusks are the 212 species of shorebirds. In any one location, shorebirds will have beaks that among them cover all the options like a plumber's or a surgeon's toolkit, with a different instrument for every task. Some have very long beaks, like stilts, curlews, and godwits, for probing deep; some have short beaks for turning over stones rather than probing mud, like the turnstones; and many are intermediate, like Dunlins. There are also extreme specialists, like the lovely avocets, whose beaks turn up at the end and which sweep the water just beneath the surface, setting up minute currents that waft small creatures between the mandibles. The related spoonbills, too, feed by sweeping movements, again setting up local currents that draw the food into the beak. This method works best when the birds move line-abreast, each one stirring up the prey for its neighbors so that they feed more efficiently in a group than they would alone—yet another example of nature's cooperativeness. Phalaropes commonly forage along the shore-line conventionally enough, searching for midges and gnats in stranded wrack; but they also, quaintly, spin in tight circles in shallow water to stir up invertebrates, including small jellyfish. Not many animals eat jellyfish, which has often struck me as an awful waste of protein.

In shorebirds we see how touch-and-go life in the wild can be—that strategy is indeed vital. For although some shorebirds live or nest far in-land, most of them feed on estuaries and mudflats—and on many days es-pecially in the autumn there are only a few hours when the tide is out far enough to expose the mud, and also enough daylight to feed by. Too much disturbance, wasting their time and energy in useless flight, can be fatal. In certain kinds of movies, the heroine expresses her freedom after her many travails by running across the beach and stirring up the shorebirds. A fine symbol, but if it happened too often it would very probably sign their death warrant. No doubt with this very kind of thing in mind, most shorebirds migrate south for the winter.

Among the favored fare of shorebirds are mollusks, notably clams. The flesh inside is succulent and sweet, but any predator that aspires to get to it must be something of a safecracker. The long, straight, typically brightly colored beak of oystercatchers is a wondrous, multifunctional

Avocets feed most efficiently in the company of other avocets.

tool, like an all-in-one Swiss Army Knife. It's a probe for prying mollusks out of the mud. It's a crowbar for levering the shells of bivalve mollusks apart—or for flicking limpets off their rocks. It's a pair of kitchen scissors for slicing through the adductor muscle that keeps the bivalve shells together. It's a hammer for peremptorily smashing the shells of mollusks that refuse to cooperate. It takes time for young oystercatchers to learn the techniques—and until they do they must feed on scraps. Different species—and indeed different individuals within species—develop their own methods (just as different groups of chimpanzees evolve different ways of using stones and sticks).

Some birds, including Carrion Crows, deal with the shells of mollusks and other beasts by dropping them onto rocks. Crows quickly learn to drop their booty from just the right height: too low and the mollusk fails to shatter, too high and other birds will swoop in and snatch the succulent morsels before the crow can get down to them. The famous Lammergeier Vulture of the Old World, also known as the Bearded Vulture, derives 90 percent of its diet from the marrow of big bones that it shatters by dropping them onto rocks—typically onto favorite rocks, analogous to the thrush's anvil described below. The Greek playwright Aeschylus was allegedly killed by a tortoise dropped by an eagle. I bet it was a Lammergeier. The Lammergeier was once known as the Ossifrage—bone smasher (*frage* as in "fragile").

Several birds of quite different kinds use stones in different ways to break their way into mollusks. Members of the thrush family, which includes the European Blackbird and European and American Robins, in general eat small creatures on the ground. Song Thrushes take some berries from trees and shrubs, but they also, famously, break into snails by bashing them on favored stones known as "thrush anvils." You cannot mistake a thrush anvil with its penumbra of sacrificial shards. In similar vein, the Woodpecker Finch of the Galápagos uses a twig or a cactus spine to winkle larvae from their burrows. Biologists used to claim that human beings were the only tool makers; then Jane Goodall showed that chimps are, too. Now, more and more birds are joining the ranks. As we will see in Chapter 9, some crows even qualify as tool makers. Human beings are still unique in some important ways—but not in this way.

Most birds that prey on other animals, whether insects or worms or crabs or fish, catch them in their beaks. But the supreme avian predators are the raptors, both falconiforms and owls, which seize their victims in their taloned feet. All of the raptors, whatever their ancestry, have big eyes, even by avian standards: the falconiforms with a huge concentration of cones, suitable for day vision; the owls with the emphasis on rods, for night vision. Most raptors hear very well, especially the owls, which have huge ears, though hidden beneath the feathers (the tufts that some owls have on the top of the head have nothing to do with ears). In owls, one ear is bigger and higher than the other, and this enables them to pinpoint the direction of sound with wondrous accuracy—owing to the asynchronous way in which the sound reaches the two ears, placed far apart on a big wide head.

Raptors, both owls and falconiforms, come in "suites" of species: groups of species that fill the available ecological niches. Big ones like the Eagle Owl or the mighty Gyrfalcon take big prey, which among Eagle Owls can include other birds of prey such as harriers. Small species, like the sparrow-size Elf Owl of the southwestern United States and the Black-thighed Falconet of Asia, Africa, and South America, among the falcons take little prey—although the feisty Falconet takes on prey its own size, as well as insects. Their feet are specialized for whatever they feed on: raptors that capture other birds in flight, as the falcons do, have long and slender toes for maximum reach, with long, sharp, very hooked talons. Fish-eating raptors have very long, very curved talons. In raptors that capture big prey, like the Harpy Eagle, which preys largely on monkeys, the tarsus (or fused foreleg) is short and robust and the toes are thick with well-curved talons. In such birds the blow with the feet at the moment of strike is what really subdues the prey. The tarsi of big eagles, too, have muscles within them to enhance the grip even further—whereas the tarsi of most birds are all skin, bone, and tendons. The beaks vary, too, to match the diet but in general the upper mandible has cutting edges, which shear the flesh as the lower mandible—which provides the power—pushes against it. Some raptors swallow their prey whole—including Bat Hawks, which with their huge gape swallow entire bats on the wing—but most tear bits off, twisting and pulling as they hold their captives down with their prodigious feet.

We have seen how Harris's Hawks cooperate to hunt. So, among falcons, does Eleonora's—positioned on the falcon suite between the Peregrine and the Kestrel. Eleonora's Falcon is beautifully equipped to catch songbirds in flight—and is also a great strategist, planning its life to be in the right place at the right time to make the best possible haul. The birds spend the northern winters in Madagascar (with a few in Tanzania), but they breed on the faces of cliffs, in colonies of from a just a few pairs up to 200, on islands in the Mediterranean. Most Mediterranean birds nest early, from March to May, to catch the early insects, but the Eleonora hen does not lay her first egg until late July. The brood hatches in late August—and from then until the chicks are grown some time in October, both parents feed them, and themselves, almost exclusively the songbirds that at that time are migrating south across the Mediterranean into Africa. In one study, a group of 150 Eleonora's males moved out to sea at dawn, spaced themselves 100 to 200 meters (330 to 660 feet) apart at up to 1,000 meters (3,300 feet) above the waves, and hovered, using the prevailing north wind to keep themselves aloft, as Lesser Kestrels do inland. Several working together would harry a single songbird, forcing it nearer and nearer to the waves, until just one of the falcons caught the exhausted traveler in its talons and bore it off to its nest. Like the Three Musketeers, their principle seemed to be "All for one, and one for all." Working in groups they proved very efficient. One male brought five songbirds to its chicks in twenty-five minutes.

There are about 6,000 breeding pairs of Eleonora's (12,000-plus individuals), and since each pair raises an average of two chicks each year, the total population by late autumn is at least 20,000. Between them they must kill around 10 million migrating songbirds. Feathers from more than ninety different songbird species have been identified among the debris they scatter around their nests, with an abundance of warblers, shrikes, swifts, small thrushes, larks, and flycatchers. But as we will see in Chapter 8, thousands of millions of songbirds migrate south each year, and alas, they face much bigger hazards than Eleonora's.

Many birds are fishers. Eleven orders of modern birds—more than one-third of the total—include at least some that eat some fish. Some of the best-known of the ancient, pre-neornithine birds were fishers, too. Birds owe a lot to water.

The bird for whom the entire kingfisher family—Alcedinidae—is named is one of the most romantic birds of Europe, or indeed of the world. It's a flash of blue on a summer's day as it dives into the stream from an overhanging branch, seizes its prey in its dagger-like beak, and zooms out again, all in one movement, to eat at leisure on its perch. But the Alcedinidae did not begin as plunging fishers. Almost certainly the family arose in tropical forest. And almost certainly the earliest types were exclusively terrestrial, and they plunged from their perches to the forest floor to grab insects or lizards. Australia's Kookaburra, alias the "laughing jackass," still does precisely this. The Kookaburra's beak is not like a dagger. It is flat, almost like a duck (although pointed), as befits an insect eater. Clearly, somewhere along the evolutionary line some ancestral kingfisher took to leaping after prey in shallow water; and from this evolved the method that we think of as the family's metier.

As a fisher, too, the European Kingfisher and its kind do not represent the end of the evolutionary line. To plunge from overhanging branches is a fairly primitive technique—clearly of use only when there are branches to plunge from. More advanced, technically, and of more use in open water, is to hover before plunging, as a tern does, with no other means of support. Some kingfishers do this—notably the Belted Kingfisher of North America and the Pied Kingfisher of Africa, which hunts sardines far offshore in Lake Kariba: a deep diver among kingfishers, plunging up to 2 meters (6.5 feet) beneath the surface. So we can see kingfisher evolution mapped out before us, re-enacted by the present generation: the Kookaburra at the primitive end, the tern-like Pied Kingfisher the most "derived." But there have been many variations along the way. Some species of kingfisher catch insects in the air after a leap from the perch, like flycatchers; some prey on other birds; some search for earthworms in the forest leaf litter.

Many raptors, both falconiforms and owls, are fishers, too. As with the kingfishers, it's clear that the fishing raptors evolved from terrestrial hunters. Obviously, in raptors, fishing has evolved independently several times.

America's Bald Eagle is an accomplished plunderer of salmon in due season, like the grizzly bear. Britain's commonest eagle until the early nineteenth century was the White-tailed Eagle (or Sea-Eagle), when it was wiped from its remaining Scottish strongholds for fear that it might kill lambs, which were the latest economic bonanza. Biggest of all the eagles—biggest indeed of all the raptors that actually kill their own prey—is the mighty Steller's[1] Sea-Eagle, which hunts for salmon in particular along the northeast coast of Eurasia. Females can have a wingspan of 2.4 meters (8 feet) and weigh up to 9 kilograms (20 pounds), although the males are much smaller, as is often the way with falconiforms. There are tropical fish-eating eagles, too, like the African Fish-Eagle, the Grey-headed Fish-Eagle of Asian rivers, and the extremely rare and Endangered Madagascan Fish-Eagle.

The supreme falconiform fisher is the Osprey, which is different enough from all the rest to have a family all to itself (the Pandionidae). Like many a seabird, the Osprey plummets into the water from a great height—anything from 15 to 60 meters (50 to 200 feet). It fishes from the surface like a tern, not diving like a gannet, although it usually finishes up submerging. Its long legs end in huge raptorial feet with spiny soles that can grip the slipperiest fish. The outer toe can be swung back like an owl's so that the Osprey can embrace the struggling fish with two talons in front and two behind, fish that are usually less than a half a kilogram (a pound), but may weigh up to a kilogram and a half (3 pounds). As the Osprey flies off it turns its prey to face the wind—which gives extra lift. Ospreys expect to catch a fish on two dives out of three—a very high success rate; and they can usually get all they want within two or three hours. But sometimes (or so it has been reported) Ospreys miscalculate the size of their potential prey and get dragged under. Sometimes, too, the prey is pinched by a sea-eagle, just as the prey of leopards or cheetahs on the plains of Africa is sometimes stolen by lions.

There are fish-eating owls, too—different groups of them: one genus from Asia and one from Africa. They eat frogs as well as fish, and whatever else looks tasty.

Of all fishing birds, only the raptors catch their prey in their feet. All the rest use only their beaks—but in a whole variety of ways. Some swoop down to the water from flight—some from a perch, some while soaring, some while hovering. Some fish from the surface, hoping (it seems) to avoid getting wetter than is strictly necessary. Terns, though they are magnificent seabirds and great hoverers, cannot float for too long without becoming waterlogged. But some of the swoopers plunge into water, sometimes to great depths. Many others from several different groups dive from the surface. Once under the water, some swim using only their feet, keeping their wings tight to their sides; others use only their wings; and a few propel and steer themselves with a judicious combination of the two. Some that hunt underwater are heavier than birds generally are or use various devices to increase their body density so they can stay submerged more easily. Some simply stand at the water's edge on long legs and spear or sometimes grasp passing prey.

The kingfisher fishing technique, as we have seen, is primitive—just swooping from some overhanging perch. Gannets take off from the shore and dive into the water from aloft from variable but sometimes great heights. For this they are highly adapted. Inflatable air sacs between skin and muscles cushion the blow of entry. Even so they need to fold their wings before they hit the water or they would break them. The calculation is fine. If they fold their wings too soon, they could lose control, hit the water at the wrong angle, and break their necks, air sacs or no air sacs. But always they fold them just at the right time. The apparent distance between the waves increases as they get closer and they seem to calculate the speed of descent from the rate at which the wavelength seems to change. It takes pages of math to show how this is done, but gannets do what is needed in a twinkling, with absolute accuracy. Olympic divers sometimes get their timing wrong, but gannets never do.

Pelicans look clumsy, but in truth are superb fliers, especially good at soaring—happy to journey more than 150 kilometers (100 miles) in a day after food, and gliding home against the evening sun. They, too, dive into the sea after fish, which they scoop up in their huge beaks: the lower mandible has a floor of elastic skin that extends like a fishing net. The beak of the pelican can indeed, as the nursery rhyme has it, hold more than its belly can. The technique is like that of the blue whale: scoop up a lot of water, prey and all, then strain off the water.

Albatrosses stay more or less constantly aloft for weeks or months at a time, patrolling the oceans of the Southern Hemisphere, riding the constant winds with supreme efficiency; and although one of the biggest kinds is called the Wandering Albatross, they do not wander aimlessly but follow well-established routes. Mostly they feed from the surface, but some of them plunge as well: the Grey-headed gets down to 6 meters (20 feet), the Light-mantled Sooty down to 12 meters (40 feet)—an extraordinary feat for such long-pinioned beasts. Sometimes albatrosses feed at night, when sea creatures come nearer to the surface. Biologists know this because they have fitted albatrosses with internal thermometers, and their temperature drops whenever they swallow a sea-cold fish.

Terns characteristically hover before swooping to take their prey from the surface. In terns, as in falcons, we find a complete spectrum of techniques. Most species hover over sea or marsh before plunging after fish (or squid and crustaceans or the occasional frog), which they seize or stab with their long pointed beaks, but they don't swim underwater, like gannets. Common and Roseate Terns are largely commensal feeders, relying on big, predatory fish to force smaller fish to the surface before they can catch them. The larger terns, like the Caspian and the Royal, are less agile and graceful, but they plunge from greater heights—up to 15 meters (50 feet). The dark-colored noddies patter the surface with their feet to attract their prey, like storm petrels—and may catch flying fish on the wing. Marsh terns plunge for crustaceans and frogs, but also hawk for insects, like swifts. The most terrestrial of terns is the Gull-billed, which swoops to the ground for lizards, insects, and small rodents. Whereas fishing kingfishers evolved from ground feeders, ground-feeding terns evolved from fishers. Most terns feed by day, but the Sooty and some others may sometimes feed at night.

Terns are related to gulls and also to shorebirds within the order Charadriiformes—and so too are skimmers, whose approach to fishing is decorous and unique. Their lower mandible protrudes far beyond the upper one—by about a quarter or a third of its length. This is not a mistake, but a fine piece of engineering: when the bill is closed, the upper mandible notches into the lower. Skimmers swim just above the water, trailing the long lower bill into the water like a plough; when it touches a fish, "snap": the beak closes and the fish is caught—and hence the alternative name of "scissorbill." Well-developed muscles around the head and neck help absorb the shock. Because of the way they fish, skimmers can fish only on smooth waters, on lakes or the laziest rivers. Often they fish at dusk—which is when I watched them from a jetty in the Florida Everglades. Their eyes have a vertical pupil, like a cat, which may help to gather more light. Usually they fish alone or in pairs, but sometimes (as they were when I saw them) they skim the waters in groups of ten to fifteen.

Cormorants, though again related to gannets, generally prefer to plunge less adventurously from a floating start at the surface to catch fish with their long hooked beaks. They go to great lengths to make themselves heavy, as befits a diver. They have heavy bodies with dense bones and little body fat, and so float low in the water. They lack a preen gland, so their feathers lack oil—and this is not an oversight but an adaptation: the waterlogged feathers are heavy and help them to stay submerged. This is why they need to dry out when they emerge, holding out their somewhat ragged wings as if in supplication. They drive themselves through the water and along the seabed with their powerful legs and big webbed feet. They keep their wings to the side and use their tail as a rudder. The sides of their mouths and their throats are highly expandable, so they can swallow large prey.

The twenty-one or twenty-two species of grebes again form an entire suite of predators, from the size of a big duck down to the size of a very small moorhen, spread over much of the world and catching everything from insects and mollusks to fish. Like cormorants, they plunge from the surface, typically submerging for 10 to 40 seconds. The Western and Clark's Grebes spear fish with their dagger-like beaks, like a heron, not seizing them like a cormorant or a puffin. To reduce the internal damage that can be done by the sharp spines and shells of their prey, most grebes ingest their own feathers and drink a lot of water, so their stomachs acquire a felt-like lining. Like cormorants and anhingas, grebes seek to increase their body density before diving—but by a different technique. They press their feathers to their sides to expel the air from them, and from the subsidiary air spaces in the body, while the flank feathers absorb water to reduce buoyancy even more. When frightened, grebes may hide underwater. They propel themselves underwater with feet set well back on the body on highly flexible ankles, so they are extremely maneuverable; the toes are not webbed like a duck or a gull but lobed, with flanges on either side of the toes. On land, with their rear-mounted legs, they tend to be fairly hopeless (but you can't win 'em all).

Anhingas like freshwater, tropical to warm-temperate: Florida suits them (I have seen them in the Everglades, too). In general form, Anhingas are like cormorants but they can float so low in the water that only their heads and sinuous necks are showing—hence their alternative name of "snakebirds." They are heavy for much the same reason as cormorants—they are not waterproof—so when they emerge they too must stand with their wings held out to dry. Their beaks, however, are not hooked like a cormorant's but pointed like a heron's—and, like herons, they often spear fish rather than seize them. They are stealth hunters: they submerge slowly, then scout like cats beneath the water. Their feet are set well back for fishing like a cormorant's or a grebe's, but their legs are long and they climb in trees with remarkable agility. Anhingas are actually increasing in New Guinea, where they feed on tilapia, imported as a food fish from the Great Lakes of Africa.

The alcids—including the auks, murre, murrelets, guillemots, and puffins—again form a suite of predators, from very small to goose-size, with a wide variety of prey. They propel themselves with their wings when they dive; as in penguins, their wings act as hydrofoils, providing lift as well as thrust, and so they can truly be said to "fly" underwater. Unlike the penguins, however, most of the alcids have retained the ability to fly through the air. Of the flightless kinds, the best-known and also the biggest at 75 centimeters (2.5 feet), was the Great Auk of North Atlantic islands; but it paid for its flightlessness and was rendered extinct in the 1840s. It is not difficult for alcids to evolve flightlessness, for the bigger kinds routinely lose the power of flight for about six weeks when they are molting. Most kinds fish on the shallower waters of continental shelves—only the puffins routinely wander farther. The auklets of the Pacific and the Dovekies of the Atlantic feed on plankton—which in due season they carry back to their young in a pouch in the throat, in the form of "plankton paste"—and in them the beak is wide and short and the palate and tongue are equipped with horny tubercles that help to manipulate the prey. The Parakeet Auklet is yet another consumer of jellyfish (as well as crustaceans) and has a unique beak like a scoop. In the Razorbills and in puffins, the beak is deep and narrow for seizing fish. The puffin's beak is large and colorful and is encased in nine distinct plates, which play a great part in courtship and then are shed at the annual molt—simply a piece of finery. Puffins have a marvelous ability to carry many small fish such as sand eels sideways on in the bill—up to sixty at a time—which means they must hang on to the ones they already have while catching the next one; which their beaks are specialized to do. Black and Pigeon Guillemots feed on bottom-living fish. Like penguins, some alcids can dive to prodigious depths and stay submerged for long periods.

But penguins are the supreme fishers. As the New Zealand biologist Lloyd Spencer Davis has commented, penguins are birds that are themselves striving to be fish. Their bones are more solid than in most birds and so the body overall is almost as dense as water; they also take stones into the gizzard to weight themselves, like the lead weights of old-fashioned deep-sea divers. Penguins cannot bob on the surface like air-filled ducks.

But the penguin way of life is hard work. Chinstrap Penguins, when feeding chicks, may make nearly 200 dives a day. On a long sea trip, King Penguins may make nearly 900 dives, expecting to catch a squid or a fish on about one in ten of them. Gentoo, Chinstrap, and Macaroni Penguins, which largely eat krill, typically submerge for half a minute at a time, and sometimes up to one and a half minutes. Jackass Penguins, which eat mainly fish that are harder to catch, have to stay down longer; they typically manage two and a half minutes but may stay down for five minutes. Emperor Penguins also prefer dives of two and a half minutes, but have been timed at up to eighteen minutes—well within the typical range of big whales. Studies with pressure gauges have shown that Chinstraps can dive to depths of up to 100 meters (330 feet), although they usually dive to less than 45 meters (150 feet). The Adélie can go down to 170 meters (560 feet). King Penguins can reach 240 meters (790 feet). But the supremo in this as apparently in all things is the Emperor, which can descend to 530 meters—about a third of a mile. Diving isn't just a matter of endurance—holding the breath and hoping for the best. When Adélie Penguins dive, the heart rate drops from 80 to 100 per minute down to 20, so that the oxygen in their lungs and air spaces lasts at least twice as long as it would on land. But how penguins such as the Emperor avoid the bends on deep dives—the curse of human divers—is unknown. The bends happens when nitrogen gas from the air, dissolved in the blood under pressure, bubbles out again when the pressure is released. This is simple physics. No creatures on earth can defy the laws of physics. But penguins, like whales and seals, get around them.

Thus penguins demonstrate a prime principle of optimum foraging strategy. On the one hand, they live at their physiological limits—and so they must, if they are to get the best out of their environment. Yet what they do must be cost-effective, and clearly it is; for despite the depredations of leopard seals and southern sea lions and sometimes of killer whales and even of sharks, penguins flourish. Or at least they would be flourishing were it not for the universal menace of humankind, wrecking their habitats, pinching their grub, and now, the *coup de grâce*, causing the entire world to warm up and melting the ice and possibly redirecting the ocean currents to which many of them are adapted.

PREDATORS MAY ALSO succeed who only stand and wait—as do the storks and the various birds that are traditionally supposed to be related to them, such as herons (although, as we have seen, some at least of these supposed relationships may be more apparent than real). The usual method, shared by storks, herons, and so on, is either to wade through shallow water or else to stand stock-still at the water's edge, and either to spear or to seize fish or frogs in a long pointed beak. But there are some interesting variations on that basic theme. The Black Heron arches its wings over its head like an umbrella to create shadow, beneath which fish apparently like to shelter; and this also reduces surface reflection, which makes it easier to see what's going on—this is known as "mantling." The Shoebill looks like a massive stork—up to a meter and a half (5 feet) tall—except that its beak really is like a Dutch wooden clog, up to 20 centimeters (8 inches) long and very deep, with sharp edges and a hooked tip. The Shoebill hunts among the floating rafts of papyrus that traditionally spanned the Nile to form the Sudd, standing stock-still for thirty minutes or more at a time. Its huge bill is especially adapted to catch the African lungfish, *Protopterus*, which is built like a huge, thick, moray eel but really does have lungs and rises to the surface at intervals to breathe. Then the Shoebill strikes, forward and down, using its whole body, helping out with its wings if necessary—for *Protopterus* is tough; it requires commitment. *Protopterus* is weird and the Shoebill is weird—here is a Gothic encounter, indeed. Despite the Shoebill's great apparent bulk, it can soar very high on the thermals with its neck tucked in like a heron. Living as it does in such a place, it is well to be able to cover huge distances with as little effort as possible.

HOW TO BE A PIRATE

Then again, a whole suite of seabirds has evolved that can and do hunt perfectly well for themselves and find carrion, but at least in the breeding season they supplement their rations by robbing others. The skuas and jaegers (from the German *Jaeger*, meaning "hunter") are relatives of the gulls, and for most of the time are straightforward predators—from insects and even berries through lemmings and rabbits to mountain hares to Greylag Geese, depending on what's around, even sometimes feeding on the chicks of their own species in neighboring colonies. But in the breeding season, all these birds turn pirate, harassing gannets, gulls, terns, puffins, what you will, to disgorge the food they have taken such pains to catch for their own offspring. Skuas and jaegers are placed in the family Stercorariidae, which derives from the Latin name for "excrement" (as in *stercoraceous*). They are stinkers, too.

From the deep south again are the sheathbills, which look a bit like scruffy white chickens, but in truth are skuas in chickens' clothing. They are bold, unfazed by humans, and take bits of whales from around the whaling stations. In due season they gorge on the placentae and stillborns of breeding seals, as farmland crows may do at lambing time. At other

Arctic Skuas are broad-based hunters—and also efficient pirates, stealing the hard-won booty of rival predators.

times they search among the seaweed for invertebrates. But pairs also build their nests close to penguins and then can be serious predators, stealing eggs and chicks or—skualike—harassing parent penguins to re-gurgitate the food they have brought for their chicks and feeding it to their own chicks. After fledging, the young pick up debris from the beach and are then known as "shitehawks"—as indeed are Black Kites in India and South Africa. More stinkers. But sheathbills are clean birds. They spend a lot of time preening.

There are pirates in the tropics, too—notably the five species of frigatebirds that among them patrol the equatorial oceans all around the world on their long, black scimitar wings. Frigatebirds are basically fishers, scooping prey from the surface, but they are also notorious buccaneers, dressed for the part in ragged black and occasional flashes of gaudy, pillag-ing food from other seabirds on the wing and also their nesting material.

Finally, animal flesh is accompanied by all kinds of debris—spines, shells, fur, feathers, bones—that is not digestible. Many birds, within their crops, parcel all this rubbish into neat pellets, which they then regurgitate. Owls are best known for this, feeding as so many of them do on mice and voles, with their small and vicious bones. But raptors and gulls often make pellets, too, and they have also been recorded in flycatchers, crows, herons, sandpipers, kingfishers, barbets, rails, and even honeyeaters—which mainly eat nectar but also arthropods. Pellets offer a grand opportunity for the field zoologist to see what the birds have been eating, which often is all but impossible to observe directly. Field studies are difficult, and naturalists need all the help they can get.

All in all, it is hard to make a living by eating other animals.

SO WHY NOT EAT PLANTS?

Despite appearances, plants are more difficult to cope with than animals. They are typically less nutritious and tend to need more processing, in a more elaborate digestive tract, once they are down the throat. So we often find that herbivorous birds have smaller relatives who are carnivores or at least are omnivores, since small animals find it very hard to be out-and-out vegetarians. On the other hand, herbivores often have a more friendly relationship with their prey than carnivores do. After all, it is hard to see how any animal apart from parasitic worms could benefit from being eaten, but among plant eaters and the plants that they eat, we see many different kinds of mutualism—mutually beneficial symbiosis. The main advantage in eating plants, for those that can manage it, is that there are an awful lot of them. In any terrestrial environment they generally outweigh the animals by at least ten to one and generally outweigh any one kind of animal by hundreds or many thousands to one.

Small birds that aspire to live substantially on plants must focus on seeds because no other part of the plant is nutritious enough—and since seeds are hard, the birds must be properly equipped. Among the "true" finches of the family Carduelidae, different species specialize in different seeds and have appropriate beaks to match, from the slim tweezers of the Siskin that probes the cones of alder trees for their tiny seeds (in Europe there are entire forests of alder) or the Goldfinch that loves the tiny seeds of teasel, to the blacksmith's pliers of the Hawfinch that can crush the stones of cherries and olives. There is far more to the finch's beak than meets the eye. On each side of the top mandible is a groove where the bird can wedge the seed and crush it by raising the lower mandible, peeling away the husk with the tongue and then spitting it out. Hawfinches could not eat what they do by power alone: the beak is fitted internally with knobs to spread the massive load of crushing evenly. The Bullfinch feeds on soft buds, berries, fruits, and the odd invertebrate, as well as seeds; its beak looks almost as heavy as a Hawfinch's but it has a sharp cutting edge. Quaintly, too, the Bullfinch carries food home to its babies like a hamster, in pouches in its cheeks and throat. Some carduelid finches can even move the lower bill from side to side, like a mammal chewing.

In the crossbills (*Loxia*), the top mandible and the lower cross over, one over the other. The crossbill inserts its beak between the scales of conifer cones, and as it closes the mandible, so the protruding beak tips on either side force the scales apart, and the bird can then scoop out the seed with its tongue. Europe's four species of crossbill have different-size beaks for different kinds of conifer—larch, spruce, and pine; natural selection can be remarkably precise. The crossbills of Scotland are particularly adapted to the Scots pine, *Pinus silvestris* (which, despite its common name, grows widely in Europe). Until recently the Scots kinds were classed only as a subspecies of the Common Crossbill (*Loxia curvirostra*), but now they are recognized as a distinct species, *Loxia scotica*, the Scottish Crossbill. In fact, it is Britain's only endemic species of bird (*endemic* meaning that it occurs in Britain and nowhere else). Thus we see how small shifts of diet, or some particular specialty, in an isolated population can produce a new species.

It seems, though, that the carduelid finches evolved from fringillid "finches," now represented by the chaffinches and the Brambling. Bramblings love the seeds of beech trees (beechmast) in winter and may turn up in huge numbers to gobble them up in a good "mast" year, but they rely on insects in summer. Chaffinches, too, eat seeds but also rely heavily on insects. Insects are easier to cope with than hard seeds, and the fringillids accordingly have much simpler beaks (although they look much the same as the carduelids from the outside).

Many birds have muscular gizzards like an annex of the stomach, which they may pack with stones. Thus equipped, fowl such as turkeys and peafowl can swallow big seeds whole and crush them, as in a mill. Until the nineteenth century (when an imported fungus infected the entire population), the eastern United States was largely forested with American chestnut trees, *Castanea dentata*. The American chestnut was, perhaps, America's most abundant hardwood. A squirrel could hop from chestnut to chestnut all the way from Georgia to Maine without ever coming to ground, or so it was said. In the Appalachian Mountains, the American chestnut fed an estimated 10 million Wild Turkeys. Passenger Pigeons, too, fed on the chestnut and were one of the most abundant birds in the whole world—the vast flocks were said to "blacken the sky." Now the American chestnut and the Passenger Pigeon are gone, and the Wild Turkey is much reduced, but the point here is merely that big birds like turkeys and pigeons can crush big seeds with the aid of gizzards. We know from their fossils that many herbivorous dinosaurs had gizzards, too. Gizzards are a fine device, but the grindstones are heavy; and although many birds with stone-filled gizzards do fly, as Wild Turkeys do, this way of dealing with the diet clearly militates against the general avian trend toward minimum weight.

Whole new nutritional vistas are open to animals that can derive energy from cellulose. Cellulose in principle is highly desirable because it is pure carbohydrate—it is compounded from molecules of sugar joined end to end. It is also the commonest polymer in nature by far since it is the main component of plant cell walls. Indeed, there are billions and probably trillions of tons of cellulose out there—the world is a cornucopia. The most obvious source of cellulose is green leaves: their total weight hugely exceeds the total weight of seeds or fruit. Animals that specialize in eating leaves are called "folivores." Folivores that specialize in grass in particular are called "grazers"—a good thing to be since grasses are so abundant, the principal plants over vast areas of the globe. Animals that eat the leaves of trees and shrubs are "browsers."

But cellulose has drawbacks. The first is that no known animal can digest it. Only some microbes, meaning bacteria and some protozoa, have the necessary "cellulase" enzymes. Many animals from termites to cows to elephants can and do use cellulose for energy, but only by harboring suitable microbes in various kinds of chambers in the gut to do the digesting for them, by the general processes known as "fermentation." They then absorb the products of this fermentation. But the microbes take their time, and the only way for an animal to get enough nutrient from their labors is to harbor a large quantity at any one time. So cows and other ruminants have a vast fermentation chamber in the foregut known as the "rumen" where the bacteria and protozoa go to work, while elephants and horses and rabbits have a vast diversion of the hindgut, known as the "cecum," for the same purpose.

Small warm-blooded animals cannot hold enough cellulose and microbes in their guts to provide enough energy to maintain their high metabolic rate (termites do okay on cellulose because they don't waste energy keeping themselves warm). So as we noted earlier, birds must be at least as big as a grouse to live full time on leaves. For an Ostrich, with its huge body and relatively low metabolic rate, vegetarianism holds no fears, although Ostriches will snap up the odd small animal when the opportunity arises. Ostriches often graze side by side with zebras, and the two seem to be doing the same thing—eating the local vegetation. But while zebras shovel back the grass as fast as they can go—they are "bulk feeders"—the Ostriches are far more selective, taking only the buds and succulent, nutritious shoots. Thus two or more creatures may feed side by side ostensibly on the same fare without getting in each other's way at all, and thus Ostriches show that it is possible to find adequate nourishment from unpromising fare provided you choose well.

Almost all the big folivorous or near-folivorous birds from grouse to rheas to the Ostrich ferment the cellulose in a cecum, as a horse does. But there are two exceptions among leaf-eating birds. One (which always seems to be an exception to everything) is the extraordinary Hoatzin, which lives in the forest trees along the riverbanks of South America and looks a bit like an unkempt chicken, but has sometimes been classed as a cuckoo, although it probably is not related to either. The Hoatzin lives almost exclusively on the leaves of trees and has a fermenting chamber in its foregut, like a cow or a kangaroo. Foregut digesters tend to belch surplus gases, including methane, and so do Hoatzins—so they smell like cows. Hoatzins, too, uniquely, feed leaves to their babies from day one, albeit in largely predigested form (and with accompanying bacteria). The other major avian leaf eaters that lack a worthwhile cecum are the geese, swans, and wigeons. They have straight guts, like a pig or a human. So it is that geese, in addition to the normal watery excreta of a bird, produce vast amounts of characteristic goose turds that largely consist of undigested grass. It seems odd for an animal to rely on food that it cannot digest very well, but nature is full of surprises. Note, too, how widely the diet varies within the duck family, Anseridae. Some ducks, after all, are fish eaters.

Parrots also demonstrate radical shifts of diet within a single family. Most parrots mostly eat fruit and seeds, plus flowers and buds. Most are fairly catholic, but some are extreme specialists. Fuerte's Parrot of the High Andes lives mostly on mistletoe seeds. Pygmy-parrots of New Guinea specialize in lichens and mosses. The Great Blue Macaw of South

America feeds mainly on palm nuts. Australia's lorikeets prefer pollen and nectar from trees. Small parrotlets and lovebirds, and the Budgerigar, are seed eaters, mainly of grasses near the ground. But some have broken wildly with convention. Cockatoos and rosellas in Australia search for grubs on the trunks and boughs of trees—like woodpeckers, for there are no woodpeckers in Australia. Hyacinth Macaws in South America have been known to eat aquatic snails. The Kea of New Zealand mostly eats insects, but has expanded into carrion and one colony was known to excavate the chicks of Sooty Shearwaters from their nesting holes. Keas also favor human company and have been known to come down chimneys to steal food, and are renowned for picking the rubber from around the windscreens of cars (there are warnings in the parking lots). They also pick greenbottle maggots from the backs of sheep, which is helpful, but then they peck the wounded flesh of the sheep itself, sometimes fatally, if the shepherd doesn't get to the sheep in time.

We find spectacular shifts of diet, too, within and among the fifty-four species of hornbill—the smaller kinds catching insects, which are concentrated fare, while the bigger ones focus on fruit, which is much bulkier.

Fruit can be fine food, but it is highly seasonal; and although trees in tropical forests may bear fruit year-round, they tend to do so spasmodically and sometimes, as far as can be seen, randomly. But when fruit does turn up in tropical forest, it tends to be superabundant. The orangutan, the world's biggest committed fruit eater, tends to operate alone or in small family groups—for abundance is no problem for a lone creature with such an appetite. But small fruit eaters, including monkeys and most fruit-eating birds, tend to operate in groups. That way they have more chance of finding the fruit in the first place, since many pairs of eyes are better than one—and when they do find fruit, there is plenty for all. Turacos, for example, tend to operate in groups of up to twelve. Vultures, in their search for cadavers, follow much the same principle.

The sociality that is encouraged by the diet tends to spill over into all aspects of life. So it is that hornbills are fruit eaters—and also, as we will see in Chapter 7, are outstandingly social breeders, with various kinds of social arrangements. But also among hornbills we see an interesting twist—where the innate sociality has in turn become adapted to a quite different kind of feeding. For among the biggest of all hornbills, and in various ways distinct from the rest, are the two species of ground-hornbills from Africa. Ground-hornbills are not mere fruit eaters; they are formidable predators. The beak is like an icepick. They can hack their way into a tortoise. The Northern species is among the biggest of all avian predators. The ancestors of ground-hornbills were presumably fruit eaters, and that, perhaps, is how they first evolved their sociality. Now, as predators, they hunt in packs. Typically they chase some hapless creature like a hare into a bush and then some act as beaters while others lie in wait and deliver the *coup de grâce*. The packs are usually family groups. They can be seen as strategic predators like wolves or perhaps as problem families, terrorizing the neighborhood.

Ecologically speaking, toucans are the New World's answer to hornbills—again, the piciform–coraciiform connection. Toucans also are long-beaked fruit eaters, able to reach the most succulent fruit without risk of falling from their perch. Like hornbills, too, toucans are often social—and sometimes extremely so, as shown in this tale from *The Naturalist on the Amazons* by Henry Bates. It was Bates who lent his name to Batesian mimicry. He was a great explorer and naturalist, but living as he did in the mid-nineteenth century, he took it for granted that killing the creatures he loved was all part of the game. In the following story he was after Curl-crested Toucans:

> I had shot one from a rather high tree in a dark glen . . . and entered the thicket . . . to secure my booty. It was only wounded, and on my attempting to seize it, set up a loud scream. In an instant, as if by magic, the shady nook seemed alive with these birds, although there were certainly none visible when I entered the jungle. They descended towards me, hopping from bough to bough, some of them swinging on the loops and cables of woody lianas, and all croaking and fluttering their wings like so many furies. If I had had a long stick in my hand I could have knocked several of them over. After killing the wounded one I began to prepare for obtaining more specimens and punishing the viragos for their boldness; but the screaming of their companion having ceased, they remounted the trees, and before I could reload, every one of them had disappeared.

Here is sociality of the highest order, each bird apparently prepared to risk all for the group. Fruit eating encourages sociality. Diet affects everything.

In the dry tropics in particular, many plants grow only after the rains and then die back or hide entirely beneath the surface until the good times come again in the form of seeds or tubers or woody roots, and if the would-be eater of plants is not on hand at the right time, it misses out. Many dry-land plant eaters are therefore obliged to be opportunist nomads, going to where the vegetation happens to be.

The queleas of Africa, which are small weavers, move in vast flocks, in the manner of forest fruit eaters, to wherever the plants happen to be flourishing on the unpredictable plains. Often they stop to feed on crops and then become major pests, like locusts. Some quails, too, are nomads—like the Harlequin Quail of Africa and the Rain Quail of Asia, which follow the rains and hence follow the food supply. In this way, it seems, quails turned on cue to feed the fleeing Israelites in Exodus.

But the world's most spectacular food-seeking nomads are Australia's Emus. Like Ostriches, Emus live in dry open spaces and yet manage to find a diet that is nutritious and succulent—the tastiest bits of plants eked out with small animals, notably insects. But they can find such pickings only after rainfall. When food abounds, they put on fat, and their weight shoots up to 45 kilograms (100 pounds). But when the food runs out in any one place, they undertake what may turn into huge treks in search of more—seeking out places where it has just rained. Other dry-land animals, such as the Arabian oryx, pursue the same kind of tactic. Arabian oryx locate new, promising places by the scent of damp earth and fresh vegetation. Emus may use scent, too. But they seem to depend mainly on sight—like vultures they go where the clouds are—and to some extent they also depend on sound: they move toward thunder. Wild creatures "read" the landscape as aboriginal people do everywhere—far more astutely than all but the most practiced Western naturalists. Emus with their great three-toed feet are prodigious walkers, keeping up a steady 7 kilometers per hour (4 miles per hour) for hundreds of kilometers, and on their great food expeditions they lose the fat they put on during the good times and by the end may be down to 20 kilograms (44 pounds). So, en route, their weight may halve. An ability to gain weight prodigiously and then lose it again is characteristic of many wild creatures, including all birds that undertake long migrations over sea or unforgiving land. Many human beings who live as hunter-gatherers in countries that have a dry and barren season are similarly adapted to long periods of fast, and their weight routinely rises and falls.

Spectacularly, and famously, Emus undertake their enormous hikes in great herds. First, various Emus in any one region find themselves meeting up with other Emus, as they all head toward what they hope will be more good times, like Chaucer's pilgrims on the road to Canterbury. This seems to get them in herding mood; otherwise, when times are good, and each can find good food, they avoid each other. A kind of collective zeal takes over until they may be 70,000 birds or more all marching together. The trek stops only when the birds at the rear start to dawdle. Then the buzz gets around that new feeding grounds have been found and the Emus disperse once more. Until the next time.

These migrations are most conspicuous in the west of Australia, where the birds trek north to catch the regular summer rains and south for the rains of winter, brought in by Antarctic depressions. Farmers regard the migrating Emus as the people of ancient Egypt regarded locusts, and have built a 1,000-kilometer (625-mile) fence to protect the cereals against them. They have made similar fences to keep out rabbits and kangaroos. Long wire fences are to the Australians what the pyramids were to the Egyptians.

When a bird does find plants, the plants may fight back. Many leaves are toxic or at least are resinous or otherwise designed to be unpalatable or indigestible—including the leaves of most forest trees. Although it costs a plant a great deal of metabolic energy to make toxins, it can be worth it when there are folivores around. But a few specialist animals have special regions of the gut, typically harboring microbes, and well-equipped livers that can cope with the toxins; for them, otherwise inedible leaves offer a niche market: a source of food that nobody else wants; an excellent

Australian farmers conducted major "Emu wars" against the sometimes vast flocks of Emus that trek in search of fresh pasture. But the wars failed.

survival tactic. The prime examples among mammals are Australia's koala, which subsists entirely on the toxic, resinous leaves of eucalypts; and the leaf monkeys, of China and Southeast Asia. Various birds adopt the same strategy. The Capercaillie, the Old World super-grouse, lives mostly on pine needles—highly resinous and fibrous. The leaves that the Hoatzin eats are often toxic, at least to most other creatures. Like the goat, the Hoatzin largely overcomes the problems by eating small amounts of a wide variety of leaves, so that no one toxin is overwhelming.

Many plants produce bright and tasty fruits specifically to attract fructivorous animals, which eat them and with luck carry them off and either discard the seeds or else allow the seeds to pass through their guts unharmed. But not all animals that are attracted to fruit are good dispersers. Monkeys tend to be pillagers rather than dispersers, stripping trees of their fruit but often eating in situ, with very little dispersal at all. So we sometimes find co-evolution between plants and the particular animals that the plant wants to attract, such that the plant attracts its favored dispersers but repels or indeed poisons the ones that are less useful to it.

So it seems to be with the turacos—bright, flamboyant, crested birds that may be seen as Africa's answer to tropical America's macaws. Turacos forage for fruit in bands of up to twelve, often in people's gardens, on all kinds of fruit, and since about 80 percent of the seeds pass straight through them, they are excellent dispersers. But among the fruits they feed upon are berries that are toxic to many other animals, including humans. So the bearers of the berries ensure that only the turacos get to eat their seeds; and so the turacos emerge as niche feeders, with a diet laid on for them that other animals leave alone. Turacos even feed their chicks on fruit, which is unusual, but they supplement the infant diet with invertebrates, including snails. Incidentally, the family name of the turaco is Musiphagidae, which means "banana eaters." But bananas originated in Malaysia and were introduced to Africa only recently; in truth, turacos eat very few of them.

Honeyguides—finch- to thrush-size relatives of the woodpecker from Africa and parts of Asia—are niche feeders *par excellence*, and their niche is most extraordinary. First, they are called honeyguides because they are just that: they feed mainly on insects but have a particular predilection for honeybees' nests—though not primarily for the honey but for the wax. Wax is chemically a fat, but it is the fatty equivalent of cellulose: it is basically a structural material, and for all but a very few creatures it is indigestible. But honeyguides possess special enzymes for digesting wax—and indeed can live on pure beeswax for a month. To satisfy their craving they literally guide other creatures—including human beings—to the wild bees' nests, fluttering ahead of their co-opted aides with all the arts of seduction. When the co-optee arrives at the nest, he or she is liable to pinch the bulk of the honey, but the honeyguide—which cannot break into the nest itself—is content to feast on the debris. In the general course of their lives honeyguides keep an eye open for anyone who might seem willing to be led to a bees' nest. They have an agreeable relationship with many African forest people. Honeyguides, then, offer a fine example of recondite niche feeding and also provide one of nature's finest demonstrations of commensal feeding. But honeyguides are not the bird world's only wax eaters. The Yellow-rumped Warbler is known in the eastern United States as the Myrtle Warbler because in winter it favors the waxy fruits of the myrtle, and also of poison ivy. Like honeyguides, Myrtle Warblers can digest the wax and process it for energy. In comparable vein, some of the larger bustards feed on gum from acacia trees.

Most plants don't like being eaten—or at least, they don't like being gobbled up entirely; many have long since been resigned to this fate and many of them have come to rely on the animals that feed upon them for their own survival. Often we find that the various kinds of herbivore and the plants that are fed upon have co-evolved to form a symbiotic, mutualistic relationship. Grasses positively need to be grazed. If they are not, then they become senescent and rank, and the healthy sward is replaced by scrub. Grasses survive grazing because they keep their growing tips very low down so the grazer eats only the waving "flag leaves" above. Most other plants keep their growing points near the tip and when they are grazed they are severely compromised. Thus grass can outcompete its rivals, provided some obliging wigeon or Emu (or cow or antelope) is on hand to graze it.

But mutualistic relationships work only if both parties benefit; if the plant wants help from animals, then it has to bribe them. Succulent fruits are sweeteners for seed dispersers (often literally, although many wild fruits are sour). Plants that seek the help of animals for pollination provide gifts of surplus pollen, protein rich; but above all, they bribe their putative helpers with nectar—a solution of sugar produced for no other purpose. Many plants in Australia in particular seem to drip with nectar—including those of the primitive family Proteaceae, such as the various

Banksias, with their foot-long, bottlebrush inflorescences that may include more than 5,000 individual nectar-bearing flowers. Nectar is serious bush tucker. Accordingly, in Australia and round about, the nectar feeders include bats and other mammals, including possums, a host of insects, and many birds.

Australia's nectar-feeding pollinators include many of the 177 species of honeyeaters, which also live in New Guinea, Indonesia, New Zealand, and Hawaii. Some are highly specialized, with long curved beaks adapted to particular flowers that are correspondingly long and hard to get at unless you have a long curved beak. Some honeyeaters also eat the sugary secretions produced by the nymphs of some insects, and the leaves and branches of some eucalypts, the tree that is most characteristic of Australia. But honeyeaters are not exclusive nectar feeders, by any means. On the whole they prefer a balanced diet, some eating insects and spiders, and some with a penchant for fruit.

Then there are the small and brightly colored sunbirds, spider-hunters, and sugarbirds from sub-Saharan Africa, much of Asia, and New Guinea. They all have beaks and tongues adapted for nectar—different beaks for different flowers—but, like the honeyeaters, they very sensibly eat quite a few arthropods as well, and sometimes some fruit. In East Africa, there is a most peculiar social arrangement among different species of sunbird. Some sunbird species dominate others, and the more subordinate ones can breed only when nectar is particularly abundant. I know no other examples of interspecies competition of this kind.

The tanagers of the American tropics—more than 400 species of them—range from thrush-size to crow-size, with a huge range of diets and anatomy to match. But many are part-time nectar feeders, and as they feed they scatter pollen—again, of huge importance ecologically, for without tanagers many a plant would die unfertilized.

Most nectar feeders are passerines, including all the kinds mentioned so far. But a few nonpasserines feed on nectar, too, including some barbets—and the world's most famous and specialist nectar feeders are related to the swifts. Again, like the tanagers, they are exclusive to the New World. Among their ranks are the world's smallest birds. They are, of course, the hummingbirds.

Hummingbirds feed on nectar with just a little pollen and perhaps some tiny invertebrates—just enough to give them the protein they need. They lap up the nectar with their extraordinarily long tongues—sheathed and protected in the long slender beak—from the depths of deep-tubed flowers: a thousand to two thousand flowers per day. They defend their own patch of flowers aggressively, as if their life depended on it, which, of course, it does. In return, the flowers are pollinated. Hummingbirds and their specialist flowers offer supreme examples of co-evolution.

But so much nectar raises problems. Nectar is sugar in solution. This means that hummingbirds, if they are to get enough energy, must take in

Hummingbirds like this Streamertail are not the only avian nectar feeders, but they are the most specialist.

an extraordinary quantity of water—equivalent to 160 percent of their body mass each day. So whereas most birds go to extraordinary lengths to conserve water, and excrete nitrogen in the form of uric acid (not as a dilute solution of urea as in the urine of mammals), hummingbirds must go to equal lengths to get rid of it. Their kidneys are adapted to produce enormous quantities of very dilute pee. Indeed, if hummingbirds were scaled up to human size, they would pee about 100 liters a day, around 25 gallons. If they made such a ceremony of micturition as humans do, this would surely take them an hour or so a day. But hummingbirds just spray as they go.

The passerine nectar feeders have strong feet and generally perch alongside the flowers they are feeding from, gripping the stem. But hummingbirds hover like hawk moths. Their feet and legs are not for perching; like swifts', they are very small and indeed the name of the group to which they both belong, Apoda, means "no feet." Hummingbirds do perch to rest, however, which swifts do not (though some kinds cling to walls to sleep).

Hummingbirds are among the finest hoverers in nature, certainly in the bird world—more adept and controlled even than kestrels or terns. To achieve this, their wings move horizontally in figures of eight, like a dragonfly, and a slight change in their pitch moves the bird upward, sideways, or backward. Hummingbirds can even fly upside down. The Giant Hummingbird flaps its wings at a mere 10 to 15 beats a second—a veritable sluggard. The smaller ones, like the Amethyst Woodstar, get up to 70 or 80 beats a second. Some North American types, like the Ruby-throated Hummingbird get up to 200 beats per second—although they don't do this every day. Like swifts, they reserve their greatest aerial extravaganzas for courtship.

Such hovering requires prodigious power, provided by flight muscles that account for more than 30 percent of the body mass—huge, but not much bigger than the bird average. These are attached to a deeply keeled and long sternum, which in turn is supported by eight pairs of ribs—two more than in most birds (and it's odd that the flight muscles of swifts are rather small). Modified shoulder joints allow all-round movement. Few animals apart from humans and other primates can move their forelimbs with such versatility.

To power these muscles at such a pace hummingbirds have the highest rate of metabolism of any vertebrate: they use about 400 times as much energy per unit body weight as a human. Commensurately, hummingbirds also have the highest throughput of oxygen of any vertebrate—twice as much per unit volume as the smallest mammals, such as shrews. At rest they take 300 breaths per minute; in flight, up to 500. All modern birds have supplementary air sacs in addition to lungs. Hummingbirds have nine of them.

But when hummingbirds are not feeding they would cool too quickly. Rather than waste energy keeping themselves warm, they simply allow themselves to cool down—thus demonstrating yet another trick that is not unique to hummingbirds but is nonetheless highly unusual, at least among birds. They can enter a state of torpor in which their body temperature drops to that of the surroundings. Hummingbirds at high latitudes routinely do this at night. In such a state they save 60 percent of their energy. But there's a downside. Torpid animals are easy prey, and if they become too lethargic they may not be able to revive.

Most of the nectar feeders also eat small insects that they catch on the wing. Even hummingbirds, the supreme nectar specialists, may catch a few. Australia's woodswallows perhaps are showing us how this transition might have come about. In truth it isn't clear that woodswallows feed on nectar at all. Like Barn Swallows (although they are in a quite different family), woodswallows catch small insects on the wing, staying aloft for hours and hours. But they catch those insects with the help of a brush-like tongue—precisely the kind of tongue that's needed to gather nectar. As hawkers of small insects, too, they are wonderfully precise fliers. Small insects are attracted to flowers, and their predators are likely to follow them there. Put everything together and we can see how some ancestral hawkers after small insects became nectar sippers, and indeed how one of those early insect chasers became a hovering nectar sipper. Hummingbirds have no relationship at all to woodswallows. Hummingbirds are related to swifts. But we can imagine how some ancestral insect-hawking swift might have been pre-equipped to be a nectar feeder just as the insect-hawking woodswallow seems pre-equipped.

Hummingbirds take in a surplus of water and their problem is to get rid of it. For others, it's hard to find enough water. Most need to keep their weight down and cannot afford to take on too much at a time. It can be very hard to balance the books. Some birds build their lives around the need to drink.

WATER

Some birds need more water than others. Seed eaters generally have a particular need for water because seeds tend to be dry. Most kinds of pigeon primarily eat seeds and some, like the Namaqua Dove of Africa and some Australian species, are nomadic—they go where the water is, as Emus go where the grazing is. Pigeons uniquely among birds have mastered the art of drinking. Most birds must lift their heads after each sip to tip the water back into their throats, but pigeons immerse their entire beaks and then, aided by special throat muscles, just suck.

Sandgrouse are related to pigeons, although they look a bit like grouse. They, too, eat mainly seeds—and for good measure they generally live in dry land, even semi-desert, in Africa, southern Europe, and much of Asia. Farming may have helped them to expand their range by providing both seed and watering holes. For sandgrouse, watering is a big social ritual. In fact, you are most likely to see sandgrouse when they gather in flocks, sometimes huge, to drink at waterholes in the morning and/or the evening like wildebeest—perhaps only once every several days in temperate weather but at least once and perhaps twice a day when it's hot. They fly swiftly to water—up to 80 kilometers (50 miles) one way, which they can do in little more than an hour—calling all the while, with others joining the party as they go. If two sandgrouse are drinking together, then one waits for the other to finish before they both fly off. But when they flock to drink they fall prey to raptors. Lanner Falcons love them.

Helmeted Guineafowl make their way in orderly fashion to the waterhole.

For the Helmeted Guineafowl of sub-Saharan Africa, too, water is the center of social life. Early each morning the birds march from their roosts to the waterhole in single file with the dominant males in the lead, apparently acting as scouts. Drinking done, they advance across the plain line-abreast—an efficient way to seek out bulbs, seeds, and insects, like hired hands flushing pheasants for the hunters to shoot, or avocets sweeping the water collectively to stir invertebrates. But the Vulturine Guineafowl is so well adapted to the semi-deserts of northeast Africa that it seems to need no drinking water at all. Among mammals, only the oryx and the odd gazelle can match its thriftiness. Camels, for all their reputation, are far less independent. Almost but not quite matching the Vulturine Guinea-fowl for thriftiness is a bird that could hardly be more different: the Scaly Weaver, smallest of all the weavers. Again it is a seed eater, which suggests a particular need for water. But it can metabolize the fat within seeds to release water, so it does not need to drink so much. Camels pull much the same physiological trick.

Chicks need plenty of water, too, and several kinds of bird have special ways of bringing water to them. Secretary-birds may bring water to their chicks in their crops, as well as food; the chicks solicit a drink by vibrating their beaks against the parents' beaks, whereupon the parent disgorges. The male sandgrouse has special sponge-like feathers on his breast, and for about two months after the chicks hatch he brings them feather-spongefuls of water. (But the chicks do have to find their own food—and indeed may find thousands of seeds a day from day one.)

In fact, for all species of birds, the feeding and watering of chicks is a big problem that shapes their entire lives.

PROVISIONS FOR THE VERY YOUNG

Many parents from thistles to barnacles simply abandon their offspring to the four winds or the tides with a metaphorical packet of sandwiches and, doubtless, a message of good luck; but many, including such unlikely candidates as earwigs and seahorses, take good care of their children. Sometimes, as in earwigs, only the mother is devoted. Sometimes, as in seahorses, the father takes responsibility. Sometimes, as in some primates and dogs, both parents lend a hand. In mammals, of course, the mother has to be involved since she provides the milk that, for the newborn mammal, is the sole provender, but sometimes the male lends a hand, too.

All birds offer serious child care—even the big, strange fowl known as megapodes, which may seem to abandon their offspring. Indeed they build their lives around it. More of what's entailed is in Chapters 6 and 7, but meanwhile we should note the special problem of provisioning.

Some baby birds can allow their temperature to drop somewhat, but in general they need to stay warm, and since they are very small—and often are born without feathers—the problems of warmth loom very large, coupled with the need to grow. Parents provide much of the warmth their offspring need by brooding—sitting on them and imparting their own prodigious body heat. But the babies also need a great deal of food. Because they are so small they need their food to be concentrated, and because they lack the powerful beaks or stone-filled gizzards of the adults, they need it to be soft and easily digested.

Different birds arrange this in different ways. Small passerines that, when adult may mainly eat seeds, take care to feed their babies on insects, and especially on soft caterpillars. Birds that migrate to northern Europe to enjoy the long summers must get the timing right: their eggs must hatch just as the caterpillars are emerging—but the first flush of caterpillars is largely geared to local temperatures. While the birds must preplan their migrations, and they do this according to day length, as the climate changes with global warming, the two can become seriously out of synch. Thus, studies of flycatchers and other small songbirds suggest that they may currently be arriving in Europe too late to take advantage of the caterpillars that emerge early in the strangely warm spring. Natural selection should ensure that they adapt, but whether they can do so before their populations are dangerously reduced remains to be seen.

Raptors feed meat to their babies—but are at pains to tear off small shreds, holding the prey down with one powerful foot and tugging with the beak. Many birds, especially seabirds that must forage far from the nest, carry the food home in their guts and so are able to feed it to their chicks already somewhat processed and predigested. Vultures do this. The Hoatzin, unique in this as in all things, feeds predigested leaves to its young—along with the microorganisms that will colonize the guts of the chicks and enable them to digest the cellulose in plants for themselves.

Diving-petrels feed their young on predigested fish or whatever they have caught, but the rest of the procellariforms—the petrels and shear-waters, the albatrosses, and the storm-petrels—go one step further. They also secrete custom-made oil from the gut wall and feed this to their chicks, *plus* predigested fish and squid—a rich stew, indeed. Thus they demonstrate that mammals are not the only creatures that feed their young on their own bodily secretions.

Indeed, three kinds of birds that are quite unrelated to each other or to the procellariforms take matters even further than this. Pigeons, flamingos, and the Emperor Penguin produce a secretion from the gut wall known as "crop milk" (although, as it happens, the Emperor Penguin doesn't have a crop).

Pigeon milk, in truth, is more like cottage cheese. The pigeons—both sexes!—start to produce it in the wall of the crop when the eggs are about half incubated, stimulated by the hormone prolactin, and then shed not only the milk but also the cells that produce it into the crop. For the first few days of the hatchling's life it is all the food it gets, or needs. Whereas the fluid milk of mammals generally contains only about 5 percent dry matter (much less in horses and humans), the cheesy milk of pigeons con-

Pigeons—like this rare Crowned Pigeon—feed their young with cheesy "milk" that they secrete from the lining of the crop.

tains 19 to 35 percent. The protein content is very high—13 to 19 percent; compare that to about 4 percent in a cow and 2 percent in a human or horse. Fat is high, too—7 to 13 percent, the upper level almost in the seal bracket. Minerals account for another 1 to 2 percent, and the milk also contains vitamin A and some B vitamins. But it contains no carbohydrates—that is, none at all of the mammalian "milk sugar," lactose. Pigeon milk is also low in calcium and phosphorus, whereas mammalian milk is high in calcium, and cow's milk has become a major source of it for human children. Oddly, too, pigeon milk is high in sodium, which in more than minimal quantities is now considered very bad for human infants. The parents (both of them) go on producing the milk in fairly constant amounts until the youngsters are well grown—but after the first few days they gradually increase the content of other food, which mainly means vegetation. Thus they wean their babies very much as mammals do. Again, we see birds and mammals arriving at much the same endpoint but via completely different routes.

In short, the business of staying alive is complicated. Just to get enough to eat, birds, like any other creature, need the right physical equipment, the right techniques, and, perhaps above all, the right strategy, which among other things means being in the right place at the right time. But creatures that focused only on feeding themselves would die out after one generation, so all must juggle the generally conflicting demands of feeding, finding mates, and raising young. Most wild animals, too, have to fit all this into a single year. Many flit from hunting ground to breeding ground as the season unfolds, generally striving to avoid the worst of the winter or the summer drought—but no animals surpass the birds in this and only a very few (like some of the whales) even begin to match them. Because birds can fly, the whole world is their oyster, but to take advantage, they must migrate. For many birds, migration has become one of the outstanding events of their working year.

Bar-headed Geese migrate almost at jumbo-jet height over the Himalayas, where oxygen is low but so is the air resistance and the extreme cold prevents overheating.

6

THE WORLD AS AN OYSTER

ACH AND EVERY AUTUMN, 5 BILLION BIRDS ABANDON
northern Europe and Asia for the easier climes, fruits, and summer insects of southern Europe, Africa, India, or Southeast Asia. Another 5 billion leave North America to while away the northern winter in the tropics of Central America or the Caribbean, or in the southern summer of Argentina. Some (though not many) migrate each southern autumn from Australia or New Zealand to New Guinea, or from Australia to Sulawesi. All in all, about half of all the known species of birds undertake some form of migration, mostly north–south but also sometimes east–west—like the Redwing, which breeds in northern Europe and Siberia but flies to western and southern Europe to escape the continental heartland winter.

Not all populations within any one species migrate, and not all members of any one population migrate, and sometimes the decision to migrate or not seems somewhat flexible. Most Barn Swallows in northern Eurasia fly south to Africa, but some of the ones that live in southern Spain stay put year-round. Similarly, North America's Killdeer—very like Europe's Ringed Plover—generally migrates south, but those that live around the Gulf of Mexico are content to stay where they are. As global warming bites, we can expect more and more birds that formerly were migrants to stay at home. Some birds hop only a few miles by way of migration, sometimes for no apparent reason at all. In Britain, some of the European Robins that breed in Lancashire fly south to Suffolk, while some Suffolk Robins take off for Kent, and a few cross the Channel into France, which all seems fairly pointless, although I do not presume to know the Robins' business better than they know it themselves.

Many birds fly prodigious distances—and most prodigious of all is the Arctic Tern, which breeds in the north and flies all the way to Antarctica to feed itself in the southern summer and prepare for next year's breeding. Some Arctic Terns follow the west coast of the Americas and some take the eastern route, hopping across open ocean from Newfoundland to Brazil. Those of the Old World follow the west coast of Europe and Africa. Whatever the route, this means that the Arctic Terns that breed farthest north—within the Arctic Circle—make a round trip each year of up to 40,000 kilometers (25,000 miles), which is almost the circumference of the globe. The oldest known Arctic Tern to migrate was at least twenty-six years old, which we can be sure about because it was ringed, and this means that in the course of its life it had migrated a million kilometers (625,000 miles), equivalent to going to the moon and back and halfway back again. In compensation for these extreme endeavors, Arctic Terns spend more of their lives in daylight than any other creature.

Birds of prey can be prodigious migrators. Swainson's Hawk is the greatest traveler of all North America's hawks, typically migrating in huge flocks from the prairies and plains of the central states down to the pampas of Argentina—and so it spends its entire life in summer grassland. Nowadays, however, fewer of the hawks travel to the pampas than in the past. Traditionally in the south they gorged on locusts, but now in these days of pesticides the locusts are not so common. Dickcissels, small songbirds in the cardinal grosbeak family Cardinalidae, used to do much the same: flock from North American grassland to South American grassland. But whereas Swainson's Hawk mainly eats small animals, the Dickcissel is a seedeater—and it is content these days to cut its southern migration short and settle on the farmlands of Venezuela. So it has become a serious pest, comparable in its impact with the queleas of Africa.

Body size is no barrier to migration, at either end of the scale. Some of the biggest flying birds—eagles, storks, cranes—migrate by riding the thermals, rising to prodigious heights on ascending spirals of warm air, then gliding for mile after mile. Since there are no thermals until the sun warms the ground, the birds that rely on them cannot take off until some hours after dawn, and they stop in mid- to late afternoon. Since these big birds take care to fatten themselves up before they set out, and soaring and gliding take very little energy, they do not generally need to eat much en route, and so can spend their resting hours roosting. For the thermal riders, then, migration becomes a leisurely affair, like nineteenth-century

Americans on the Grand Tour, straight out of Henry James. I love the thought of this—who would not?—but when you think about it, the realities of high-altitude gliding are not quite so alluring. The sun is surely relentless at those unshaded heights while the wind, presumably, can be devilish cold. But then, we who contemplate such matters are merely human, bound to the earth. Birds are built for what they do.

But birds that ride the thermals cannot fly far over water, for thermals form only over land; if they want to cross from continent to continent, they have to follow whatever corridors of land there may be, and cross the water where it is most narrow. So big birds migrating out of Europe or Asia into Africa either go to the west of the Mediterranean Sea, crossing the Straits of Gibraltar, which are a mere 14 kilometers (9 miles) across; or through the Middle East, with only the Suez Canal to cross. Those that migrate from North America to South America generally follow the Isthmus of Panama.

Because the big birds must follow such narrow corridors, their numbers en route can build up spectacularly, like traffic compelled to follow the only freeway. So in the autumn of 1992, Hawks Aloft Worldwide reported that more than 2.5 million raptors had gathered over Veracruz, in Mexico.[1] There were more than 900,000 Broad-winged Hawks, 500,000 Swainson's Hawks, and 12,000 Mississippi Kites. There is an informal law of ecology that says that "big, fierce animals are rare" because each big predator needs entire populations of smaller creatures to prey upon. But this was like a world conference of wolves or an international convention of leopards. The only mammalian predators I can think of that assemble in such numbers are the crabeater seals of Antarctica.

Some big birds of prey, however, are not soarers and gliders, but flappers. They can and do cross large stretches of water, and they do not need to squeeze along narrow corridors of land. These are accordingly known as "broad front migrants." The Osprey is one such, happily migrating right across the Baltic and the Mediterranean Seas. Young Ospreys from the same brood may migrate in quite different directions. In one study, one Osprey born in northern Europe took a month to fly the western route to West Africa, while its sibling and erstwhile nestmate first flew east and then south, and so took three months to find its way to South Africa.

The smallest birds may migrate, too—and, by contrast, since they rely on powered flight, they are just as happy over sea as over land (provided the distances are not too excessive) and so they simply tend to take the straightest route. So it is that many hummingbirds fly up and down the west coast of North America between Canada and New Mexico and back, while on the eastern side of the country, the Ruby-throated Hummingbird flies 6,000 kilometers (3,750 miles) south, including a huge leap across the Gulf of Mexico.

So how do the birds find the energy for such exertion? Some take their time and stoke up at judicious intervals, like old-style stagecoaches. The advantages of such a strategy are obvious—but of course there are serious drawbacks, too. The birds may arrive at a staging post too early in the year and find that the insects or buds they expected to find have not yet emerged. In these days, when humanity rules and speculators often seem to have carte blanche, they are liable to find that their essential feeding ground has been drained, or built over, or fished out. Even in good years there will be heavy competition at the stopping-off points for what there is. A few migrating birds—like swifts—can refuel at least up to a point by picking up insects in their flight path, like long-distance war planes refueling. But all to some extent rely on body fat, accumulated before they set out. Migration cannot be a sudden, spontaneous decision; it takes weeks of preparation. Many birds put on so much fat that they double their weight before setting out: a Ruby-throated Hummingbird swells from a minuscule 3 grams to a formidable 6 grams before venturing forth across the Gulf of Mexico. This may not seem much, but it is equivalent to a human being adding about 60 kilograms (130 pounds) of fat. But then, as we have seen, fat provides an enormous amount of energy. One kilogram of adipose tissue (fat plus its accompanying water and membranes) provides around 7,500 kilocalories, which would be enough to keep a 130-pound person going for at least three days, even if he or she was reasonably active. Sixty kilograms of fat tissue should provide enough for half a year. These calculations can be applied only roughly to birds, which on the one hand generally have a far higher metabolic rate than we do, but on the other, generally expend far less energy than we do in getting around. If an animal can put on enough fat, then in principle it should have enough fuel on board to perform prodigious feats for weeks at a time with no further sustenance.

However, the niceties of physiology cannot be reduced quite so glibly to simple physics. It is one thing to accumulate fat as a store of energy, but it is quite another to metabolize that fat in the short term and turn the energy into useful action. If riders on the Tour de France adopted the strategy of hummingbirds or warblers, they would begin their exertions at more than 120 kilograms—around 260 pounds. They wouldn't find it at all easy to get up the hills (and would need heavier bikes), and if they simply starved as they went, they would get sick because human beings just aren't equipped to live on undiluted, unsupplemented body fat. Instead, long-distance cyclists start out at fighting weight, but then must gobble 7,000 calories or so's worth of pasta each evening to keep their strength up. For this, they stop at pre-arranged feeding stations. Birds have to make do with what geography provides.

Birds that rely on thermals must, of course, migrate by day. So do some small birds that do not soar, such as swallows and Sand Martins. But there are several advantages in migrating by night: it's cooler, the air is less turbulent, and if there is no cloud many birds can navigate by the stars. A few migrate exclusively by night, but many fly both by night and by day, even those birds that are diurnal, or active only during the day. Land birds (as opposed to seabirds) must fly by night as well as by day when they migrate long distances across the sea—including, for instance, thrushes, Goldcrests, and starlings. North America's Blackpoll Warblers fatten themselves up along the Massachusetts coast before heading boldly and apparently suicidally off into the Atlantic—but there they meet the northern trade winds, which bear them almost effortlessly to their winter quarters in the Caribbean or along the north coast of South America. The whole trip takes them about four days. Terns, too, prefer to fly nonstop over sea because, although they can rest on water, they soon become waterlogged if they do so. But some small birds that could stop en route if they chose to opt instead to fly nonstop, like Europe's Sedge Warbler, which doubles its weight before it starts its annual migration down to Africa and then flies up to 3,000 kilometers (1,875 miles) nonstop. At an average of 40 kilometers per hour (25 miles per hour), the minimum for such a hike is around seventy-five hours; in practice, the trip takes three to four days. Some swifts simply don't bother to land. Young swifts that leave their nests in northern Europe in the autumn may not land again until they return to breed nine months later, having tripped down to Africa in the meantime.

Snow Geese are among those birds that do stop en route to stoke up. They fly north to breed, in Arctic Canada and Alaska, Siberia, and western Greenland. They need to arrive early because the northern summer is short, and they have to incubate their eggs and raise their young to the point where they can fly before the cold weather closes in. But in practice they arrive before there is enough vegetation to feed them, so they need to be fat when they arrive to get through the first couple of weeks. They need to feed as they go, and on their way north they stop at various regular staging posts—farmland, coastal marshes, and other wetlands. Birds like the Snow Goose show why conservation needs to be a global affair. It isn't enough to conserve the breeding grounds or the feeding grounds. All the essential places in between must be maintained as well.

Birds on migration fly at very different heights: the Chaffinch at a mere 200 to 300 meters (660 to 1,000 feet); the Snow Goose at 1,000 to 2,000 meters (3,300 to 6,600 feet); swifts at 200 to 3,000 meters (660 to 10,000 feet); White Storks soar up to 5,000 meters (16,500 feet); Bar-tailed Godwits at 6,000 to 7,000 meters (19,800 to 23,000 feet); Whooper Swans at more than 8,000 meters (26,400 feet). But the Bar-headed Goose is the champion—recorded at around 9,000 meters (29,700 feet), almost jumbo-jet height, over the peak of Everest. In general, birds on migration are forced upward by the altitude of the ground beneath—birds that migrate through one high pass in the Alps, the Col de Bretolet, fly a mere 3 to 4 meters (10 to 13 feet) above the valley floor. To some extent the higher the better, because at great heights the air offers less drag—although eventually it becomes too thin to fly in at all and supplies too little oxygen. The cold at great heights is an advantage; Bar-headed Geese typically encounter temperatures of −30°C (−22°F), but cooling is vital with such exertion. Yet the birds must avoid clouds, for when fat is burned for energy, water is produced as a by-product, and if the air is too humid they cannot get rid of the surplus quickly enough. They vary their height, too, according to the wind, which tends to be stronger as height increases. But the direction of the wind varies at different heights, and it is essential to be in the right airstream. For small birds migrating at around 40 kilometers per hour (25 miles per hour), winds even of 15 kilometers per hour (10 miles per hour) can make a huge difference, whether they are tailwinds that help them on their way, or headwinds that may stop them in their tracks, or sidewinds that may blow them horribly off course and turn them into "vagrants," the birders' delight. Many birds leave and return to their breeding grounds by different routes, to take advantage of particular winds each way.

Geese, of course, famously fly in a V-shape: one lead bird at the apex, with the others trailed out diagonally on either side. Each bird flies with one wing tip just behind and beneath the wing tip of the bird in front.

Theory has it that turbulence from the wing of the leading bird causes a reduction in air pressure, which lifts the bird behind and helps to pull it along. Only the lead bird takes the full brunt of flight—and the geese alternate the lead so that all take their share. In this cooperative share-and-share-alike fashion, geese can fly half as far again as they could do if they flew alone, or so it's been calculated. Of late, however, some aerodynamic experts have suggested that geese in reality fly slightly too far apart for this slipstream effect to work. The V-formation may benefit them mainly in the way that soldiers are helped by running in formation—each picks up the rhythm from the other and pounds along almost in a trance-like state, more than any one of them could do alone. For my part, however, I like the traditional explanation—each bird slipstreams the one in front. It is far too plausible to abandon without much more evidence.

For the birds that do migrate, the world truly becomes their oyster. But to take advantage of this particular oyster, they have to do everything right. Timing is vital. Leave the northern breeding grounds too early, and the chicks will not be ready for the exodus. Leave too late, and the weather will catch them out. Set out too late for the breeding grounds in spring, and they will find the best nest sites gone, the potential mates all spoken for, or the first essential flush of caterpillars already past—or there is not enough time (for those breeding in the far north) to fit in the complete cycle of mating, laying, incubating, and fledging before the fleeting summer is ended. Set out too early, and they will find the feeding grounds still encased in ice or snow, and the essential feeding grounds still dormant. Early arrivers sometimes turn around and fly home again, if they have enough fat in reserve to manage this; those that have stayed at home then realize that they must wait a few more days before setting out. Thus, those that jump the gun act as scouts for the rest, though that is probably not their intention.

But how do they do it?

HOW BIRDS MIGRATE

There are several conceptual problems. First, there is the matter of timing. How do birds know when to migrate—or for that matter, when to fatten themselves up, or molt from eclipse plumage into breeding fig, and back to dull feathers again, and expand the quiescent ovaries and testes that produce the gametes and help to fire their desire to mate, and contract them again when the season is over?

Then—among the greatest of all nature's puzzles—how do they find their way? This problem has two elements to it. The migrating birds have to know where they are at any one time—which on the face of things can seem well-nigh impossible. How, for example, does a Wandering Albatross contrive to patrol over the same area of the Southern Ocean week after week when the sea stretches featureless to the horizon on all sides, and the waves themselves—the visual clues—are constantly on the move? Wandering Albatrosses and their like face such issues even in the usual course of their lives, when they are not migrating. But any bird that crosses a forest or a desert or an ocean when it is migrating must face it, too. But, then, even when migrating birds do know where they are, how do they know what direction to fly in?

Finally, we must ask, how do they know that migration is necessary at all? Some young birds on their first migration must follow their parents, so we merely (merely!) have to explain how the first of their species first learned the journey, for we can assume that all generations since have simply followed in their metaphorical footsteps like the pageboy of Good King Wenceslaus. But some young birds, including Barn Swallows, set out on their first migration in the autumn *after* their parents have already left for warmer climes. How on earth does a baby swallow hatched in Surrey know the way to Africa? And what prompts it to set out on such a journey that, on the face of things, seems suicidal whatever the eventual rewards may be and in practice is so hazardous?

Definitive answers to these obvious questions are not yet forthcoming. There are many theories, for some of which there is good evidence; and if you put all the theories and the evidence together, it is possible, in the early twenty-first century, to tell a story that sounds coherent and can be made to seem convincing. But actually, the overall picture is a hotchpotch: a small area of near certainty here, a solid-looking fact there, all stitched together by narrative that still relies heavily on invention and supposition. I do not mock. Progress in understanding has been impressive, and the fieldwork and many of the experiments have been immensely ingenious and painstaking. But nature is difficult. I feel in my own bones (as many do) that there are still fundamentals to be unearthed.

Meantime, here is some of the progress so far.

First, how do birds know when to migrate? Equally to the point, how do they know when to start preparing for migration by fattening themselves up? Here, we must invoke the concept of an innate "circannual" rhythm. Birds—like bears, or trees, or indeed like all of us, up to a point—know what time of year it is. Obviously for day-to-day purposes we judge this by external cues: how cold it is, what time the sun rises, whether the leaves are bare or in full leaf. But we can judge passing time to some extent even if we are locked up in cages without a view of the natural world, at uniform temperature and with artificial daylight of unvarying length. Some creatures can judge the passage of time far better than we can, including birds.

So it is that when birds that normally migrate are caged in days of constant length, twelve hours of light and twelve hours of dark, they still put on weight some weeks before the time that they would normally migrate. When the time for migration comes around, they show signs of restlessness, fluttering to the top of the cage. They also point themselves in what would be the appropriate direction. Furthermore, the time for which they remain restless corresponds, at least roughly, to the length of the normal migration. Such experiments have been carried out with many species, especially songbirds such as European Robins and Blackcaps. (This premigration restlessness is known by its German term, *Zugunruhe*. This, unsurprisingly, is because some of the fundamental work in this field was done by German ornithologists—and also because Germans have a particular penchant for words that express very particular and complicated states of mind, as in *Weltschmertz* and *Schadenfreude*).

Superimposed on this circannual rhythm, which keeps track of entire years, is the circadian rhythm: the in-built ability to judge the length and time of day. At least in some species, the length of the days and nights is registered by the pineal gland. This is a bead of tissue at the top of the brain that seems to be all that is left of an organ that, in primitive vertebrates, served as a "third eye" on the top of the head. Some lizards, and the lizard-like tuatara of New Zealand, still possess such an eye. In some birds, including starlings, the light that impinges on the pineal has been shown to enter not via the eyes but through the feathers and thin bone of the top of the skull. We might suppose that it is dark inside an animal's skull, and so it might be in a human's, but in birds apparently this is not the case. Our skulls are thick and solid (relatively speaking) while bird skulls, though wondrously strong, are thin and spongy—translucent enough. The pineal, responsive to the alternating light and dark, in turn produces melatonin, which interacts with the in-built circadian rhythm by which animals judge the length of the day. Modern pills for fighting jet-lag are based on melatonin: they reset the circadian clock. On migration, the ability to judge the time of day helps at least some birds to navigate by the sun (whose position of course changes as the day advances).

But how do they find their way?

First—working from primary principles—what possible clues are there? If a bird is flying over land, where there are features below that are distinct and stay the same for year after year—rivers, roads, forests, coastlines—then, of course, they can use their eyes. There is plenty of evidence that birds do just this. Many, for example, as already intimated, follow coastlines and thread their way through isthmuses and mountain passes.

When they get very close to where they want to be, many use their sense of smell. Homing pigeons give a clue to this. ("Homing" is not the same as migration. It alludes to the fact that pigeons can find their way home when taken by train or truck to some far-distant place and then released. But homing surely partakes of some of the same mechanisms as migration does, and so can give clues to how it works.) It seems that as pigeons get fairly close to their home loft, they first pick up general smells that tell of avian dwellings—perhaps the general alluring stink of ammonia. As they get nearer, the smells become more specifically pigeonlike and less generally feral. Finally, as they get very close, they recognize the very particular odor of their own flock in its own space. More and more evidence (not least by the manufacturers of toiletries) is revealing that humans, too, have a wonderful awareness of odor, even if they do not consciously recognize it, such that they find particular men or women attractive or repellent according to their primitive exudates: no doubt a sobering thought for those who like to suppose that human beings have risen above such things. We do not normally think of birds as creatures that put great store by smell, but as we have already seen, many of them do, in many contexts.

But what use are visual clues when a bird is above some apparently boundless ocean? What value is smell when it is a thousand miles from where it wants to be? What else is there?

Quite a lot, is the answer. On the visual front, there is the sun by day and the moon and stars by night. These are hard to make good use of unless the bird also has some sense of time, so it knows where the sun or the moon ought to be at a particular time; but as we have seen, birds do have a sense of time.

Human beings navigate by the heavenly bodies, too, but we make a great song and dance about it. The skills of the navigator were among the most intricate and prized in all the world's navies until well into the nineteenth century, when sailors in extremis could find out where they were by radio. Traditional long-distance sailors needed telescopes and sextants and astrolabes and chronometers and charts, and pages and pages of tables, to help them work out where they were. Birds have to do all this in their heads, in their bird brains, on the wing. The problem conceptually is the same as we met in discussing the diving of gannets (how they always fold their wings at exactly the right time). In each case the math is immensely complicated, once you spell it out. But presumably birds on the wing, not versed in mathematics and with no slide rules about their persons, don't spell it out. They must have some simple rules of thumb that instantly translate the cues that are offered by the sun and stars and moon into directives for purposeful action. The task for ornithologists is to discover these rules.

Again, there are clues and stories that seem to be throwing some light. For instance, many kinds of birds—including Indigo Buntings, much favored for such experiments—are known to use star maps. In the early weeks of life the baby birds sit in their nests and study the night sky—and are somewhat thrown if those early weeks are too cloudy. But they do not, as human amateur astronomers might do, spend their time learning the individual constellations—how to recognize Orion or trace the fanciful outline of Taurus, or whatever. Instead, they focus on the bit that does not move as the night progresses, which in the Northern Hemisphere means the North Pole or North Star. They can see, if they look at it long enough, that as the night progresses, all the stars in the sky, including the mighty Orion and the notional Taurus, seem to revolve around the Pole Star, which sits in the middle like the hub of a giant cartwheel. Once they recognize the hub—the bit that doesn't move—the most fundamental problem is solved. The creature that can do this knows where north is and everything else can be inferred. I don't know what the equivalent would be in the Southern Hemisphere, but undoubtedly there is one. Navigation simply does not seem to need the details of astronomy.

What if it is cloudy? Many birds seem to lose their way, is one answer.

But there are other clues to whereabouts and direction that do not require a clear view of the heavenly bodies. It has been claimed that birds can make use of polarized light—that they can see what direction the light is coming from even if they cannot see the source—so they can judge where the sun is even when it is behind clouds. This seems somewhat implausible since clouds, one might suppose, would scatter the light from the sun hither and thither, making it well-nigh impossible to judge exactly where it is coming from. When I was at school in the 1950s, much interest attached to the pecten, a strange, pleated, sail-like structure found in the eye of birds, attached to the retina and protruding into the vitreous humor—the jelly-like fluid that gives the eyeball its shape. Did the pecten help the bird to distinguish polarized light? But this was among about thirty theories put forward to explain this curious device. None of them is fully convincing, and some suggest that the pecten has no visual function at all. It simply helps to provide the retina with nutrient.

Then there is the earth's magnetism, which indeed is reliable, if not quite totally consistent, since the poles wander over time and every few thousand years the North Pole and the South Pole reverse: magnetically speaking, the North becomes South and the South becomes North. But magnetism could obviously be very useful nonetheless—magnetic compasses kept all the world's navies on track until the days of satellite signals—and some birds clearly make use of it. European Robins and pigeons are among the many birds that have been shown to harbor magnetite at the top of their beaks—magnetite being an oxide of iron (ferric-ferrous) that is magnetically sensitive. The magnetic field gives two kinds of clues. First, the orientation of the field shows the general north–south orientation. Second, the "dip" of the field—how steeply the lines of force are heading away from the earth or ducking back into it—gives a clue to longitude. Experiments at Manchester University have shown that human beings have some such ability, too. Students were fitted with headgear that scrambled the earth's magnetic signal while others, serving as controls, had equally weighty headgear that had no such effect. They were taken on a convoluted bus ride, then they were discharged at intervals and invited to find their way home. The controls, whose magnetic wits had not been scrambled, fared much better. There is some hard evidence and a great many anecdotes to suggest that people living in a state of nature, such as the Aborigines, who still live by hunting and gathering in Australia, have a very powerful magnetic sense. Indeed, it is hard to see how they could make their way around such vastness if they did not. People who live in Western societies, equipped with maps, compasses, signposts, and now personal satellite tracking systems, seem largely to have lost such native skills.

More generally, perhaps we should not be too surprised that birds have such a skill as this, even though it is far beyond our own feeble abilities, and so far at least is beyond our ken. Many animals do things that beggar belief. Bumblebees, according to a great deal of perfectly respectable physics, could not possibly fly with their great fat bodies and tiny wings—but fly they do and very well. To catch insects on the wing at night by means of reflected sound waves seems beyond all imagining—but this is how bats earn their living. On the purely athletic front (given that athleticism is not just a matter of muscle), our own ability to walk upright is really quite extraordinary. A physicist would say that a person standing upright is "metastable," like a pencil stood on end: give it the tiniest shove and it falls over; it does not rock back into position, like the toys that litter the floors of parakeet cages. Yet all of us can stand upright, unless suffering from some special pathology. Indeed, we can walk downstairs while carrying piles of books, and whistling a happy tune. We don't even need to think about it. Many other animals are bipedal, but none matches our skill—not even birds, since they are balanced fore and aft as if with a tightrope walker's pole, and have big spreading feet for good measure. We need no such aids. Yet the complexities of standing upright are horrendous. It would take volumes of math to show what is entailed. The point is not that we defy the laws of physics—nothing on this earth defies the laws of physics—but that we have fabulous reflexes built into the fine nerves that coordinate the tendons and ligaments in our ankles. If we thought about these reflexes, or consciously dwelled on the math involved, we would surely fall over. It would be far too hard for us. By the same token, the nervous systems of all creatures—or at least of creatures like us and birds—are innately equipped to make use of the minutest clues and to respond appropriately; somehow or other all the complexities are taken care of, built into the system, all operating by rule of thumb—just lie back and admire. More of this in Chapter 9.

Finally, in addition both to their innate timing and to the clue of changing day length, birds time their migration according to conditions. If it is too cold, or stormy, or the wind is in the wrong direction, they wait for better times. If the signs are good, they may shift their schedule forward. So in a warm spring, birds that have arrived at the breeding grounds early, or who never went on migration at all, may breed seven to ten days sooner than usual. The external clues that allow fine adjustments to be made are known, in ornithological German, as *Zeitgebers*, or "time givers."

Still, though, we have to ask—how come? Whatever the mechanisms of navigation, why does a European swallow, say, set off for Africa in the first place? We can guess how it finds its way south, but what tells it, in due season, to start flying? And what tells it, once it is flying, why it is a better idea to fly south rather than north, or west or east? To be sure it is warmer in the south. But when spring comes again, it will go in the opposite direction—through the hot tropics and then out of the tropics and into the temperate north.

There are various kinds of explanation—none of which is really an "explanation," merely an observation that seems pertinent. First, it is obvious that swallows that do head south in the autumn are more likely to survive and leave offspring than those that stay put in the north, for if they tried simply to endure the winter they would starve from the lack of flying insects; any swallow that failed specifically to fly south would certainly perish. In other words, natural selection favors the birds that favor the south.

But what tells them to go? Here, we have to resort to the same general term that has been around for centuries and in truth is just a label for a whole swatch of behaviors that we don't really understand: "instinct." Clearly, all animals (and if we stretched a point we might say, "and all plants") have a repertoire of behaviors encoded in their genes that predispose to their own survival. As animals (and even plants) grow older they supplement, augment, and sometimes override these inbuilt, instinctive behaviors with behavior that is learned. But, clearly, a great deal of what they do is not learned—it is just "innate," acting out the predilections that we are born with. Thus newborn calves have to learn how to drink milk from a bucket if the farmer wants to take them from their mothers. But the calves know how to suck from the udder without any prompting at all. Human babies, too, fresh out of the womb, do the same—with a little assistance, but no instruction. By the same token, as the days grow shorter and cooler, baby swallows feel the *Zugunruhe* and, if the *Zeitgebers* are auspicious, they take to the air—and, by whatever collection of cues it is that enables them to stay on course, off they go. If they follow the right route, they live to tell the tale and leave offspring who will do the same. If not, not.

Of course, such descriptions are not entirely satisfying. One vital step that is obviously missing is: how does the code in the DNA translate into a feeling that some action must be undertaken, and an innate understanding of what that action actually is? DNA, when you boil it down, is just a chemical—a string of nucleotides cobbled together from the everyday elements of carbon, nitrogen, oxygen, hydrogen, and phosphorus. How does chemistry translate into psychology? Here, whether scientists like it or not, we are into metaphysics; and science, on the whole, does not do metaphysics. So there we must leave it.

But although it is clearly better for swallows to fly south than to stay put or to fly north or west or east, the hazards are nonetheless enormous. At least half of all Barn Swallows that begin their very first unguided trek to Africa perish along the way. On subsequent migrations those that do survive the first trip will have more strength and should make fewer mistakes because they have some experience, but even the experienced birds have a one-in-five chance of perishing.

Young Barn Swallows prepare well for their first migration. They practice hunting and get to know the layout of the area where they have been hatched and reared, which is where they will return to when they in turn are ready to breed. But alas they lose weight in this preliminary period, which means they lose precious body fat—sometimes up to 10 percent. This, presumably, makes it easier for them to fly, but it also means that some are likely to starve to death en route. Just before they are ready to leave, they roost together in local reedbeds or, if there aren't any, they gather on telegraph wires.

If the swallows are setting off from Britain for Africa, the first major hazard is the 35 kilometers (22 miles) of the English Channel. If the weather is too bad, they simply stay put. When it is calmer, some of the birds fly out to sea but then return. Sea breezes may still be a problem; some birds just can't get across. Those that make it to France stop off at various points for two or three days at a time to rest and feed, although bad weather may put an end to the feeding.

But the journey across France leads to the Pyrenees. Many skirt around the edge—to the west of them or to the east—but some run the gauntlet of the mountain passes and so may perish in thunderstorms.

Those that reach Spain then have an easy run (relatively speaking) for the next thousand miles or so.

Then they must cross into Africa—and there is no easy way to get there. The narrow crossing of the Straits of Gibraltar seems no problem at all, but sometimes there is an easterly wind that blows them out into the Atlantic Ocean. Others, instead, set out across the Mediterranean, but that means 500 to 600 kilometers (300 to 375 miles) of open sea. Some stop to rest on the various islands, but the men of the Mediterranean have a traditional penchant for small, migrating birds. In Malta I assumed in my naïve and effete British way that the small, walled enclosures along the cliffs to the north must be for birdwatching. How quaint. But then I saw the cartridge cases, dozens of them, in each strategic enclave. The birds that are not shot may be netted. On other islands, and along the shore of North Africa, battalions of Eleonora's Falcons lie in wait.

But still the journey is only half over. Peace reigns again as the swallows cross north Africa, for the Altas Mountains are far more agreeable and hospitable than the Pyrenees. But then comes the Sahara Desert—more than 1,500 kilometers (1,000 miles) across in places. Perhaps there are more oases en route than we generally imagine where the birds may stop and conceivably may catch insects, but many perish nonetheless. There is little or nothing they can do in the face of duststorms.

Then comes the enormous stretch of semi-arid land known as the Sahel—but in truth at the time they cross, in autumn, it is the rainy season, so these again are easy times, with plenty of food on the wing.

Then there is more trouble: the forest of the Congo Basin. Many of the migrating swallows fly to the west or the east of it, but some try the straight route over the top—and many of those that do are killed by thunderstorms.

Finally, about half of those that started out reach South Africa. Still the birds' troubles may not be over, for there may be late frosts in the early southern spring of November and early December, which means they will find few insects to feed upon. But if they do survive such snaps, then at last they can relax. The South African summer can hardly be improved upon.

But then it is time to go north again.

One last and obvious question of an evolutionary kind: how did birds develop such instincts and such prowess in the first place? When birds simply fly a little way south by way of migration, or even a long way, with no wild forays over sea, the answer seems simple enough. Presumably the ancestral types simply headed to where it seemed warmer and brighter, and their descendants inherited this general propensity. The tendency to migrate is inherited, and the required techniques are partly inherited and partly learned, so in principle at least migration seems flexible. We can imagine how each generation built on the practices of the ones before. Clearly, too, the bird's-eye view really does count for something. From a few thousand feet a bird really can see where it is going.

But sometimes the migrations seem too elaborate and too adventurous to have evolved simply in this step-by-step, precautionary fashion. How can so many European songbirds launch themselves so intrepidly over the Mediterranean, en route to Africa? Or over the Sahara, to get to the balmy latitudes of the south? How did America's Blackpoll Warblers ever get hold of the idea that it would be useful to fly out from New England, deep into the Atlantic to catch the offshore tradewinds?

Earth history has undoubtedly helped to shape these wild excursions. Birds have been around for at least 140 million years, and recognizably modern birds have been patrolling the skies for up to 80 million years; in that time, the continents themselves have shifted significantly. The Mediterranean, for instance, wasn't always the vast sea it is today. Through much of the Late Miocene, around 6 million years ago, the Mediterranean was dry land; it did not start to take on its present form until about 4.5 million years ago, when water from the Atlantic began to cascade through what are now the Straits of Gibraltar (in what must truly have been the mightiest waterfall of all time). We can imagine how ancestral birds once simply flew from Europe into Africa without running any risk of getting their feet wet, let alone of drowning—and then as the sea filled up, gradually extended the journey year by year. At first, the migrating birds could see the other side. Later, they were driven to head south by a general urge that had evolved to become a rule of thumb that they followed, even though the other side was now hundreds of kilometers away. Contrariwise, 50 million years ago, South America was still an island, 1,125 kilometers (700 miles) to the south of North America, and the two great continents did not finally meet until about 2.5 million years ago, in the Late Pliocene. The migration routes between the two Americas are less various and bold than between Eurasia and Africa. It seems that ancient Eurasian birds found the trip to Africa fairly straightforward, and simply lengthened their stride as it became more difficult. But ancestral

North American birds simply knew nothing of South America until recently (where "recently" is measured on the evolutionary scale), and they are still growing used to its proximity.

The ice ages undoubtedly helped even more. There have probably been around twenty alternating cycles of ice and warmth over the past 2 million years—not so long in biological history—including two in the past 150,000 years. There would be times when some routes were easy, over lush land that offered plenty of resting places to stoke up, and other times when that same land offered nothing but snow and ice. In ice ages, too, the sea level falls, and many present-day seaways were once dry land even in fairly recent times, Britain was simply a wing of mainland Europe until about 10,000 years ago. So again, we can envisage that ancestral birds established their migration routes in part to avoid the main ice fields; and they continued to fly in the same direction toward better climes even when their established routes were flooded, as the route from Britain to France and the Lowlands was flooded by the English Channel when the last ice age ended around 8,000 years ago. The ice ages, too, put a premium on migration. So long as northern winters are merely cold, then birds may sometimes tough it out, as many do nowadays. But if the winters are fierce to the point of wipe-out, then the need to migrate becomes absolute.

Many other creatures migrate: some insects, including many butterflies; some mammals, including many of the big herbivores of the African plains like the wildebeest and zebras, and some whales; many fish. But migration is unusual in most classes of animals, whereas in birds it is commonplace; no other animal even begins to emulate the feats of, say, the Arctic Tern or the Bar-headed Goose. No other animal can claim so convincingly, as birds might do, that the world is their oyster. But then, as Orwell's pigs declared, some animals are more equal than others.

7

Idyll and Mayhem: The Sex Lives of Birds

THE STORIES OF OUR CHILDHOOD AND EVERY CHRISTMAS
card with robins on it tell us that birds are sweet—their love life
and their family life a model for us all. Mummy bird meets Daddy
bird. They fall in love, and live together happily ever after. Swans, beautiful, powerful, and serene, stay together till parted by death, then mourn
till reunited on some higher plane. The small birds of the garden build a
nest together, line it with feathers, search all the daylight hours of spring
for caterpillars to cram into their offspring's clamorous beaks, then heave a
giant sigh of relief and fatten themselves for the winter so they can do the
whole thing again next year—or else pack themselves off to some warmer
clime where the going is easier, each with its own private air ticket.

Well, broadly speaking, this is how some birds do manage their affairs.
But many do not—and only in recent decades have biologists begun to realize how wildly and absolutely they may deviate. We know that we ourselves don't always behave like this, partly because we may stray from the
ideal but mainly because, for many societies through all of history, simple
monogamy, fidelity, and shared parental care were never the ideal in the
first place. Males and females have to cooperate to produce babies, and
the babies need looking after, but there are hundreds of ways to go about
it with endless variations and people and birds explore all of them—but
birds as a whole are even more various than we are. Among them, acted
out in every swamp and hedgerow, are all the great themes of literature—
which, of course are the themes of nature: true love, betrayal, jealousy—
Romeo and Juliet, Antony and Cleopatra, Troilus and Cressida, Othello,

One of the world's most conspicuous nests—but its creator, the Hammerkop, is flourishing.

Hamlet, and a prime theme of many eighteenth- and nineteenth-century novels, of babies born into one family yet cared for by others, with or without their new host's blessing. All this has become apparent largely through new approaches in biology—and notably by continuous, round-the-clock study of individual birds, made possible by extraordinary patience but also by a host of new technologies. These include radio-telemetry to track each individual; video recorders that turn themselves on with great fiscal prudence only when things start to happen; and minute studies of DNA that enable biologists to see who exactly is related to whom and who has consorted with whom—and in particular to discover the true parentage of eggs within each innocent clutch.

At times the data have been so at odds with our preconceptions as to beggar belief. Scientists have been wont to lose their dispassion and write about the goings-on with ill-concealed shock, or sometimes, as the fashion has changed, with lip-smacking relish. The facts of bird life have given a new spin to human sexual politics, as if they weren't spinning fast enough. But through the confusion of facts has come revelation: that nature can be cruel, to be sure, and full of deceit and lies—that it can be, as Lord Tennyson said, "red in tooth and claw." But it can be noble, too, and benign, inclusive, and self-sacrificing. These qualities are as much a part of life and of the universe as are the many and various forms of viciousness. Birds on the whole are not sweet; they do what they have to do. But if they—or living creatures as a whole—were simply vicious, red in beak and talon, then there could be no life at all; and life there very clearly is, in superabundance.

Yet legends are often rooted in truth, and the legendary fidelity of swans is indeed the case. Bewick's Swans, the European equivalent of America's Tundra Swans, have lived at the reserve in Slimbridge, Gloucestershire, ever since Sir Peter Scott founded the Wildfowl and Wetland Trust in the 1940s, and in all that time no pair has ever separated unless, as has occasionally happened, the partners failed for whatever reason to raise young. In the wild, when one of a pair of Bewick's Swans dies, his or her surviving partner may take two or three years to find a new mate, and some take as long as six—half the average life span of a wild swan (although some are known to have lived to more than twenty). For swans, fidelity clearly pays. If there is a dispute over territory, then established pairs almost invariably repel singleton invaders. If different families are in dispute, then the females (pens) and youngsters (cygnets) are content to let the males (cobs) slug it out between them. The longer the partners stay together, the better they are at raising chicks. At least for the first eight years or so, their success rate rises season by season.

The same goes for many geese, which of course are relatives of swans. The great mid-twentieth-century Austrian naturalist Konrad Lorenz kept a huge variety of animals over many years, not as pets but as house-guests, and Greylag Geese were among his favorites. In his classic *King Solomon's Ring* he describes the "courteous chivalry" of male Greylags, evident in the devotion of Martin the gander toward his partner, Martina:

> In Greylag geese, the bridegroom follows literally in the footsteps of his bride, but Martina wandered free and fearless through all the rooms of our house, without stopping to ask the advice of her bridegroom who had grown up in the garden; so he was forced to venture into realms unknown to him. If one considers that a Greylag goose, naturally a bird of open country, must overcome strong instinctive aversions in order to venture even between bushes or under trees, one is forced to regard Martin as a little hero as, with upstretched neck, he followed his bride through the front door into the hall and then upstairs into the bedroom. I see him now standing in the room . . . shivering with tension, but proudly erect and challenging the great unknown with loud hisses.

Many a seabird forms lasting relationships, too, including the procellariforms in general. Albatrosses are as faithful as swans, and giant petrels have model marriages even though in other ways they are mobsters: general niceness has nothing to do with it. But often, though, we find marked contrasts between similar birds. Gannets are faithful from year to year, but frigatebirds, which look vaguely gannetlike and are related at least distantly, find new mates after each bout of breeding. Why the difference? An answer comes from the brilliant seabird watcher Bryan Nelson. Northern Gannets nest on rocks in the north. They contrive to lay their eggs within a critical few weeks, beginning in late March. Too early, and the eggs will perish; too late, and the chicks will not be ready to face the following winter. Space is key—the nests are crammed, as Nelson says, to within jabbing distance of their neighbors' murderous beaks. So when a gannet pair does lay claim to a good nesting site, it pays both partners to return to the same spot at the same time the following year. They don't want to waste time either in finding a new place to nest or in finding a new partner. Faithful monogamy suits them both perfectly, and they have developed elaborate rituals to bond their relationship—pointing to the sky with their beaks, clacking their beaks together.

But frigatebirds nest on islands in tropical seas to the south. There are no clear seasons, the food supply is uncertain, and it takes a frigatebird pair more than a year to raise their chick to the point of independence. So the option of returning to the same spot at the same time each year to breed is simply not open to them. It is better just to breed when conditions look right and to find a new mate each time. Their courtship rituals are astonishingly elaborate: the males sit on the nesting sites they have commandeered and puff out the bright red pouch of skin beneath the throat (the gular pouch) like a great red balloon to attract the females wheeling overhead; and the pair then go through a breathtaking aerial routine where the males throw nesting materials for the females to catch. But once the contract is signed, no further bonding is visible. At the nest, the frigatebird pair seem to ignore each other. We may choose to be romantic, but birds must be pragmatic.

Male frigatebirds that have claimed desirable real estate flash their inflated, scarlet throat pouches to attract the females cruising overhead.

Ducks are related to swans and geese, and some treat their partners and families much as they do—soberly, and on the whole faithfully. But some decidedly do not. America's Wood Ducks practice what has emotively been called "gang rape." In one "assault" observed in the Muscadeck National Wildlife Refuge in Indiana, 50 to 75 males chased one female for twenty minutes, trying to mate with her. The Mallards of North America and Europe are just as unruly. There were plenty of Mallards on the river that ran past my university, and I remember one springtime two old ladies chasing the mobbing drakes with their umbrellas like a couple of hockey fullbacks. I can hear them to this day: "You horrible birds! Get away! Get away!" One afternoon in my medieval courtyard the room was suddenly filled with glass; a duck, panicked by pursuing drakes, had flown straight into the window. I found her stunned on the lawn outside. She recovered and flew away, I am happy to relate, but female Mallards and Wood Ducks are sometimes killed in such forays. Sometimes, since copulation takes place in the water, they are drowned.

In truth, "rape" is proving remarkably common in nature, and for a whole variety of reasons. Among young orangutans, it is the norm—basically because they have no obvious alternative. Orangs are monogamous and, as big primates, they invest an enormous amount in gestation and child care. Their choice of mate is very important to them. Female orangs don't like to mate with males who are not already mature and possessed of a convincing territory. Mature males, for their part, want to know that their chosen partner is fertile, so they prefer mature ladies who already have a youngster at heel. For the virgins of both sexes, it looks very like a catch-22. The young females prefer mature males but are rejected by them, but in their turn they reject the virgin males. Unsolicited assault offers the only opportunity that either will get to reproduce. At least the female knows that the male who overcomes her is not a complete wimp. The arrangement is not nice, but we can see how, in those very special circumstances, natural selection would have favored it. No "rape," no offspring, and the lineage of orangs would come to a halt. The basic reproductive strategy of orangs seems sensible: go for partners with a proven track record. But it has forced them into a strange corner. Despite the traumas of their first encounter, once the pair have formed a bond, they stay together. The male protects his partner and his offspring.

With Wood Ducks and Mallards, the outcome is quite different. They, too, are ostensibly monogamous—but most of the would-be "rapists" already have partners and, for them, impregnation of unwary females is just a bonus. If the female conceives and already has a partner, then she and he together may raise her assailant's baby as one of their own, and then it would have as much chance as any other duckling of living to maturity—another bonus for the adulterous male. The behavior may seem unsavory, but again we can see why natural selection has favored it. Yet hooligan behavior has a downside. While the male is away, philandering, his own partner may be gang-banged by others like himself. So it is that male Mallards (and other birds, too) commonly copulate with their mates immediately on return after a separation. To put the matter anthropomorphically, the male wants to make sure as far as he can that any eggs laid subsequently are his own offspring and "hopes" that his own, fresh sperm will override the earlier efforts of passing strangers. Clearly, though, this tactic doesn't always work. DNA studies show that in up to half of all of Mallard broods there is multiple paternity. On the other hand, this may not always indicate adulterous relationships. Many ducks and geese practice "egg dumping"—laying eggs in the nests of others of their own species. More of this later. (Some female birds take matters into their own hands, forcibly ejecting sperm from males they would rather not have mated with. Domestic hens, social climbers that they are, will eject the sperm of a subordinate cock if the boss cock starts to pay them attention.)

There is a further twist. Males who play away from home clearly benefit because they have more babies than they otherwise would, some of which may be raised free of charge in another family. But then, the same is true for females. Sex is risky genetically as well as physically: each partner links his or her genes to another individual who is something of stranger and may turn out to be a ne'er-do-well. The risk of marrying into an unknown gene pool is, of course, reduced if animals mate with their own brothers and sisters, so that they already know each other, and each other's family. But incest carries risks of its own, and most animals (and plants) go to great lengths to avoid it. The safest strategy both for males and for females is to choose the best possible partner; establish a good safe home where the offspring may be raised; and then practice a little judicious adultery, so that some of those offspring have different genes. It's a question of spreading options—not putting all the eggs in one basket, or indeed in one nest.

Ducks, though, make life easier for themselves by being monogamous—which in theory gives their partners less opportunity for extramarital relations (or "extra-pair copulations," as the technical expression has it). For polygynous species (*polygynous* meaning "many wives") the problems are compounded. The males of some polygynous species such as pheasants and other gamebirds avoid raising eggs and chicks fathered by other males by taking no interest in child care at all—they leave it all to the females. But males of other polygynous species do contribute to child care. The Red-winged Blackbird of North America (one of the many icterids) seeks to maintain several nests within his territory—and DNA studies show that one in five Red-winged Blackbird babies has been sired by a male other than the one that is helping to look after him, usually by the male in the neighboring territory. Even after a male Red-winged Blackbird has been vasectomized (this being the kind of thing that scientists do in their quest for knowledge), his mate continues to lay fertile eggs. The male behaves as if he does not realize that he has been cuckolded and continues to help out at the nest, helping to raise babies that are not his own. It may look very bohemian, but in truth everybody gains. Both the males and the females spread their genes among several partners, and all the babies get looked after, whatever their parentage.

You might well feel that for animals who raise only one offspring at a time, like an eagle (or an orangutan), adultery is not a sensible option. Either partner would surely be at least as well advised to produce a singleton offspring with his or her own well-chosen mate as with some outsider, however forceful or dashing. But this isn't necessarily so. For instance, male penguins of many species are happy in due season to copulate with any female who comes within range—and since most penguin species nest in colonies as seabirds often do, there are bound to be quite a few females about. The ideal for the male is to raise one chick with his regular partner, while ensuring that as many females as possible produce more of his chicks—and that they raise them in partnership with their own chosen partners, at no cost to himself. For females to do something comparable, they would need to produce several eggs by several partners and dump them on other couples. In a crowded penguin colony, with everyone watching everyone else, this surely would not be possible.

But in birds that lay several eggs, females may gain just as much by playing around as the males do. If one or more of the eggs in any one clutch contains a different subset of genes, then this increases the overall variety and so in general should raise the chance that one or several of the emergent chicks will survive. So it is that while female Wood Ducks and

Mallards can suffer horribly at the hands of marauding males, females of many species seek extra-pair copulations at least as eagerly as their mates, and sometimes more.

Knowledge of infidelity in birds or other creatures has not derived entirely from modern research. Past writers often seem to have known far more about our fellow creatures than is usual today—as witness King Lear's passing comment: "The wren goes to't, and the small gilded fly Does lecher in my sight" (Act IV, scene vi, lines 112–13). But the extent to which extramarital relations are routine is only now becoming clear. Indeed, adultery by one or both partners has now been reported among at least some species within *every* bird family, although there is huge variation even within bird families and it certainly is not the case that every known species is adulterous. Sometimes, too, presumably, adultery might simply be an aberration: in every population of all creatures, there will be some individuals who behave unusually.

More generally, right through nature, there is tension between the sexes—another great theme of literature. Sometimes the interests of both sexes coincide; sometimes they do not. They rarely coincide exactly. In penguins, it seems, males have more to gain from adultery than females do. For ducks, adultery may pay equally for both sexes. Different species, and different couples within species, overcome that tension to varying degrees. Some species (and couples) live permanently with significant tension. Natural selection can produce sweetness and light, but survival is its only inescapable condition.

Let us look again at the mythology. We saw in Chapter 5 how far the European Robin deviates from its Christmas card image—how feistily it defends its feeding grounds. Indeed, the only time that adult robins can stand each other's company is when a male and female establish a breeding territory together. He does most of the courting, in a no-nonsense, workmanlike manner; he shows her his territory, sings, and performs a rudimentary dance. Well, what more is a fellow supposed to do? If she likes what she sees, she moves into his territory and from then on, apart from their extreme unsociability, they are the model couple. The female builds the nest while he does most of the defending. He warns intruders by singing and strutting, but if anyone does intrude he is more than prepared to fight. In general he will attack anything with red on it, even red rags—anything that looks like one of his own, in fact. As Charles Darwin recorded in 1871, in *The Descent of Man and Selection in Relation to Sex*, one of his bird-keeping acquaintances, Mr. Weir, "was obliged to turn out a robin, as it fiercely attacked all the birds in his aviary with any red in their plumage, but no other kinds; it actually killed a red-breasted crossbill, and nearly killed a goldfinch." Once the eggs are laid, Ma and Pa share all the duties between them. This approach—male stakes out territory, sings to repel rivals and attract a mate, and the pair settle down to raise a family—is a fairly standard pattern that we also see, for example, in cardinals, finches, warblers, and wrens. Though there are differences in detail (sometimes both sexes build the nest together, sometimes the male does it), many birds do indeed conform to their nursery-rhyme image. But by no means all.

Take the Dunnock, formerly known as the Hedge Sparrow but in truth from the passerine family of the accentors. What, one asks, might Beatrix Potter have made of her? Miss Potter pulled no punches in her immortal children's stories. Peter Rabbit and Squirrel Nutkin were both exceedingly foolish, Jeremy Fisher the frog was outstandingly self-satisfied, and all were almost killed for their pains. If she had written of robins, then surely she would have portrayed their bellicosity at least as frankly as their prettiness. But even she would surely have censored the Dunnock and its fellow accentors.

For the sex lives of accentors is prodigious. The male has no penis (few birds do), but in due season his cloaca swells and protrudes. In the Alpine Accentor, from the high mountains through Europe and Asia, the testicles swell until they account for 7.7 percent of the male's body weight—equivalent to 6 kilograms (13 pounds) in a 78-kilogram (170-pound) man—more than fifty times more, weight for weight, than is the norm in men. Yet there is a twist in this. Huge testicles, of course, reflect a prodigious output of sperm, but this is necessary only if the sperm has to compete with the sperm of other males within the womb. In other words, male animals with the biggest testicles are the ones most liable to be cuckolded. Among gorillas, the alpha males have a remarkably small penis and testicles—smaller in absolute size than a human's, even though gorillas are far bigger—and much smaller than a chimpanzee's. But that is because alpha gorillas really are alphas—cuckoldry need not enter their heads. Chimps have to be more careful. As for human beings—well, think of Othello, for whom betrayal was a constant nightmare, or Troilus, for whom it came true. Almost the entire oeuvre of August Strindberg is devoted to the fear that men might be cuckolded and raise children who are not their own.

Sexually, the female Dunnock is highly charged. She is the one to initiate sex, turning around before the male, bending over, lifting her tail to expose her own cloaca, also swollen. The male may peck at her cloaca until it extrudes a drop of fluid. Then he mounts her—and as soon as the deed is done (barely a second), she may immediately approach him again—or another male, and then maybe a third. A Dunnock may copulate up to thirty times in a day, and at the end of the mating season will probably have had sex with every male in her circle. The drop of fluid that she extrudes after his pecking is the semen (or at least some of it) from the previous copulation. The males, in their turn, aim to cover every available female. Yet Dunnocks and the other accentors are monogamous up to a point. Males and females do raise chicks together like any nuclear family. It's just that a fair proportion of the chicks do not belong to the resident male—although he has plenty more dotted around other nests in the neighborhood.

Dunnocks and accentors in general are not the only birds in which the females initiate sex. Females of America's Black-capped Chickadee have been seen to approach neighboring males, settle nearby, and crouch with wings quivering in the submissive-solicitous gesture that invites copulation; and since copulation in these birds is over in a second, she can return to her mate before he has had time to turn around.

Even Dunnocks are nothing compared to the Cliff Swallows that nest in spring and summer throughout North America. Indeed, the sexual and social lives of Cliff Swallows are so complicated that biologists are still trying to make sense of them, although few birds are studied more intensively these days. Again, in essence, Cliff Swallows are monogamous, or at least they live in monogamous pairs. But the pairs group together to form colonies, in nests shaped like small pitchers, side by side, up to 5,000 at a time. The result, as Stanford University biologist Joan Roughgarden puts it in *Evolution's Rainbow*, "amounts to a city of mud huts"—although some Cliff Swallows are content simply to be villagers and live in groups of twenty nests or so. Some males, too, just hang about, out of town, with no nests to call their own. In short, their living arrangements overall are as varied as those of human beings. Indeed, says Professor Roughgarden, socially speaking, "Cliff Swallows are perhaps our closest cousins."

The city of the Cliff Swallows, as Professor Roughgarden puts the matter, features "a hot real estate market, trespassing, robbery, hanky-panky with the neighbors, plus presumably some compensations"—just like, say,

Los Angeles. The nests are very close together—so close that they may block each other's entrances—so that residents sometimes find themselves trapped, and starve. Or their home may be engulfed in droppings from above. The whole urban stew is a hotbed for pathogens, a veritable shanty.

Sometimes the swallows—usually males—trespass on each other's property, either barging in or following the owner home and blocking the entrance. The single most common reason for this, so scientists have shown—accounting for 14 percent of all trespass in one study—is simply to steal grass to line their own nests—in human terms, popping next door to steal a blanket. In 7 percent of intrusions, the trespassers steal mud that is not yet dry to build their own nest. Do the birds know what they are doing?, one might whimsically ask. Do they feel the *frisson* of forbidden fruit? Or do they simply feel that grass is grass and mud is mud, and next door is as good a place as any to get it from?

But sometimes the visits would, if this were a human city and we were being anthropomorphic, attract the attention of the social services or even the serious crime squad. In one study, in about one in ten visits a male trespasser forced himself on his neighbor's "wife." In about one in thirty intrusions, a female laid an egg in her neighbor's nest or carried one in her beak, ready-laid—the strategy known as "brood parasitism." Usually the nests she favored with her surplus eggs were close to her own—not more than five doors away—and usually it was when the neighbor was not at home. Sometimes she would seize the opportunity to throw out one of her neighbor's eggs, but often she would just add one to the existing clutch. Sometimes the female Cliff Swallow would bring in a chick, ready-hatched, and add it to the resident family. But in one in a hundred visits a visiting female would also toss one of the resident chicks out of the door, like Judy, the lovely wife of Punch, as if to keep the numbers even. In one in every 300 visits the trespasser would simply evict the owner.

Of course, the term *rape* has vile connotations, but it is sometimes used in zoology in a neutral way simply to describe what is observed: one individual forcing its sexual attentions on another who is clearly unwilling. Usually males rape females, but all other possible combinations are observed from time to time. Rape, thus conceived, is a common part of Cliff Swallow life. The females are vulnerable when they flock together to gather mud and grass for their nests. Then males—mainly apparently the unattached ones—circle overhead and leap upon them. Yet the female Cliff Swallows do not always seem to resist. Again, from the female's point of view, the assault is not all bad. The male who overpowers her at least has physical prowess, and if she has a chick by him to raise alongside those of her regular partner—well, she has spread her genetic options. The "rape," however, can be indiscriminate. Sometimes mud-gathering males are raped, too. On the other hand, the regular male back at the nest would miss out if his partner were impregnated by a third party, and like a Mallard drake, he generally copulates with her every time she returns to the nest after an outing as if to ensure that his own sperm overlies any that she might have acquired elsewhere. In busy times the pair may copulate dozens of times in a single morning.

For their part, some but by no means all of the "married" male Cliff Swallows seek extra-pair copulations. This seems yet another example of mixed strategy: some males seek to impregnate as many females as possible; some are content to focus on their own offspring, and do their best to ensure that their partner does not play away from home; and many are somewhere in between. Each strategy makes some genetic sense, and it also makes sense if, in any one population, some individuals do one thing and some do another.

Because of all the extramarital transactions, about 15 percent of Cliff Swallow nests wind up with one or more eggs that were not laid by the resident female. In one study, 38 percent of nestlings were the fruits of extra-pair copulations. In one brood, none of the six eggs belonged to the resident male who was helping to look after them. Yet, by the standards of passerines, such figures are not outrageous. In a study of Superb Fairy-wrens in Australia, 98 percent of nests contained at least one baby that had not been sired by its mother's regular mate, and 75 percent of all their babies had been fathered through extramarital copulation.

Sometimes, to the neutral observer, Cliff Swallows simply seem confused. We can see why a female might want to leave a few eggs with her neighbors. We can see why males might want to impregnate passing females when the opportunity arises, and vice versa, why a married female might not always reject the advances of an outsider. But sometimes a male in his nest will allow a strange female to enter and lay an egg, and then depart. What's in it for him, in such a case? Perhaps, Professor Roughgarden suggests, the female may be dropping off an egg that the accommodating male has previously fathered (although this does raise the question of how she knows that the particular egg is his, and why he should take her word for it). Still, it is a possibility.

Not all the females in any one colony of Cliff Swallows lay eggs in others' nests. In one study, only about a third were "brood parasites." But again, we must ask if this is simple miscreancy. The females who lay eggs elsewhere must leave their own nests to do so and are more likely to be parasitized in their turn. So they, too, finish up raising chicks that are not their own. The picture begins to resemble the arrangements that were common among the eminently respectable English middle classes of the late eighteenth and early nineteenth centuries, who when the family was large, sometimes asked or allowed richer relatives or friends to bring up at least some of their children. Overall, each nest winds up with the same number of chicks, and each couple raises the same number of chicks, and the number of strangers that any one couple finds itself raising must, in the end, equal the number that one or both of the partners have sired or laid in other birds' nests. So the overall picture is fair, and all the birds in the colony have a more or less equal chance not only of raising their own chicks, but also of ensuring that somebody else raises some of theirs.

It would be at least peremptory to view the Cliff Swallows as unreformed bandits and sex fiends. They might just as soon be seen to be sensible and humane, helping each other to spread their genetic options, with net benefit for all. In practice, too, a female that transfers an egg may be transferring it to the nest of the real father—in which case it wouldn't be "brood parasitism" at all—or at least, if it is called brood parasitism, then it need not be seen as a competitive pursuit.

There is one further twist. After fledging, the juvenile Cliff Swallows gather in crèches. Within them, adults seek out their own young and continue to feed them, rather as Emperor Penguins do when they return to their famished youngsters after a long forage. But some juveniles do not join the crèches, but return to their nests, block the entrance, and intercept food that (it is generally assumed) was intended for the new clutch of babies inside. These apparent miscreants have been labeled "kleptoparasites." But this term, too—like brood parasite—is judgmental; perhaps it is simply the case that the parents are happy to feed the juveniles. Certainly they never throw out their adolescent children, as they would usually throw out visiting adults.

Apart from the Red-winged Blackbird, all the birds described above are monogamous, up to a point. So indeed, are 90 percent of birds—in diametric contrast to mammals, 90 percent of whom are polygamous. Clearly, though, different bird species interpret monogamy differently.

Some are what zoologists call "reproductive monogamists," meaning that the two partners do indeed share their genes. Others are merely "economic monogamists." They may share a nest, and up to a point share the parental duties, and perhaps help to feed each other as well as the young, but either or both partners routinely produce youngsters through extra-pair copulations. Cliff Swallows, as Professor Roughgarden puts the matter, are extreme examples of birds that "have decoupled economic monogamy from reproductive monogamy."

What else do birds get up to? And, above all—why such variation? Why don't all birds behave like Bewick's Swans, or robins? Why, for that matter, don't they all behave like Cliff Swallows—if it works for them, why not for all? It will never be possible to answer such a question definitively. But we can address it sensibly, and for this we must return to first principles.

WHY REPRODUCE AT ALL? WHY SEX?

None of us, taken individually, actually need to reproduce. Life would usually be far easier if we didn't. Maiden aunts and bachelor uncles, provided they are well cared for, often outlive their fecund siblings, surrounded by fishing rods and wool, smiling to the end. As for sex—ridiculous! If you really must reproduce, then split down the middle, in the time-honored manner known as binary fission; or do as some lizards and fish do and produce eggs that develop to become exemplars of their kind, but asexually, without the mess and angst and danger of union with males. Why do we make life so complicated?

The first question—why reproduce?—can be answered only in retrospect. In principle, any creature might live forever. There is no biological rule to say that senile decay is inevitable. Among trees, some bristlecone pines and probably some yews are flourishing after 5,000 years. If 5,000, then why not 500,000? Even among birds, a fulmar may be as sprightly at forty as at four—and if at forty, then why not 40,000, or forever?

The answer lies in time and chance. Sooner or later, life is bound to catch up with each individual creature, no matter how comfortable it may feel. The mountains where the bristlecones grow will be dust one day, and they must die when the mountains die. Sooner or later, the most skillful of fulmars will catch its beak on some fishing line, or die like Jonah in the belly of some whale (an orca is the most likely candidate) or at the jaws of some sportive sea lion. The creatures that die without issue take their genes with them, never to be recovered. We cannot say that any one creature *needs* to reproduce in order to pass on its genes—for any one creature that preferred an easy life might simply say, "So what? What difference would that make to *me*?" Many creatures in the history of the earth may well have taken such a view, or been forced to take it, but if they did, then they have left no trace, no descendants to continue their dynasty and to inherit their solipsism; and those who took that view now would leave no trace in the future. It is absolutely certain, then, that all the creatures that do exist, and that will exist in the future, must be the scions of ancestors who did take the time and trouble to reproduce.

Reproduction is something that living creatures are stuck with, like it or not. Incidentally, bachelor uncles and maiden aunts—if they really are uncles and aunts—do leave a genetic legacy. So do the elder daughters in Victorian novels who looked after their ever-growing train of siblings and then took care of their old Dad. One in eight of the genes of an uncle or an aunt is shared by each of his or her nieces and nephews, and a girl or a woman has as many genes in common with her siblings as she does with each of her own children—so caring for younger relatives is reproduction by proxy (although, biologically speaking, looking after your old Dad is a waste of time).

So why sex? We speak of sexual reproduction, but if we think of reproduction as multiplication, then sex is its precise antithesis. If a female produces an egg that needs no fertilization, then for every parent there is an offspring. But if the egg needs first to fuse with a sperm, then it takes two parents to produce each baby. Head for head, sex halves the possible rate of multiplication. If sexually reproducing creatures are competing with asexually reproducing creatures, then the asexual reproducers should surely prevail. They reproduce twice as fast. Sex gets in the way.

In truth, sex is *not* about reproduction—or at least, it is not primarily about multiplication. It is about mixing genes, trying out new combinations. When genes are recombined, there is constant variation, generation to generation. Ducks produce ducklings, which indeed are baby ducks, but no two ducklings are alike and none is exactly like either of its parents. In the long term, variation is the stuff of evolutionary change—but that can't be the reason creatures go to such lengths to generate variation, because natural selection is geared to the here and now. It does not look to the distant future. There must be a short-term advantage—so what is it? There have been various guesses, and one of the most favored came from the English biologist Bill Hamilton in the 1960s: that it is harder for parasites and pathogens to cause epidemics among populations of creatures that are genetically various than it is if they are all the same. Many a farmer has discovered this, as his fields of identical corn or grapes succumbed to some virus or fungus that just happened to hit the right note—and what did for one plant, did for them all. It is chastening to contemplate that sex, and all the passion and courtliness that go with it, should be driven by the need to avoid invisible worms, but that is how it seems to be. Parasites and pathogens are a constant fact of life. Disease is serious, so sex is serious too, and for all its obvious drawbacks, it has been favored by natural selection.

So what does sex entail? At the strictly mechanical level, not much. A big sex cell—gamete—of the kind known as an egg, produced by an individual that by definition is a female, must fuse with a small sex cell of the kind known as a sperm, produced by another that is defined as male, and nature takes care of the rest. But the more you look at it, the more problematic it all becomes. This is why the intricacies of sex account for at least half of all literature, from Homer and Shakespeare to Mills and Boon; and what is hard and complicated for us is hard for birds, too. It is the nature of things.

WHY SEX IS DIFFICULT

First, creatures that aspire to reproduce by sex must ensure that their mates are of the appropriate kind. Unions between partners who are too unlike—duck on crow, for instance—come to nothing, and so at best are a waste of time and at worst are extremely dangerous, since an uninvited sexual advance can hardly be distinguished from simple aggression. Unions between partners who are unlike but not *too* unlike may produce hybrids—and the fortunes of hybrids are very mixed. Hybrids between different races of the same species sometimes exhibit what Darwin called "hybrid vigor"—they may be more robust and accomplished than either of their "pure-bred" parents. But unions between members of different species—interspecific hybrids—are usually in some way compromised, as we have seen. Thus, creatures that aspire to reproduce by sexual means must be sure to identify a partner that is indeed appropriate. Recognition is crucial—and it isn't always easy to get it right.

Production of sperm and eggs is a great physical strain, because reproduction is important and so natural selection has in general favored those creatures that invest as much in it as they can afford. The paired testicles of a bird sometimes account for 10 percent of body weight—a big investment in a flying animal for whom weight is critical. In most species the testes dwindle when breeding is over, but this means they have to be grown again the following year; and so it is that at the start of the breeding season the testes of a mature Japanese Quail grow from 8 milligrams to 3 grams—3,000 milligrams—in just three weeks, an increase of almost 400-fold. Eggs, produced usually from a single ovary (the left one of the original pair), seem an even greater strain. Of all birds, the kiwi lays the biggest egg relative to body weight—kiwis are about the same size as farmyard hens, but their eggs are about six times bigger. But then, kiwis lay only one or at most two eggs at a time. Jungle Fowl, the wild ancestors of domestic hens, can lay up to twenty, at one-day intervals. In some birds the total clutch, laid over a few weeks, may weigh as much as the mother. This takes prodigious energy. Snow Geese do not arrive at their breeding grounds in the American Arctic until late May, and so to save time, the females must provide at least some of the energy for their eggs from body fat that they have accumulated in the previous few months, while they sat out the northern winter in the southern United States and Central America. But most birds provide the energy for their eggs simply by feeding as they go. The females of some species do all their own feeding, but many rely on the males for supplementary food. In those species it is very much in the males' interest to be good "husbands." No help, no offspring. The crude idea that it must always pay males to philander—restlessly seeking new mates to impregnate—is simply not true. Often the male's best strategy is to help the female he has already impregnated to raise the eggs and chicks that they have produced between them. Here, at least, their interests do coincide.

A bird's egg is one of nature's finest creations: a superb survival kit: a self-contained unit providing the embryo with nourishment and safe housing, in some cases until it is able to fend for itself; a brilliant exercise in engineering; and on the part of the female, an extraordinary feat of physiology.

An unfertilized egg leaves the ovary, deep in the female's body, to enter the trumpet-like opening to the top section of the oviduct, known as the "infundibulum." There it meets the sperm, which may take as little as half an hour to swim and to be carried up from the cloaca. The eggs are usually fertilized within a few days, although the females of some species can store the sperm for up to ten weeks in the oviduct before transferring it to the infundibulum, and by such means domestic chickens and turkeys can lay fertile eggs from thirty to seventy-two days after copulation. If the female mates with more than one male, then the sperm of the last one usually does the fertilizing—though why this is so is not properly understood. Some females, like the Dunnock, actively eject any remaining sperm from previous copulations before accepting a new male, but this does not seem to happen in every species.

After about twenty minutes in the infundibulum, the fertilized egg passes down the oviduct at just a few millimeters (½ inch) a minute. First, it is coated in four layers of albumen secreted from the oviduct wall; these form the white of the egg. Then the white is coated in threads of keratin—hard protein—which forms the thin inner membrane of the egg. Then a second, thicker layer is added—a felt-like mat of protein fibers cemented together with albumen. Finally, in the uterus, the pigments are added—porphyrins, which are created from breakdown products of hemoglobin from the blood, and the shell itself, formed from calcium carbonate in the form of calcite crystals (with small amounts of magnesium and phosphorus).

With the egg complete, the female commits it to the outside world by voluntary contractions of the vagina—usually in the early morning, which perhaps reduces the risk of damage to the almost-finished egg that may result from daytime activity. The whole passage from infundibulum to vagina typically takes only twenty-four hours, although in some species it takes up to a week. Many birds—particularly small birds, and a few bigger ones such as domestic chickens—can and do lay an egg every day until the clutch is complete. Many others take two days between eggs, while in boobies and hornbills the interval between successive eggs is a leisurely seven days; in megapodes it is four to eight days; and ratites and large raptors need three to five days. In general, big birds obviously lay bigger eggs than small birds, with hummingbird eggs down to 0.2 gram (.007 ounce) apiece and the eggs of the extinct elephant-bird weighing in at 9 kilograms (20 pounds)—a 45,000-fold difference between the smallest and the biggest. But each egg of a big bird typically weighs only 2 percent of the mother's weight, while each egg of a small bird commonly weighs up to 11 percent. Each Brown Kiwi egg weighs a quarter as much as the mother—and she usually lays two of them, or even three. The only other warm-blooded animals I know that produce such relatively huge off-spring are some small bats, whose single baby is half as big as its mother at birth—which it needs to be because it is usually required to fly while still young, and it could not do this if it were too tiny.

Even the simple act of fertilization—bringing eggs and sperm to-gether—can be extremely risky. It may not seem too risky for those aquatic creatures that simply release their gametes into the passing cur-rents—sometimes just the sperm, as in swan mussels, which must then find its way to some sedentary female, ensconced in the riverbed, and sometimes both eggs and sperm, which must meet up in the plankton. It may not seem too physically dangerous for grasses that simply shed pollen into the air or wait for pollen to be wafted in. But it is risky for animals like birds and mammals (and insects and spiders and most others) that practice copulation, and so must make direct contact, one individual with another. If an animal approaches another animal that is not in the mood for sex, then the approach is liable to be construed as a threat and the would-be seducer could be badly hurt (and, indeed, male spiders and praying mantises are regularly gobbled up by their intended paramours). So courtship is vital: seduction; foreplay; the cultivation of appropriate mood in the potential partner—a mood, furthermore, that runs ab-solutely counter to the cautious standoffishness that is usually advisable in day-to-day socializing. Courtship requires serious effort, usually but by no means always on the part of the male, and it also requires that the female should find the male's advances truly seductive, for nothing is more ridiculous, irritating, or threatening than sexual advances that are unwel-come. Courtship accordingly plays a huge part in the lives of birds. Some males devote their entire working year to it.

In general—not universally—male animals of all kinds have to com-pete with other males for the attention of females. Sometimes the compe-tition is indirect: the males strut their stuff and the female chooses between them. Sometimes there is direct physical strife. This is more con-spicuous and typically more violent in mammals than in birds—elephant seals, bulls, stags, billy-goats, lions—but birds may fight, too. Farmyard cocks may fight to the death, although their semi-captive state exacer-bates the tension. Gannets, jostling for sites on their crowded rocks, may seriously damage each other. Usually—as in gannets, or penguins, or robins—the fight between males is not ostensibly over a female, but about territory. Yet it amounts to the same thing, because the males who have the best territory attract the greatest selection of females, and the males who have no territory are not likely to find a mate at all. So male rivalry is another hurdle to overcome—and again, the overcoming of it requires tremendous investment of time and energy, and can be very dangerous.

Copulation itself can be risky. We have seen the hazards of Mallard mating: females sometimes drown. In gannets, there is plenty of room for confusion since the males and females look virtually identical, and their ritualistic signals of mating are very like the ones that signal aggression. So it is, says Bryan Nelson, that the male may invite the female to approach, but when she does get close he may switch to vigorous attack, her male-like appearance tipping the balance that way. She always tries to weather the assault by facing away, but in this she may fail even if she endures the punishment for ten or twenty minutes—a long time to endure such a battering. Many a time I marveled that she, who would attack an intruding female with a total commitment not a whit inferior to the male's, nevertheless meekly accepted such harsh treatment just because it was handed out by a male. But the stakes are so high that natural selection has ensured that she can and does. In the gannet's case, no pain, no offspring.

For many other birds, the main hazard in copulation comes from predators: what better time to strike? (I once saw copulating House Sparrows in London run over by a car.) Gannets take about thirty seconds to copulate, the male standing on the female's back, trampling with his feet and gripping the nape of her neck, hard, with his lethal beak. But many other birds take less than a second, and then they are off and away. It's as if they feel it is just too dangerous to hang around. (Not all, however; Fiery-throated Hummingbirds may copulate for more than fifty minutes, and Vasa Parrots for around 100 minutes.)

Copulation is often frequent—partly for the obvious purpose of fertilizing eggs, but also, as we have seen, to ensure or at least increase the chances that eggs laid by the female have indeed been sired by her ostensible partner. In many species, too, including gannets (and of course in humans and some other primates), copulation is a form of bonding—particularly important when the partners must stay together for a whole season, or even for many seasons. The record, according to some accounts, is held by Smith's Longspur, a North American bunting of the genus *Calcar-*

ius. Female Smith's Longspurs solicit copulations around seven times per hour and are mounted for their pains about three times an hour, and so they finish up with around 365 copulations per clutch. The record apparently stands at 629 times in 6.5 days.

Just to complete the list of hazards, copulation is a prime route for the transmission of disease—an irony, given that one probable reason sex occurs at all is to outwit parasites.

Then there is the business of staking territories, building nests, protecting and incubating the eggs, and protecting and feeding the chicks. It all takes tremendous effort—and, as always, it needs sound strategy. Good territories for nesting and feeding are limited, and although the world as a whole is large, it is nonetheless finite. In the north and deep south, cold succeeds hot succeeds cold. In the tropics, dry succeeds wet succeeds dry. Everywhere, timing is all. This is why half of all bird species migrate. It takes enormous effort and is hazardous. But it means, at least in principle, that they can treat the whole world as their habitat.

MAKE SURE YOUR PARTNER MATCHES UP

To reproduce sexually you must first make sure that the individual you mate with belongs to your own species—which, on the face of things, seems simple enough. Weekend birders may find it hard to tell a female Mallard from the female Gadwall, but the males of nearly all duck species are as distinctive as rival football teams. Yet just about any two species of duck can hybridize—and often do, more usually in captivity or on the local pond, when the different types are thrown into each other's company, but also in the wild. The mistakes are even more striking among the forty-two species of birds-of-paradise, sicklebirds, and riflebirds, in the family Paradisaidae. They vary so much in appearance that they are divided among no fewer than seventeen genera. Some are plain, but many are metallic or iridescent, and some have plumes of red or yellow or pastel blue, and some have "wires"—modified from feathers—protruding from their heads like pompoms or from their tails; some have bright wattles and some have special feathers in their wings that whirr or rattle as they fly, like a snipe. Yet thirteen different hybrids are known between different pairs of species—and seven more between members from different genera. As we will see later, Darwin's friend and rival Alfred Russel Wallace tried to explain the bright coloration of many birds entirely in terms of species recognition. But often, it seems, the color coding does not work.

How can birds make such mistakes? I do not presume to know, but it seems that some birds at least are not born with a clear knowledge of who they actually are, and may be confused by their early experiences. Konrad Lorenz records that the Greylag Geese and Mallard that he raised by hand became convinced that he—the first living creature they encountered apart from their siblings—was their mother. Lorenz being Lorenz humored them in this and would squat and then waddle around his garden quacking with the ducklings in train while they in their turn called "peep-peep." They would not follow him if he stood upright (for then he was too tall for their search image) or if he stopped quacking (for they clearly were born with the simplistic notion that if it doesn't quack, it can't be a duck, and if it does, it is). At other times, various of the jackdaws he kept in an aviary on his roof took him as one of their own kind and courted him, dropping fat caterpillars into his mouth as if he were a young female. One of his hand-reared jackdaws took a special shine to the maid, and would thrust miscellaneous invertebrates into her ear. (It must have been fun to live in the Lorenz household, but it clearly required forbearance.)

To be sure, the folklore of many cultures seems to take mating between unlike creatures for granted. In medieval England, robins were apparently assumed to be all male and mated to wrens. Their common names reflect this: robins were originally known as "Ruddocks" or as "Robin Redbreast"—and the nickname took over. Wrens had a feminine cognomen as in "Jenny Wren"—little heard these days, which is a pity. Wrens were also thought to be married to Great Tits (also known as "Ox-eyes"). In reality, robins make very sure that they mate only with robins, and wrens with wrens, but many birds have trouble with this. The need to look or sound distinctive is part of the burden that sexual reproduction imposes.

HOW TO ATTRACT A MATE—AND KEEP ONE

Birds seek to attract mates in many different ways, but there are a few main patterns. In the species in which the male helps with child care, and including most monogamous species, the male typically begins by carving out a territory that is suitable for nesting. He sings or calls or otherwise signals to keep other males at bay, and those same signals, generally, lure the females in. But in other species, the males simply seek to attract as many females as possible and then leave them to find territories, build nests, incubate, and raise the young by themselves. Some males from various orders group together to form "leks" for the purposes of seduction: they all stand together, or at least within close range of each other, and make the calls or strut the stuff that is characteristic of their species. Lekking is more or less cooperative, since the females are attracted to the virile males en masse. But it is mainly competitive, since each male is at pains to ensure that he does most or preferably all of the mating. Sometimes, as in peafowl, the males aim their display purely at the females. Sometimes, as in Sage Grouse, the display seems aimed at the other males, and the females choose the winners.

In all species, the males have to be attractive to the females, and in many, they look very different. The males typically have far brighter feathers—this is "sexual dimorphism." Often the males grow special crests and quills in due season and sometimes both sexes do, as in egrets and herons. Many males have areas of skin devoid of feathers but brightly colored, sometimes dangling to form wattles, as in the turkey or the cassowary. Sometimes the skin is perched on the head to form a comb, as in the Jungle Fowl; and sometimes it is inflatable, as in the frigatebird, with its

Many male gamebirds, like these Sage Grouse, display en masse to the females in collectives known as "leks."

throat pouch. Commonly, the colors and plumes are shown at their best by strutting, dancing, and aerobatics. All may be enhanced or largely replaced by calls and, often, by songs of marvelous elaboration. In general the motto is, "If you've got it, flaunt it." In some species, as the male sings and cavorts, the female looks on with studied indifference. But in some she joins in so that male and female build the courtship ritual between them, in dance or acrobatics or song, and this is called "duetting." Sometimes the roles are reversed: the female is the brighter and/or the larger of the pair, and sometimes she hands over child care entirely to the male. If she does this, then she may simply clear off and leave him to it, or she may take on the role of guardian, like a male lion. Endless variations are possible, and birds seem anxious to try as many as they can.

The seductive deployment of plumage, combined with posing or dancing or aerobatics or song or some permutation thereof, is taken to its peak—and sometimes to human eyes to its absurd extreme—in birds-of-paradise.

Some birds-of-paradise are beautiful to human eyes. The tail of the Black Sicklebill is a meter long (3¼ feet). The long plumes of the King Bird-of-paradise—the smallest of the family—trail behind it like fire, or the mane of some winged horse, or the tresses of some northern goddess. Wallace, in the jungles of the Malay Peninsula, lamented that such creatures should be "living and dying amid these dark and gloomy woods, with no intelligent eye to gaze upon their loveliness; to all appearances such a wanton waste of beauty."

Others, frankly, seem ridiculous. I confess I have not seen birds-of-paradise in their native jungles, and now that gratuitous air travel is known to be so damaging, I don't suppose I ever will; perhaps their feats and antics lose something when translated onto film. But while the courtship flights of an eagle or a tropicbird can take the breath away, birds-of-paradise come across, for the most part, like a series of music-hall acts: the contortionist, the illusionist who pulls a bouquet of flowers from his inside pocket and flags of all nations from his sleeve, the dog in a boater that stands on its hind legs and plays the trumpet. A favored bird-of-paradise trick is to thrust out the feathers to change the shape and hold the pose as in a tableau. One species, with one flick of its feathers, turns itself from a bird into a perfect jet-black rectangle, with two painted eyes with a pale blue crescent at the bottom like a mouth. The Twelve-wired Bird-of-

Birds-of-paradise—this one is a Greater Bird-of-paradise—are often beautiful to the human eye and sometimes absurd. Either way, the males' apparent extravagances certainly attract the females.

paradise fluffs out the bright yellow feathers of its body and the bright green feathers of its neck, with its beak pointing straight up, so it resembles, well, two fluffy pom-poms, one on top of the other. The White-plumed Bird-of-paradise likes to hang upside down, with the wispy feathers of its tail pointing up like a vase of flowers. The Parotia Bird-of-paradise is half dervish and half old-fashioned stripper, its feathers held out like a tutu, whirling about the forest floor with legs as far apart as they will go, and with four bright blue pom-poms on its forehead, shaken to a blur. It is all very amazing, but it invites the cry that all auditioning actors dread the most—"Next!"

Yet, for this, birds-of-paradise prepare the whole year, as people who live in carnival towns may prepare for the Mardi Gras. Six-wired Parotias prepare their whole lives. The dominant male displays on a special perch that is used generation after generation; the younger males cannot breed until the dominant male, the incumbent of the sacred perch, finally retires. So long as the boss male holds his position, the younger ones remain in juvenile plumage, and this may be for seven years—evidently their hormonal development is suppressed by his presence. We see the same phenomenon among mammals—for example, in the blackbuck of India. Only the dominant male blackbuck is actually black; the rest remain in juvenile brown until the alpha male is finally displaced, whereupon one of the younger males is liberated, hormonally, and allowed to mature.

But most famous of all, and far more familiar, is the peacock. Its wondrous fan is not a tail but a train, formed not from quills that grow from the stumpy tail (the pygidium) but from the back. Most attractive to the female are the hundred "eyes," each with its bright blue iris that stares from the shimmering background of greeny-bronze: "eyes" in nature are a universal icon, frightening or alluring to all creatures that have vision, from insects to fish to lizards to birds to mammals, including us. As if the visual display were not enough, the peacock adds sound effects. As Gilbert White observed in a letter to Thomas Pennant in 1771, "By a strong muscular vibration these birds can make the shafts of their long feathers clatter like the swords of a sword-dancer."

Cranes are compulsive dancers. They all do it, more or less at any time of year, for all kinds of reasons. They make their first experimental bounces when only a few days old. But they dance most enthusiastically when courting—and then emerge as outstanding duettists. Most accomplished, graceful, and rightly celebrated are the Japanese Cranes, which have inspired whole schools of human dancing. The Sandhill Cranes of North America are less balletic, but possibly more athletic. In March they gather in groups of up to half a million on the flats of Nebraska to feed, fatten, and pair off before flying north to claim their territories and raise their families. Each pair begins its courtship with a few cautious jumps, first one partner and then the other, though sometimes both at once.

Then as the mood takes hold, they leap higher and higher, up to about 2 meters (6½ feet), and sometimes right over each other's heads. Other pairs join in until the whole group is at it in one huge garage party. Then the first pair stops and starts calling instead—trumpet blasts, first one and then the other. Then they resume feeding, still perfectly synchronized, first one partner with its head down and the other partner's head up, and then the other way around. So they continue until they are rested and fed and continue their migration north to breed.

The courtship rituals of grebes are wondrous. At the end of the preliminaries, in all species, both sexes combine to build a mating platform—a raft of weed, sometimes resting on the bottom, sometimes on the submerged roots or stems of reeds or whatever, on which to copulate. Unlike ducks, grebes cannot copulate in water and only rarely copulate on land (on a shallow bank or island). They prefer to build a love nest. The proper nest, where they lay their eggs, is very similar, but is a quite separate structure. Grebe eggs tend to get wet, but they have a coating (of calcium phosphate) that allows the free flow of gases despite this.

In the smaller species of grebe—twelve of the twenty-three species—courtship tends to be noisy rather than spectacular, but the larger kinds are terpsichoreans supreme.

The courtship rituals of the Western Grebe, which lives in the west of North America from Canada down into Mexico (with another, southern race on the Mexican plateau) begin in spring. Prospective mates dip their bills into the water and flick their heads up again or point their bills at each other with necks held stiffly, either bent up or at a strange angle. Suddenly the male whirls to one side, thrusts with his feet, lifts himself out of the water, and tears off across the surface—running on the water, his feet pounding. Almost immediately—at least within half a second, and sometimes within a tenth of a second—the female runs after him, quickly draws level, and so they pound along in tandem for 20 meters (65 feet) or so, necks curved, heads thrust forward, and wings held out stiffly, certainly to provide stability and perhaps to provide some lift. Then, abruptly, they both sink to the surface. This is the "water dance." Some other grebes do something like it, including the Horned Grebe, but the Western Grebe is the master.

Sometimes two males perform the water dance—perhaps cooperating, up to a point, to attract the attention of a female, or perhaps because they mistake each other's sex. Sometimes two males and a female set off in a trio, and when this happens the female may stop suddenly and leave the males to rush on by themselves.

After the mad dash, the birds dive and shake and bob their heads and then, like the Great Crested of Europe and some other grebes, they go into the "weed ceremony." One of the pair first stretches its neck upwards, and dives, sometimes repeatedly, emerging eventually with a beakful of weed. The other quickly follows suit. When they both have weed, they approach each other, rise out of the water breast-to-breast about a body's length apart, holding themselves aloft with frantically paddling feet, and point their weed-filled beaks upward at an angle of about 45 degrees. Typically they hold this pose for about 20 seconds, but have been known to stay in this position for a minute and a half. Then they sink down again, bobbing their heads and preening, as if to signify that the show is over. Eventually, they build a mating platform in the usual manner of grebes.

It is easy to see how all this ceremony might have evolved. In the most ancient grebes, the mating rituals evolved out of rituals of aggression—as still evident in the smaller grebes of today. As time passed, the mating displays became more ritualized and less like naked aggression. Although Great Crested Grebes and Western Grebes are not thought to be closely related—at least they are placed in separate genera—we can assume that both evolved from a common ancestor that had already evolved an elaborate mating ritual. It seems most unlikely that the two could have evolved such similar routines independently. Of course, it could be that the first ever grebes had an elaborate mating ceremony, and that the smaller grebes of today are simply decadent and have simplified the ancestral routines. Comparable simplifications are common in evolution. Studies of fossil grebes and of DNA could clear this up—reveal whether the small grebes or the large grebes of today are the more primitive. Perhaps such studies have already been done (and I should know about them, but don't). My money, however, would be on the smaller grebes being the more primitive—still locked, like gannets, in mating displays that are hardly a whisker removed from naked aggression (but a whisker is all that is needed).

Many birds—especially the males, seeing off other males or seeking females—are fabulous musicians. As we observed in Chapter 1, 97 percent of the air that they force with such vigor through the reeds of the syrinx contributes to the sound, and they make full use of it.

SONGSTERS

European Robins sing beautifully year-round for one purpose or another, but as Edmund Spenser observed in the late sixteenth century, many are even more impressive:

> *The merry Larke hir mattins sings aloft,*
> *The Thrush replyes, the Mauis descant playes,*
> *The ouzel shrills, the Ruddock warbles soft,*
> *So goodly all agree with sweet content,*
> *To this dayes merriment.*

The "Thrush" I think must be the Mistle Thrush; the "Mauis," or mavis, is the Song Thrush; the "ouzel" is the Blackbird; and the softly warbling "Ruddock" is of course the European Robin. The most famous songster of all, though not necessarily the most accomplished, is the Nightingale, one of which generously sang for Vera Lynn in Berkeley Square. Or so she tells us. After all, Dame Vera was a mere Londoner, and as England's peasant poet John Clare had observed a century earlier, "Your Londoners fancy every bird they hear after sunset a Nightingale . . . when I was last there in the fields of Shacklwell we saw a gentleman and lady listning very attentive by the side of a shrubbery and . . . heard them lavishing praises on the beautiful song of the nightingale which happened to be a thrush but it did for them such is the ignorance of nature in large Citys that are nothing less then over grown prisons that shut out the world and all its beautys."

Besides, Clare added, "the Nightingale sung as common by day as night." Nightingales in due season fill half of Europe with their song, but they are rarely seen partly because they are small and pale brown, but mostly because they are wont to address the world from the heart of some favored shrub. The only Nightingale I have ever been privileged to watch at length was in Spain, in full song on a low branch one lunchtime.

We should distinguish songs from calls. Calls are very specific—like the words of human languages. Each species has its own vocabulary, with which to express particular ideas—not deep ideas, about the general state of the world, but particular notions of immediate importance in the manner of exclamations or indeed of expletives: "I am here!" "Clear off!" "Run for cover!" "Where's my lunch?" and so on. The information is terse—telegraphic indeed—but, like a telegram, it is of huge import. It makes the difference between life and death, and in a minimalist way it provides the signposts needed to give shape to a life and to keep us all in tune with day-to-day requirements.

Birds differ markedly in the size of their vocabulary. Fowl are among the leaders, with up to twenty different calls, each with a specific meaning. In the further interests of economy and efficiency (or so it seems), various groups of birds have certain calls in common. Thus in Europe, the Reed Bunting, Blackbird, Great Tit, and Chaffinch all make the same *seee* call when a hawk flies overhead. Why not? Hawks are a threat to all of them, and we can see how natural selection would favor some commonality of language. All look out for hawks, and all benefit from the alertness of the others. So it is that natural selection sometimes favors creatures that make the same kind of sound as all the others around them. In other contexts, however, natural selection favors distinctiveness. The sounds that baby birds make to attract their parents tend to differ significantly not only from species to species but also from individual to individual within species. Hence, each parent Emperor Penguin, returning to its growing chick on the thick ice after weeks at sea, can recognize its own particular baby by its own particular call.

It used to be argued, as a matter of dogma, that the basic calls—warning cries and so on—simply did not vary within a species, that they were genetically endowed, and that all birds of a given species had the same basic vocabulary that they pronounced in the same way. But this turns out to be less true than was supposed. Different populations of the same species may call in slightly different ways. They have different dialects.

Most birds have their vocabulary of calls, but only a minority have the gift of true song. Most of the kinds who do sing are specifically known as "songbirds," or more formally as "oscines"—a subgroup of the passerines that includes around fifty families. The suboscines, including tyrant flycatchers, pittas, and so on, can often sing, too, but the oscines are the supreme exponents. Calls and songs are complementary; they are not an either-or. Birds with true song also have their vocabulary of calls.

Songs are far more elaborate than calls. Sometimes they include scores or even hundreds of repeated elements, or "strophes"; the Brown Thrasher is the recordholder, with 2,000. Songs are typically open-ended, lending themselves to great variation from individual to individual, and in some cases added to year after year.

Songs, in fairly sharp contrast to calls, are not intended to contain highly specific information. The different strophes do not correspond to human words. The Nightingale, in its four- or five-hour peroration, is not laying down the rules for a better life in the manner of Martin Luther or Fidel Castro. The information in a bird's song is of a more general kind. If the song is long and highly various, this shows that the singer is mature and has learned from his experiences—and is in good health, for singing is exhausting and only the fit can keep it up. The singer needs others of its kind to know this—potential rivals, and potential mates. Yet some specific information is embedded.

In general, songs are high-pitched, meaning that the sound is high-frequency; and high-frequency sound travels well through unobstructed air, but is quickly enfeebled (attenuated) in dense forest. Sounds of low frequency travel very long distances, even through dense vegetation. Indeed, as whales and some fish show, low-frequency sounds travel very well through water—and elephants and rhinos communicate in infra-sound, which travels through the ground like the vibrations of a big bass drum. Calls are often high-pitched when intended for close-range contact, but birds that aspire to call over very long distances, or to make their presence felt in forests and reeds, commonly call in the range of the baritone or bass—like the mournful boom of the bittern or the Capercaillie and some of the big ratites, which travel literally for miles, and the monotonous, endlessly repeated tenor notes of America's Mourning Dove or of Australia's bellbirds. Thus, the call or the song tells the listener whether the caller or singer is nearby or far away: the greater the distance, the greater the loss of the high frequencies, and the more the overall sound is biased toward the low registers. For birds staking territory, the distance between callers or singers is crucial.

Broadly speaking, each songbird begins life with the template of the song that is characteristic of its species and is embedded in its genes. The genetic element is evident in domestic canaries: they can be bred to sing better than their wild ancestors, just as they are bred with brighter feathers. But, although the proportions vary from species to species, much or most of the song has to be learned—largely, in practice, from the baby bird's own father. If songbirds are raised in captivity, away from their own kind, then many songbirds (and probably most) can and do learn and sometimes adopt the songs of other species within earshot. I am told that even European Robins, which in the eighteenth century were sometimes raised in captivity as songbirds, could learn the songs of others, such as Linnets. When songbirds are raised in isolation, with no one to listen to, they tend to produce at least part of the general themes and rhythms typical of their species, but they never quite get it right. Thus in the 1950s, the Cambridge animal behaviorist W. H. Thorpe showed that Chaffinches raised apart from their own kind could never sing a complete Chaffinch song—they always left a bit out. In Chaffinches, as in most but not all songbirds, most of the learning is done within the first few months. After that, the sensitive period is gone forever. The same is surely true of humans. All of us learn effortlessly to speak impeccable Bostonian or Geordie or Mandarin within the first four years of life, but very few people indeed can ever learn another language in later years without at least some trace of accent and of general foreignness. By the same token, some music teachers claim that all children can learn perfect pitch—they can hear that a random car horn is an E-flat, for instance—if they are taught before the age of two. I have never seen this confirmed, but it is certainly plausible; if it is true, then it's a great pity we aren't all trained as toddlers. We might all be Ella Fitzgerald.

Because so much of the song is learned, there is plenty of opportunity for dialects to arise in different populations of the same species. I once stood in a field in what was then East Germany, with an extremely accomplished local ornithologist, who pointed out that the song of the Nightingale in the woods to our left was ever so slightly different from that of the Nightingale in the woods to our right. We were standing at the boundary of two dialect zones.

Songbirds in captivity can and do learn the songs of other species, and even adopt them as their own—and many also do this in the wild. Some are reluctant mimics; Chaffinches, for example, may learn bits of other species' songs, but they soon lose interest and revert more or less exclusively to their own. But other birds are committed to mimicry and for some it is a supreme virtue.

For Europe's Marsh Warbler, mimicry is all. It has been known for many years that about half of the song of the Marsh Warbler, sounding forth from thickets and wet meadows, is made up of elements from the songs and calls of other local birds. In one study in Belgium, blackbirds,

House Sparrows, and Tree Sparrows were the most often copied, with hints of linnet, skylark, stonechat, and magpie. In other regions you might catch references in the Marsh Warbler's song to Greater Whitethroat, Barn Swallow, or Great Tit. But the other half of the Marsh Warbler's song, so most birders agreed, sounded like nothing that was to be heard locally. This half, they concluded, was peculiar to and characteristic of the Marsh Warbler itself: the intrinsic bit, imprinted on its genes.

But then came Françoise Dowsett-Lemaire of Liège, Belgium, with her doctoral thesis on the Marsh Warbler. Marsh Warblers spend their summers in northern Europe, but like most warblers they are migratory and spend their winters in Africa. Perhaps, she thought, the half of the song that European birders fail to recognize is not imprinted ready-made in the Marsh Warbler's genes, but is learned from the African birds it meets on its travels. So it turned out. She visited Africa, stopping off along the Marsh Warbler's route, and found that in the north of Africa the bird picks up snatches of the Boran Cisticola and the Vinaceous Dove, and in its final destination, in East Africa, it learns elements from the Black-eyed Bulbul, the Blue-cheeked Bee-eater, the Fork-tailed Drongo, and the bleat of the Bush Warbler. For a song to consist entirely of mimicry is most unusual. So too is the propensity of the Marsh Warbler to go on adding to its song even after the first few months of life. Einstein claimed that he never stopped thinking like a child and neither, it seems, does the Marsh Warbler.

Female Marsh Warblers can sing the same kind of song as the males, but they rarely do so. Most female birds sing less than the males. Perhaps they, too, are perfectly able to sing, but choose not to. Perhaps the where-withal is in their brains but it needs a blast of testosterone to set it free. Dr. Dowsett-Lemaire notes, too, that male Marsh Warblers from neigh-boring territories sometimes sing together in a relaxed kind of way. Whether they do this just for the pleasure of it, or with some deeper agenda, I will explore a little later. We should not dismiss out of hand the idea that birds, or any nonhuman creature, might do things purely for pleasure, or dismiss the idea that beneath the pleasure lies a deeper agenda. As for why birds imitate—well, there have been many suggestions, of which one of the most attractive is the "Beau Geste" hypothesis: that by singing the songs of lots of birds, the mimic gives the impression that his territory is crowded and that every rival bird should stay away, though this has not been critically tested. It does seem that the birds who are mimicked do not generally respond to the imitations of their own calls. They can tell the difference even if nobody else can. But then again, some country people are able to call birds down from their perches by imitating their calls, and then stuff them into the pot. Why should a bird fall for the wiles of a human mimic—and not for a bird mimic, who presumably is far more adept? In the absence of critical studies, it can be hard to know what to believe. But critical studies may take years and can be done only one species at a time, so for much of the time we have to make do with anecdotes whether we like it or not—and always will, because there is not enough time left in the universe to do all the critical studies that might in principle be done.

Most mimics mainly imitate other birds, and usually other songbirds. Some, though, can imitate a huge range of sounds with uncanny accuracy—chainsaws, tractors, dogs, people. Some of the most adroit mimics are not songbirds. Parrots are the most famous of all, so that "to parrot" means "to imitate." Vultures, too, according to Konrad Lorenz, can speak in human tones. Eerie. Among oscines, the most versatile mimics are the crows, mynahs, starlings, and male cowbirds, while America's mockingbirds seem to know no bounds. They sing beautifully—others' songs and their own. As Maudie put the matter in Harper Lee's *To Kill a Mockingbird*, "They don't do one thing but sing their hearts out"—which is why, in the novel, it is a sin to kill one. But along the way mockingbirds may imitate up to 150 other birds in their neighborhood plus frogs, human voices, and even pianos. In the good old days there was ragtime from every bush.

None, however, is more dedicated to mimicry than Australia's two species of lyrebird. The male Albert's Lyrebird finds a clearing in the wood with a loop of vine like a trapeze, and there he sits and imitates all the other birds in his vicinity—Satin Bowerbirds, rosellas, Yellow Honeyeaters, kookaburras, and anything else that takes his fancy. Albert's are long-lived—up to thirty years—and it takes them about six years to string all the elements together into the complete rendition. Thus the females can tell, just from the variety of the song, whether the singer is mature; they prefer the mature ones, for they have proved by the undeniable fact of their survival that they have survivability. The males show their stamina at the same time: they accompany their song by displaying the magnificent tail, curled into a lyre in times of excitement, and with dance, and repeat the whole performance every day throughout the winter, tireless as an end-of-pier chorus line in a wet and windswept English summer (of which, in my childhood, there was no other kind). But then, female Albert's Lyrebirds are especially choosy. They lay only one egg every two years and cannot afford to waste time on duds. They need all the information they can get. The males restore their strength in the summer, when they shed their tail plumes and focus on eating. Phew.

The Superb Lyrebird, the second of the two lyrebird species, performs his routine on a mound of debris, a meter (3¼ feet) across, which he builds himself over about a week. A pleasant anecdote from New South Wales illustrates his skill as a mimic—and the way that birds can learn from each other. So it was that in the 1930s, a farmer kept a Superb Lyrebird as a pet. The farmer played the flute. After some time, he released his bird back into the wild. Thirty years later, in a nearby wildlife reserve, the strains of a flute could be heard through the songs of the local lyrebirds—themes not to be heard in lyrebirds elsewhere. They included those

Albert's Lyrebird of the Australian bush shows its brilliance with a visual display and with an endlessly inventive stream of mimicry.

much-loved tunes from days of yore, "Mosquito Dance" and "The Keel Row." After another forty years—seventy years after the initial release—those same songs were being performed by lyrebirds 100 kilometers (60 miles) away. David Rothenberg, who tells this tale in *Why Birds Sing*, asks why birds seem to pick up some tunes and not others. Why do we, for that matter? As Noel Coward remarked in *Private Lives*, "Extraordinary how potent cheap music is."

What lies behind birdsong? Why are some birds true singers and others not? The Harvard philosopher Noam Chomsky, in his revelatory discussions of human language, spoke of a "language module": a special faculty in the brain that enabled us to speak, but was absent in chimps and all other animals. This module should not be seen simply as a physical center, a particular and discrete collection of neurons, because it is clear that different areas of the brain contribute to the overall ability. Yet some areas of the brain do make specific, necessary contributions, such that people who suffer specific brain damage—for example, through a stroke—may lose very particular components of their speech. So it is with birdsong. It makes perfect sense to speak of a "song module," which some birds possess and some do not. As with the hypothetical human speech module, we should think of the song module as a faculty rather than as a discrete area of the brain; but again, as with human speech, we can identify at least some areas of the brain that clearly contribute to that faculty and form the notional module between them.

Birdsong would be easier to come to grips with if it had a form that human beings could easily comprehend. Sonata form would be convenient, as in most of the music of Haydn or Mozart—primary theme, secondary theme, development, recapitulation (roughly speaking). But alas it does not. The form of birdsong is far looser, or so at least it seems to the human ear. Neither does birdsong follow the diatonic scale that underpins Western classical music—a series of twelve tones and semitones, mathematically fitted into an octave. Birdsong can be properly represented only as frequency against time—and it did not yield to formal study until scientists were able to present it visually, first on smoke drums and now, far more satisfactorily, by oscillograph. On the other hand, most forms of human music do not have sonata form, either—the Indian raga, for instance, that lovely discursive music that unfolds over hours or days, does not and yet it has a deep structure. Many forms of music, including Indian and Chinese, are not rooted in the diatonic scale. Sitar players are happy with quarter-tones. Many reject the simple rhythms of Western music—two, three, four, or occasionally five beats to a bar. Flamenco is based on units of eleven beats, almost impossible for the nonaficionado to get to grips with; you can see the contempt in the faces of Flamenco dancers as the tourists try to clap along.

Clearly, there is structure to birdsong, and the female thrush who listens out for the most musicianly males surely appreciates that structure, the meta-music, just as the human musician hears the unfolding harmonies and the cross-references even in the deceptively simplest songs of Schubert and the easiest passages of Mozart. But to human ears, the structure of birdsong is elusive. It is very hard for humans to hear and will surely be very difficult to pin down, even with all our modern instrumentation. It is a mistake to underestimate the things that other creatures do just because we cannot understand them. Actually, that might be the most important mistake of all.

For some years, whimsically, I have nursed the question: is there a Mozart gene? I certainly do not want to suggest that animals are mere expressions of their genes or that genes exhaustively "determine" our lives. But as the sociologist Christopher Badcock of the London School of Economics is wont to say, "Genes make brains!" And without the right genes a bird won't have the kind of brain that enables it to sing, and we wouldn't have a brain that enabled us to talk and write books like this one. Genes are in there somewhere. So—is there an identifiable gene that enables birds or people (and possibly insects) to put structure into sound and build music from it? Clearly, the structure of birdsong is different from that of a Mozart sonata, yet it might in principle be the same gene nonetheless, expressed in different ways. The question is whimsical not least because the science of genetics is a million miles from answering it. But it seems to me worth toying with.

Human language, too, at least from our Western perspective, seems word based—vocabulary based. But, as comedians are fond of saying, "It's not what you say, it's how you say it." Tone and inflection matter as much to meaning as do the words themselves (one reason it is so hard to write convincing dialogue). In some languages, such as Chinese, the meaning itself is largely contained in the inflection: a word that goes up at the end may have a different meaning from the same word that goes down at the end. This has led many to speculate in the history of human language, did words (vocabulary) come first or did inflection? Was the first human language in truth a line of music? It is remarkable how much information any of us can convey by wordless sounds—"Uh-huh," "Ooh!" "Oh-oh!" and so on. If you add the odd specific word—vocabulary—to the musical thread, then most of us could convey most of what we need to say to get through a day. It works wonderfully in Paris. Chomsky revolutionized the study of language in the 1960s by pointing out that what really matters is not vocabulary—the one-to-one correspondence between sound and meaning—but syntax: the underlying rules by which words are rearranged in an open-ended way to mean anything we like. Is syntax rooted in music? Leonard Bernstein speculated roughly along these lines, and surely he was on to something. Are birds revealing to us the origins of human language—calls on the one hand (vocabulary) and song on the other (syntax and emotional drive)? Put the two together and indeed we have "meaning."

By any reasonable definition, birds have language—not human language, but language full of meaning nonetheless. But how human can they get? Parrots and other birds can imitate our voices. Are parrots really "almost human," as their owners like to claim? Or are they achieving similar ends by different means? Or is the resemblance purely superficial? The answers to such questions are pertinent not only to birds but also to the nature of language itself, and the nature of human intelligence, and to the human claim that we are different from all other creatures, unique and special. Indeed, it throws light on the nature of intelligence. Even if we didn't like birds, it would be worth studying them just to provide insights into these matters. Since we do like birds, such cogitation is icing on the cake.

Finally, do birds enjoy singing? Was it reasonable for Shelley to write to the skylark, "Hail to thee, blithe spirit!"?

Well, the skylark may not feel particularly "blithe." It has serious issues to contend with—staking territory, finding a mate. But whatever it feels, it surely feels something. It certainly isn't an automaton. Modern creators of intelligent robots have found that robots won't work unless they have artificial "emotion" built into them. Without emotion, they lack motivation. Animals, too, prefer not to do things that cause them pain, and they positively seek out pleasures. Indeed, as often shown in the laboratory, animals of all kinds will endure a lot of discomfort in order to secure pleasure; the drive for pleasure is stronger than the aversion to pain. Natural selection requires male skylarks to establish a territory and find a mate. Those that don't bother simply die without issue. But the skylark would not do the things it needs to do unless it gained pleasure from doing so. So we might suppose that natural selection would ensure that a skylark that rises from the field and fills the valley with its song does indeed feel huge exhilaration. *Blithe* is probably not the right word. *Ecstatic* is surely closer. If their hearts weren't filled with joy they would not rise and sing and threaten and seduce, and if they didn't find mates they wouldn't be here. As is so often the case, the poets have got it right. (Although they do tend to refer to avian choristers as "she," when more often it is "he.") Do not be fooled by the soprano trill of the male canary or the alpenhorn boom of the female Emu.

While some entice by their songs, others press their suit with gifts.

BEARING GIFTS

The gaudy plumes and elaborate songs that males employ to attract their mates seem for show (though perhaps never purely for show, as we will see). But other mating ploys are more obviously practical. Again, it is tempting—and surely perfectly valid—to draw parallels with humans. Some men seek to attract with medallions and muscles (used strictly for handstands and parallel bars), while others put up shelves and cook the dinner.

Courtship feeding has elements of both. By bringing food to the female the male demonstrates both that he is capable of finding food and also his generous nature. In some species, feeding really is for courtship— the male offers food to his intended early in the relationship—but in others it begins after laying, serving both to bond the pair together and to raise her nutritional status as she incubates. In some species the male delivers his offering with a flourish like a cocktail waiter, just to show he can; and in some the female responds in like manner, so the exchange of food becomes a choreography, mutually bonding. In birds, as in people, in the simple act of presenting a seed or a battered and still-warm mouse, there are infinite nuances. In some species, either sex may feed the other. In those in which traditional roles are reversed, she feeds him.

In the United States, Blue Jays pass tidbits from beak to beak—or sometimes they just touch beaks, symbolically. The male Hen Harrier drops a bloodied vole from a great height and the female, 5 meters or so (15 feet) below, flips over and catches it in her talons in mid-flight.

Frigatebirds perform similar feats with nest materials, although, of course, since they are not raptors, throwing and catching it with their beaks.

Sometimes females trade sex for food. This has been observed in hummingbirds—and not only in the breeding season. Harrison and Harrison (1997) report one occasion in which a male kept a female away from the flowers he was carefully guarding for himself—until she offered herself for copulation. Then he let her have a few sips, at least enough to send her on her way. Female roadrunners squat in the mating position when their mates turn up with a lizard and not, usually, unless they do so—three out of four roadrunner matings involve the transmission of food. Female Ospreys incubating on the nest, reliant on their mates to bring them food, have been known to attract passing strangers if the mate is late, trading copulation for a trout.

Other male birds demonstrate their usefulness by offering real estate. Many offer territory: no male European Robin could hope to attract a partner unless he had already carved out an area to live and feed, and shown he could keep out intruders. Seabirds that nest in colonies on rocks and islands have no room for feeding territories—and of course do not need them since they all feed at sea—but they have to be able to show their prospective mates that they have a site of their own, within the colony.

Yet others offer nests ready-made. The wren of Britain offers his mate a choice of three or four nests that he has built, but if she decides she likes him, they may build yet another one to lay the eggs in. Male wrens are serial monogamists and may get through three mates in a season, so they may build up to a dozen nests. Weaver birds build fabulous nests, artfully woven, hanging from branches and safe (or relatively so) from predators, often with elaborate side entrances. Again, in some species at least, the male builds a nest before he attracts a female—and if he does not, then no female will give him a second glance.

But the male of the Southern Masked-Weaver, most widespread of the African weavers, from Tanzania to the far south, outdoes even the wren. He begins by stripping the leaves from a thin branch, preferably on a tree overhanging water—an acacia, palm, mesquite, or even the odd barbed-wire fence—all to make it harder for snakes and other predators to invade. Then he brings long leaves and stems of grass, and weaves them into a ring, attached at the top to the branch. He builds out from the ring to form a doughnut, and so on until, within about five days, he has a complete hollow sphere, sometimes somewhat elongated. For good measure, to make it even harder for predators, he might also fashion a vertical entrance tunnel below, 8 to 12 centimeters (3 to 4.5 inches) long. Then he hangs underneath and sings, hoping to attract a female. But he may not succeed the first time. Indeed, he may have to build up to five nests before any female acknowledges his architectural prowess. Despite such lengths, predators may still find a way in. Eggs and chicks may be stolen—and for the Dideric Cuckoo, the masked-weaver is the principal host. Masked-weavers nest communally, the colony sometimes spreading over several neighboring trees—a common feature of the southern African landscape.

Again, though, the act of nest building is always liable to become an end in itself and to become symbolic. This is seen most spectacularly in bowerbirds. There are nineteen species of bowerbird in Australia and New Guinea. Three are monogamous with a conventional family life, but the other three are polygynous—and attract their succession of mates by showing them around what are veritable love palaces. (Note, in passing, the enormous difference in mating strategy even within one taxonomic family.)

These palaces come in three styles. In the Satin Bowerbird of Australia and others, the male builds an avenue, a corridor between two walls. Others favor the maypole—more like a tepee, up to 2.7 meters (9 feet) tall, with an entrance. Yet others build a bower proper, where the sides are brought in at the top to form a chamber.

Whatever the style, the basic idea is the same. The male bowerbird decorates his apartment with all manner of objects, the brighter the better, with the color blue much favored. As with magpies, their capacity for theft has acquired the status of legend. One allegedly stole a glass eye.

They arrange their booty with what looks like true artistry. When native Australians first showed bowerbird bowers to European explorers, the Europeans simply did not believe that the aborigines had not made them themselves. The male bowerbirds are quite shameless in their acquisition. Often they steal from each other. Some avenue-builders, including the Satin Bowerbird, paint the walls with charcoal, berries, and saliva.

Once the abode is made, the male bowerbird calls from its threshold to the world at large—not necessarily melodiously to human ears; sometimes more of a croak—but tuned to the tastes of the females. When one finally puts in an appearance, he beckons her eagerly inside and shows her his accumulated riches, picking up each object in his beak like an aging millionaire out to dazzle a starlet. If she is properly impressed, they mate, quickly—and with that out of the way he can't wait to get her out of the house, to resume his renovating and prepare for the next visitor. The bower is sometimes referred to casually as a "nest," which implies the raising of chicks, but in truth it has nothing to do with child care. It is a bachelor pad, for the sole purpose of seduction. She, once serviced, flies off to build a proper nest, for eggs and babies, which she will raise alone.

Sometimes males help each other to find mates—sometimes wholeheartedly, it seems, but much more usually with a strong edge of competition.

MATING COOPERATIVES

Mating is an intrinsically competitive pursuit, yet sometimes, males cooperate to achieve it. There are examples among mammals—notably the gray whale, the males of which find it difficult or impossible to copulate at all without help from a male friend. The prime example among birds is the Long-tailed Manakin of Costa Rica. The males—basically black with bright blue backs, red crowns, and long tail streamers—first form a partnership. In the following year they combine forces to seduce females, but only one of the males actually mates with any of them, while the other, the junior partner, merely assists.

The males first take up their positions side by side on the courting perch and call in synchrony until a female arrives. Then male A rises up, flies backward, and hovers, while male B moves to the spot he has previously occupied. Male A then lands where male B had been, and male B rises up in his turn, hovers, and so they change places again. They continue this faster and faster, round and round like a child's windmill, up to a hundred times. The female, hardly surprisingly, grows more and more excited, flicking her wings and hopping up and down. Finally the dominant male calls time—flutters his wings and calls—and his assistant flies off. He then performs an adagio dance from one perch to the next, slowly circling her. Then he mounts. She flies off, builds a nest, lays the eggs, incubates them, and feeds the young when they emerge, all by herself. The father of her brood, with his ever-faithful sidekick, calls again and hopes to repeat the performance as many times as possible before the season ends.

The two males continue their partnership, with the same arrangement, year after year until the dominant male dies—he is generally the older one. Then the erstwhile subordinate takes over the courting site, finds himself a new junior partner, and at last enjoys the fruits of his lifetime's endeavors. Other manakins have comparable mating arrangements, albeit different in points of detail.

The males of many birds—and of many other creatures, too, including some insects such as midges and mammals such as the kob antelope—gather in mating groups known as leks to attract the females. Lekkers among birds come from at least ten families and from very diverse orders: various fowl (Galliformes), Great Bustards (Gruiformes), cock-of-the-rock (Passeriformes), Ruffs (Charadriiformes), and so on.

The relationship among the various males within any one lek is ambivalent. To an extent it is cooperative: by congregating, making a collective show and fuss, they can draw in females from miles around. But ultimately, the assembly is competitive. Males in leks are typically polygynous and each wants to inseminate as many females as possible, preferably without any of the others getting a look in. In practice, in any one lek, very few males (the dominant ones) get the majority of the matings. All the females want from the males in lekking species is their sperm. They don't expect any help. They go off by themselves when they have got what they want to make the nest and raise the young while the male tries again and again until the season is over, when he can focus once more on regaining his strength.

You might suppose, then, that no individual female should care what the other females get up to—so long as they get inseminated, their mission seems accomplished. In practice, though, at least in some species, the females try to keep other females away from the most desirable male or males. It is as if each female wants to ensure that no other female produces offspring with superior genes that could compete with her own superior offspring. She wants everyone else's offspring to be wimps. On the other hand, if two or three hens show special interest in some particular cock, then others gather round. What has he got that they should know about?

In some species, too, the males seem to spend more time trying to boss the other males in the lek than on the seduction of females, even though seduction is the whole point of the exercise. Quarrelsome though they may be, though, the males stay in the lek—and presumably, in those species that form leks, any male that decided to set up shop by himself would do badly. The whole setup is like a street market, in which the stall keepers all gain from the proximity of the others because we all love a market, and we wouldn't necessarily bother to seek out a lone trader. But once the customers are in the market, the traders try very hard to outdo each other—sometimes in a friendly fashion and sometimes decidedly not.

Many fowl form leks, including the Greater Prairie Chicken and the Sage Grouse of the United States, the Black Grouse and Capercaillie of Europe, and the peacock originally from India. The males of the Greater Prairie Chicken gather on the ancestral breeding ground forty at a time to practice their dance steps and establish the dominance hierarchy before the females arrive. In full mating dress they are most impressive, with long feathers on the head and neck that they can raise like the ears of a rabbit, golden eyebrows as bushy as a Soviet president's, and bright orange air sacs like beacons on either side of the neck. When the females finally arrive and gather around, the males droop their wings, lower the head, raise the fanned tail aloft, and drum with their feet—faster and faster, whirling round and round. Every now and again they force the air from the neck sacs and boom their three-syllable boom out across the prairie. Like song, the routine is partly innate—male prairie chickens practice a few steps when only a few hours old—but to a large extent it must be learned.

The female Greater Prairie Chickens, smaller and altogether more modest, do not on the whole look terribly impressed by all these exertions. They peck the ground in front of the males, as if they were really interested in seed. But animals in general are deceptive, especially at times like this. All the time they are sizing up the talent. Some females visit several times before making a choice, but others mate with one or several males on each visit. The dominant male shows his dominance by his position—right in the middle of the booming ground—and he attracts most of the matings. As each female approaches, he bows, she comes closer and crouches, and copulation is over in a second. Mission accomplished, she clears off to scratch a nest and raise her family. The males begin their ritual each day at dawn and keep it up for several hours. Each has plenty of time to demonstrate his prowess, and the females have plenty of opportunity to see who's who. Soon all the females are fertilized. Some lose their eggs and come back for a second go, but eventually the frenzy stops and the males go their own way for another year.

The Helmeted Guineafowl, quaint and social as they are in the non-breeding season, undergo a change of personality as the breeding season approaches. The males become more aggressive, the flock starts to break up, and then begins one of nature's more picturesque mating rituals. The males goad each other. The goaded one gives chase, but takes care not to catch up, and so the two run line ahead at breakneck speed. Others may join until we find up to eight males, in a line, pounding along in a mock race—like Greek athletes, hurtling in silhouette around a vase. The females look on. Then each female pairs with the male that takes her fancy. Once paired, the males become very aggressive indeed toward other males. Fights may ensue. In captivity, where there is no escape, the contests may end in death. So the marriage may last, well, for two or three weeks—until the female starts incubating. This she does all by herself so that the male guineafowl, like the male mammal with a pregnant "wife," finds himself at loose ends. So he goes off and "rapes" any unmated hens he can find—the term apparently appropriate, since he offers no courtship and merely imposes his strength. Then he returns to his original mate and helps her to raise the chicks—which is essential, for the chicks do not survive without his additional input. If he returns to the fold too late, they die.

Reproduction, then, is a serious business, and so sex is serious, too. But serious need not mean dour.

SEX FOR BONDING, SEX FOR FUN

Christians, on the whole, have had a lot of trouble with sex. (Other people, too, obviously, but I am most familiar with Christians.) Jesus himself had little to say on the matter—on the whole he seemed fairly laid back about it—but St. Paul certainly did. Sex, Paul declared, was designed by God for the purposes of reproduction, and that's the end of it. Any other indulgence was a misuse. Some practices he said were "unnatural," and the Roman Catholic Church in particular has picked up on this. Catholic lore condemns unnatural practices of all kinds, especially those of a sexual nature that are not expressly intended for procreation. (St. Paul did not suggest, though, that what is natural is necessarily good.) At the other end of the Christian spectrum, the Puritans tended to condemn anything at all that was enjoyable, lest it detract from our awareness of our own unworthiness, the sin into which we are all born. John Calvin was especially hot on this. In large part sex has been seen as a necessary evil. The memsahibs who built the British Empire were advised to lie back and think of England.

You may feel that such notions are long since past, but in many a guise they live on—not least, until very recent years, in biology. Sex is obviously difficult and dangerous, natural selection is a hard taskmaster, and so it seems self-evident that nonhuman animals must be single-minded in their approach to all things, including sex. Many animals (including most birds) tend to reinforce this notion by confining all sexual activity to a very well-defined season: they reject all sexual contact until and unless the time is propitious for breeding. Birds in general are geared to various environmental signals—day length is the main one in high latitudes—that cause their sex organs to dwindle except in due season. Out of season, gonads are so much baggage. Animals in general often seem fairly expressionless or, at least, their expressions can be hard to read. Many *in flagrante* look bored or positively irritated. Lions have to move pretty sharpish after copulation to avoid a formidable forearm swipe from their partner (although this doesn't stop them copulating hundreds of times over a few days, when the lioness is in season). Birds commonly copulate in a second or less, as if they can't wait to get it over. They all seem to agree with St. Paul. Sex is for procreation, and anything else is deviation, and indeed is deviant; and animals, out in the harsh wild world, have no time for deviance.

Modern natural history, however, suggests that all is not so simple. Some mammals—especially many primates, and especially our closest relatives, the chimpanzees and bonobos (previously known as pygmy chimpanzees)—are extremely sexual. Bonobos copulate at the drop of a hat, often with members of their own sex. We have observed that many birds that form pair bonds indulge in bonding ceremonies that may well include copulation. Scientists tend to be minimalist in their interpretations and so tend to see these "casual" copulations in strictly functional terms. Thus, as noted above, they suggest that male Mallards mate with their "wives" when they return to the nest to override any sperm left by other males in their absence. But it is far from foolish to interpret this in psychological terms. The copulation reinforces the bond between them. Bonding, too, is functional: without it the pair breaks up, the ducklings are often unsafe, and the whole reproductive endeavor is jeopardized. But why should copulation serve to reinforce the bond? It surely would not, if neither partner enjoyed it; if it was entirely neutral, or distasteful, copulation would surely weaken the bond or destroy it.

I am sure, too (as most modern biologists agree), that natural selection is a key player in evolution, but this does not mean that every aspect of every creature can be tidily explained in functional terms. There is plenty of leeway. Natural selection may favor a particular organ or a particular behavior, but once the organ or behavior is in place, its possessor may use it for all kinds of other purposes—including self-indulgence.

So of course birds enjoy sex. If they didn't, they wouldn't do it; and if they didn't do it, they would have no offspring. It's the same with the skylark singing. The ultimate reason for doing it is that it promotes reproduction, and so was favored by natural selection. The immediate reason for doing it is that it makes them feel good. And the feeling good was favored by natural selection, too, because it promotes the thing—the singing—that is necessary.

As I will argue again in the final chapters, birds are not like us, and we do them no favors if we suppose they are. But they are more like us than they are like robots. Like us, they are driven by emotion. If they were not so driven, they would simply fade away.

SEXUAL SELECTION

Darwin always knew that his theory of evolution by means of natural selection left loose ends. For one thing, at least in its rawest, crudest form, "survival of the fittest" seemed to suggest that all creatures would be more or less obliged by nature simply to knock any other creatures aside that were weaker than themselves. Some politicians and tycoons continue to suppose that this is what natural selection does imply, and use what they imagine to be "Darwinism" as an excuse to be vicious. But Darwin, who was a liberal gentleman and far from vicious, also knew that nature is not completely ruthless. He acknowledged full well that many animals and plants cooperate with others of their own kind and of other kinds; and that at least in their dealings with their own kind, they often behave "altruistically"—helping others even at expense to themselves—even, indeed, at risk to their own lives. But he found such altruism hard to fit into the simple, crude version of natural selection.

He was also worried by nature's apparent fripperies. Natural selection has a hard-nosed, hard-bitten air to it. It is about survival and how many offspring a creature leaves. Life in the wild seems unremittingly harsh— "It is a jungle out there," as tycoons say (although Darwin didn't). It seemed to follow that nature should shape its creatures as a trainer hones his or her athletes—to be hard, swift, and sharp. Yet we are surrounded by animals of all kinds that are apparently content just to strut their stuff, gratuitously gaudy, weighted down with ornament. Birds are the chief offenders—quite ludicrously beautiful and sometimes, it seems, bound to suffer as a result. Darwin was a naturalist above all and he delighted in life's caprices. Yet, he said (as recorded by his son, Francis), "The sight of a feather in a peacock's tail makes me sick!"

Of course, to some extent, the qualities that help males to attract mates are the same as those that help them in everyday life: the gander woos the goose with the same show of strength and voice and general swagger that he might also deploy to repulse the fox. But the peacock's tail? The bright and often iridescent hues of parrots and cardinals and drakes? Who do they frighten? By declaring their owners to be popinjays they almost invite attack. For the natural selectionist, they are indeed a puzzle.

So in 1871, twelve years after *Origin of Species*, Darwin wrote *The Descent of Man and Selection in Relation to Sex*, a book with two quite different subjects, somewhat perversely bound into the same volume. Part II floated the idea that in addition to natural selection, running parallel with it and often in opposition to it, was a second shaping force, which he called "sexual selection." Natural selection, as he saw it, focused on the simple business of staying alive, day by day—a healthy mind in a healthy body and the ability to fight your corner. Sexual selection was about attracting mates. A peacock's tail does indeed get in his way and must make it harder to escape from leopards. But the hens like it, and other things being equal, the cocks that can attract the most hens leave the most offspring. So the cocks dress up, and the hens choose.

Expressed like this, sexual selection seems a very straightforward idea. Of course, animals need to attract mates if they are to leave offspring. So, of course, it pays them to smarten up. Of course, a female must choose carefully, and if there is a choice between scruffy versus smart, then, of course, she would go for smart. But the peacock and the mating egret and the Sage Grouse and the birds-of-paradise aren't simply smart. They are dandies. Their mating outfit is their obsession. It weighs them down, and their sons, who inherit their genes, must be weighed down too. So, of course, it makes sense for a hen to choose a cock that looks in fighting trim. But why choose one decked out to the point of absurdity, whose offspring—who will be her offspring, too—will be similarly encumbered?

Darwin proposed a disarmingly simple solution. He declared, "a great number of male animals . . . have been rendered beautiful *for beauty's sake*." He also said: "the most refined beauty may serve as a charm for the female *and for no other purpose*." For good measure: "that ornament is the sole object, I have myself but little doubt."

Somehow, one feels, this isn't really an explanation at all—more an exercise in arm waving. As we will see, his conjecture received some powerful support in the twentieth century, but at the time he put it forward it had very few takers indeed. One of the principal opponents of sexual selection was Alfred Russel Wallace, who was a prime advocate of natural selection and in the 1850s had come up with the idea of natural selection independently of Darwin.

For example, said Wallace, the gaudiness of birds—particularly male birds—can readily be explained in purely materialistic terms. Everyday metabolism produces all kinds of chemical by-products, some of which are colored. Even the insides of animals are colorful—the blood is bright red, the brain is creamy, and the liver is a fetching autumnal brown. So we might expect the outsides to be colorful, too. Gaudy, in short, is the default position. The real problem, he said, is to explain why many creatures—including many birds, especially female birds—are not brightly colored. But this, he said, is accounted for by natural selection. For instance, a male Mallard is beautifully colored in winter and spring because that is the color all Mallards would be if left to themselves. But the female is streaked in shades of somber brown because she spends long periods on the nest, incubating, and natural selection has ensured that she is properly concealed. The males have nothing much to do when the eggs are being incubated and can safely retain their bright feathers (although in fact in the slack season they molt and adopt the drab plumage known as "eclipse"). Sometimes female birds are brightly colored, too—but the females that are bright, said Wallace, almost invariably nest in holes, like parrots or puffins, so they don't need to be concealed. Overall, he argued that cryptic coloring—general dullness and specific camouflage—does have a function, and is specifically selected for. But bright colors, as are most common in males, are just a kind of metabolic side effect.

More specifically, Wallace suggested that the patterns of animals—apart from those fashioned by natural selection to provide camouflage—take the forms they do simply because the pigments follow the tracks of the underlying nerves and skin. This is an odd point for a serious biologist to make since it is so obviously untrue. If pigments were laid out along blood vessels, then we'd expect closely related animals (two kinds of ducks, say), with similar insides, to have similar external markings. But closely related animals are sometimes highly contrasting. Besides, the mechanics of coloring tell us next to nothing about the possible function of that coloring. Our bones are hard and strong because they contain calcium phosphate (the chemical explanation). But that doesn't mean that they don't also help us to stand up (the functional explanation).

Wallace did concede that conspicuous colors might have some function, but only to help animals to recognize their own species. This is vital, of course, because congress between different species generally comes to nothing and if anything is even more dangerous than usual. Birds rely heavily on vision and it pays them to be clearly marked. Closely related birds may often look very similar—but when they do, he said, they often bear small but conspicuous insignia that show what species they belong to. He offered the example of three African plovers that look virtually identical except for the stripes around their eyes and necks (although the average birder would still have problems telling them apart). This is a strong point. Recognition matters.

But however much we concede to Wallace, none of his suggestions quite explains why the peacock has such an extravagant tail or why the Sage Grouse puffs up its cheeks like beacons, or male birds-of-paradise dress themselves for burlesque and take such pains to practice their poses, or why bowerbirds build love palaces. You just can't explain such theatricals as expressions of surplus energy, or as metabolic side effects, or indeed as species recognition.

So in the end Wallace did concede that sexual selection is a force—females do indeed choose males who are in some ways especially conspicuous. But, still, he would have none of this "beauty for beauty's sake" nonsense. The fine colors and plumes of the male bird, up to and including the extraordinary train of the peacock, must denote some extra, purely utilitarian quality. Evolution is driven overwhelmingly by natural selection, and natural selection is tough-minded, and could not possibly favor feathers that were intended merely to thrill. As Wallace put the matter in 1891 (some years after Darwin's death), the "most vigorous, defiant, and mettlesome male . . . is as a rule the most brightly coloured and adorned with the finest developments of plumage." In the late twentieth century, in Wallacean vein, Bill Hamilton suggested that the male's bright colors and elaborate plumage—not to mention his strutting, dancing, and acrobatics—show that he is healthy, which means well-fed and, above all, is free of parasites. It was Hamilton who also suggested that the need to avoid parasites is the root cause of sex—genes must be mixed and stirred to create variety. So it is the unremitting arms race with our parasites that in the end has ensured that human beings are clever and peacocks have wondrous tails that rattle in Gilbert White's opinion like crossed claymores. Hamilton's is not a savory idea, but it is certainly salutary.

So what is the truth of the matter? As Helena Cronin puts the matter in *The Ant and the Peacock*—a definitive account which everyone should read—do females choose on the basis of Darwinian good taste or of Wallacean good sense? Well, for most of the twentieth century biologists took very little notice of sexual selection at all—they seemed to write it off as one of Darwin's crackpot little fancies (of which he had one or two). Dr. Cronin herself did much to bring it back into the serious scientific arena. Now it is a leading topic for evolutionary research.

Yet even in the doldrum years of sexual selection, there were some intriguing contributions—and one in particular came out firmly in favor of Darwin. In the early twentieth century, Sir Ronald (R. A.) Fisher used mathematical reasoning to show how "beauty for beauty's sake" might come about. Suppose that in general hen birds do indeed look for signs of vigor in the cocks, as Wallace proposed—which we can all agree would be sensible. A smart, stiff, well-groomed tail might be as good an index as any; a disease-ridden creature could hardly manage such a thing. So if a female has a choice of several males, she might well home in on the one with the longest and most flamboyant tail. If vigor were her target, this would be perfectly reasonable.

But birds can grow long and luxuriant tail feathers only if they have the right genes. And if a feature is gene-based, that means it is heritable. This means that the sons of the long-tailed male are likely to have long tails, too. If the long-tailed male has the most success in mating, then he would have the most sons, which means that long-tailed males would become more common with each generation. But this would raise the ante. A long-tailed male would not stand out from the crowd if all the males were long tailed. To be especially attractive, males would need especially long tails.

Then again, it takes two to tango. Long tails are dross if the female doesn't like long tails. The kudos of the male's long tail depends absolutely on a predilection, among females, for long tails. By the same token, there is nothing more deflating to the male owner of a BMW than a woman who prefers a nice sturdy bicycle, and the thighs to go with it. Psychological preferences can have a genetic basis, too. So the daughters of females who like long tails are likely to inherit their preference.

So the long tail and the preference for long tails co-evolve. They are locked together in what systems analysts call a "positive feedback loop." The tail and the preference for tails increase, generation by generation, presumably until the tails become so long that their possessors really cannot get off the ground, and could be caught even by leopards that have only three legs and have lost all their teeth. This mechanism, entirely logical (and backed by mathematics, which is always a good thing in science), is called "Fisher's runaway." We can see how in a thousand or a million generations Fisher's runaway might well produce the peacock's tail; and at least in spirit the whole process is indeed driven by Darwin's "beauty for beauty's sake." Darwin may seem to have been out with the fairies when he proposed it—but not at all. He was just more imaginative than most, and science needs imagination.

Veering more toward Wallace, and dating from the 1970s, is Amotz Zahavi's "handicap principle." He pointed out that Wallace's idea—the idea that big tails, say, might indicate underlying vigor—could not work unless the tail was costly to its possessor. Again, we might draw a parallel with humans and BMWs. BMWs are a sign of material success for the very simple reason that they are expensive. If BMWs cost only the price of a bicycle, then they would betoken very little because many people can afford bicycles. By the same token, if any scruffy old peacock could grow a train a meter long with a hundred iridescent eyes, then so what? Unless it is obvious that only the most vigorous individuals *can* grow mighty tails, then mighty tails signify nothing.

Clearly, however, the BMW is not simply a sign of wealth. It is also a very good automobile, and automobiles in general are useful. So the

BMW does not merely *symbolize* its owner's prowess. In a practical sense, it exemplifies that prowess. In the same way, the voice and the general pizazz of the gander do not merely reflect hidden depths—they are of obvious practical use, not least for seeing off foxes. Yet the peacock's tail really is just a symbol. In itself, it contributes nothing to day-to-day survival. Its flamboyance is entirely gratuitous. Yet it costs a great deal of energy to grow such a tail and maintain it—and neither Darwin nor Wallace realized just how costly such adornments are in metabolic terms; that was left to the twentieth century to find out. The point is, though, that the peacock's tail would not serve as a symbol of hidden strengths *unless* it was costly. The peacock, and the Argus Pheasant—and the male Mallard and Bullfinch—are literally squandering their wealth, as a potentate might scatter his gold, at least on feast days, just to show that he can.

An organ that is costly to its owner and yet serves no utilitarian purpose can properly be defined as a "handicap." So Zahavi's idea that signals of prowess must be costly if they are to be effective is known as "Zahavi's handicap principle." The word *handicap* has connotations of infirmity, and so it seems odd to suggest that males attract more females by handicapping themselves. But if we take *handicap* as a neutral term—meaning the possession of something gratuitously costly—then Zahavi's handicap principle makes perfect sense. Professor Alan Graffen of Oxford University, in the spirit of R. A. Fisher, has shown mathematically how it might work.

The handicap principle seems to put a quite new spin on many aspects of human behavior. A recent exhibition in London of seventeenth-century portraits from England's civil war showed various relatives and supporters of Charles I (the "Cavaliers") and their republican, Puritan opponents (the "Roundheads"). The Cavaliers were dressed in brocade and frills and lace with pom-poms on their pointed shoes while the Roundheads, in studied contrast, wore leather jerkins and crash helmets. Obviously, the Roundheads were bound to win. The Cavaliers, so the critics seemed all to agree, were clearly decadent, not to say louche.

In truth, though, the Cavaliers were just as tough, brave, and determined as the Roundheads were. Their frills were not foppish, but arrogant. They were not saying, "Look at me, for I am rich and can afford to be idle," but "Be afraid! I don't need to dress like a warrior because I can wipe the floor with you even though I can hardly move my arms in this jacket and I am wearing this ridiculous hat!" The swagger is sublime. Soldiers and modern prizefighters seem to me to follow the same pattern. Either they dress like natural-born killers, like the American Marines or the SAS, or as Mike Tyson was wont to do; or as dandies, in the Cavalier mode, like Colonel Custer of Little Bighorn fame, or among prizefighters, like America's Max Baer and Britain's Chris Ewbank. To be sure, it is hazardous to apply such notions too simplistically to human beings. But the handicap principle is more than plausible, and it certainly seems to apply to birds.

So with Fisher in Darwin's corner and Zahavi in Wallace's, it's a close call. How can we tell who is right? Are hens inspired by good taste or by good sense? Wallace knew full well that there was a deadlock, and declared that the two notions could not be pried apart "until careful experiments are made." Now, a hundred years on, some experiments have been carried out, together with observations in the field intended specifically to throw light on this question.

Take, for example, the widowbird. The males are polygynous, and each one aims to persuade as many females as possible to build their nests in his territory and raise the chicks that he has sired. The males seem to attract the females by their tails—which at 50 centimeters (20 inches) are long by any standards and relative to the bird's body size are prodigious. But is it really so? Are the tails really so seductive? To test this, in the early 1980s, Mark Anderson caught some male widowbirds and cut off their tails—a radical chop, down to 14 centimeters (5½ inches). Then he glued the 36 centimeters (15 inches) he had purloined onto the tails of other male widowbirds, so that they were half as long again as normal. He also had two sets of widowbirds as controls—one set he left unmolested and in the others he cut the tails and then glued them straight back on again, to see if the glue itself had any effects, one way or the other. This was done in the field, and he tested the mating success of his various groups by counting the number of new nests in each male's territory and the number of eggs that they contained. The males with the added tail feathers were the clear winners. Because numbers were small, the differences among the other three groups were not statistically significant. But there was no doubt that females preferred super-tails.

At first sight this is a victory for Darwin's beauty for beauty's sake and for Fisher's runaway. A long tail cannot possibly help a male widowbird in day-to-day survival. But the issue is not clear-cut. Perhaps, after all, a long tail in a widowbird simply indicates freedom from parasites, as Hamilton suggested—which would be in line with Wallace's idea. The long tail is not directly utilitarian (as the vigor of the gander is utilitarian), but it does reflect an underlying quality—a good immune system—which certainly is utilitarian. So Anderson's was a good experiment, of the kind Wallace wanted to happen, yet it does not give a conclusive result.

What of Australia's Satin Bowerbird? His palace, beautifully adorned with knickknacks, has no function at all apart from the seduction of females. Surely this reflects the Darwin–Fisher view of things. Yet this, too, is far from clear. Male bowerbirds do best when they decorate their palaces with rare objects—things not readily found in the area. This shows they can travel far and are discriminating—qualities of great value when it comes to finding food. Bowerbirds also decorate their bowers largely by stealing from others, and often stay to wreck their rivals' nests. The female visits many bowers in her district and can see for herself when this has gone on. So the males with the best bowers are not just those with the keenest aesthetic sense. They are obviously tough—the pick of the local hoods. This is supported by other studies showing that when the male birds interact directly, the ones with the most impressive bowers are also the most dominant. Of course, a bird may be an aesthete and a thug, just as a man may smile and be a villain; and in each case a female may be attracted to both. In such cases, beauty doesn't seem simply to be for beauty's sake.

There is also direct evidence for Hamilton's specific notion that brightness and vigor demonstrate freedom from parasites. When meddlesome scientists spattered the cheek pouches of male Sage Grouse with red paint, the birds did badly with the females: the red splashes resembled the lesions of lice. Similarly, male Barn Swallows in the wild that had short tails had far more blood-eating mites than those with longer tails, and males who had no mates were often more heavily parasitized than those that did have mates. You might think this was all too obvious—because no female would want to mate with a male that was simply sick. But it seems that resistance to parasites, and consequent freedom from them, is hereditary, and this is what makes it evolutionarily significant. When swallow chicks from nests (and parents) that had few parasites were placed in nests that were heavily parasitized, they still had few parasites. But when chicks with a lot of parasites were put in nests with few parasites, their parasites continued to plague them.

By now there are many more examples. Yet the picture is still unclear.

Usually, the results can be interpreted either way—beauty for beauty's sake, or beauty that signifies some extra, and useful, quality. Perhaps, most of the time, both mechanisms operate in tandem. Beauty in general surely does reflect vigor—but Fisher's runaway works, too, so the males wind up with far more decoration than one might suppose is strictly necessary.

Overall, though, Wallace showed remarkable prescience when he commented, in 1890, that "this most interesting question . . . will not be answered by the present generation of naturalists." As Helena Cronin comments, "one century on, matters are still far from settled."

A SOCIOLOGICAL FOOTNOTE

Equally intriguing, I feel, is the response to Darwin's sexual selection idea. Wallace at least took it seriously—although at first he rejected the idea altogether that the need to find mates could drive a creature's anatomy and behavior so profoundly, and was appalled to the end by Darwin's apparently non-Darwinian lapse into "beauty for beauty's sake." But most biologists, apart from an initial flurry, simply ignored the idea for the next hundred years. Why so negative?

The reasons, I suggest, have to do with the morality and the sociology of Darwin's time, and with the sociology and philosophy of science. For although scientists seek to be dispassionate, and generally think that they achieve this, in reality science is a human pursuit and human foibles will out, in scientists as in all of us.

In both England and America this past few hundred years a certain Puritan, tough-minded quasi-militarism has prevailed: a feeling that the proper state of man (as opposed to woman) is to take arms and fight the good fight against—well, whoever or whatever happens to be around. This was reflected by the chorus in Shakespeare's *Henry V* as the latest war with France got underway: "Silken dalliance in the wardrobe lies," he tells us with obvious relish. Instead, he proclaims, "Now thrive the armourers." *Dalliance* just means what in these prosaic times would be called "leisurewear," but it also has connotations of decadence. Later in the play, Henry proves totally inept as a lover, although with a certain raffish charm with which (together with the promise of a couple of kingdoms and a fair chance of not being killed) he wins over the fair Katharine. In similar vein in *Richard III*, during a little time-out in England's Wars of the Roses, the

Duke of Gloucester (the future Richard III) despises this "weak, piping time of peace." The lover, in feeble contrast to the soldier, merely "capers in a lady's chamber to the lascivious pleasing of a lute." Courtship requires him in a most degrading fashion "to strut before a wanton ambling nymph." Women come out of this badly and so too do the processes by which they may be won over. Their role is to fill in the time between wars and give birth to more soldiers.

The initial outline of natural selection seems to be in line with this tough-mindedness. So, too, do the imperialism and the straitlaced Christianity, whether High Church or Evangelical, that prevailed in Victorian times. Staying alive in the face of adversity—the enemy, the Devil, or life's vicissitudes in general—was serious and noble, while the mere attraction of a mate was at best a necessary diversion, and vaguely reprehensible. Darwin's suggestion that sex could in large measure determine the way a creature lives its life, and indeed how it is physically structured, was mildly shocking and indeed seemed frivolous—like dalliance itself. So, in general, his idea was either condemned or ignored. Biologists seemed to wish he hadn't come up with it, and to hope it would go away.

It also seemed to offend one of the cardinal guiding principles of science—that of Occam's razor. William of Occam (or Ockham) was a seriously intelligent fourteenth-century cleric from Surrey who proposed, during the tortuous theological discussions of his day, that those who truly aspired to get at the truth should not drag more ideas into the argument than were strictly necessary. Scientists in the interests of objectivity have not merely acted on Occam's idea but have extended it well beyond his intention and taken it to mean that all arguments should be as simple as possible (which is more or less true) and also that simple arguments are more likely to be true than complicated ones (which is far less true, and may not be true at all). Natural selection after 1859 was already on the table—and as far as Wallace was concerned it did the job; it could explain more or less everything. His innate, religious Puritanism biased him against the idea of sexual selection (or so I suggest)—but so, too, did his scientific Puritanism. He was reluctant to accept Darwin's idea of sexual selection before he had thoroughly exhausted the idea that natural selection itself could explain all that needed explaining. Occam's principle does apply in science, of course. But there can be no progress in science unless someone, sometime, is prepared to push the boat out and suggest something different, and even outrageous. Darwin, who was not a professional academic and was independently rich, was prepared to say what he thought—his income did not depend on academic conformity; and although some of his speculations have turned out to be wrong, most of the time he was triumphantly correct and, wild though it may seem, he was surely right in principle about sexual selection. Occam merely advised philosophers not to be gratuitous. He did not tell them to stop using their imagination.

I feel, too, that there was a deep-seated sociological objection to Darwin's idea. In its simplest form, sexual selection suggests that males display, while females choose. This seems to give females the upper hand. Darwin was a feminist, by the standards of his day—he certainly had enormous respect for his wife, Emma, and treasured his daughters at least as much as his sons, and apparently had no problem with the idea that females could and should do the choosing. But the notion of female choice went against at least one of the threads of his day. So it was that in *Northanger Abbey*, Jane Austen's hero, Mr. Tilney, tells the heroine, Catherine Morland, "I consider a country-dance an emblem of marriage . . . in both, man has the advantage of choice, woman only the power of refusal."[1] It is the man who does the choosing. Obviously. This notion is reflected, too, in the terms of the traditional Christian marriage ceremony—the bride's father "gives her away." To be sure, the novels of Jane Austen and many more from the nineteenth century protest the simplistic notion that women are commodities, or at best passive players with "only the power of refusal," as of course do modern feminists. But the notion dies hard. In much traditional society it is taken to be self-evident that men should make the decisions, even if in reality this is not the case (even in the most traditional societies). I suggest many scientists rejected the notion of sexual selection largely because it seemed to give too much responsibility to the females, though this, of course, is speculation.

Finally, there is a close relationship between the sex life of birds and their social lives; and clear connections between both and their intelligence. More of this in Chapter 9. Meantime we should look at the consequences of successful union: how they raise their families.

*As perfectly synchronized as Rogers and Astaire—the mating sequence of the
Great Crested Grebe.*

8

FAMILIES AND FRIENDS

BIRDS ON THE WHOLE SEEM FAR MORE EGALITARIAN THAN mammals: far less sexist. Ninety percent of mammals are polygynous ("many wives"), while 90 percent of birds are monogamous. On the other hand, birds are far more inclined to be bohemian. Many female birds are outstanding moms, but others are as feckless as most male mammals—happy to leave all baby care to their partners (who may not always be the fathers of the chicks they are rearing). As for male birds, some are just as chauvinistic as the most shameless of male mammals while others are model single dads. Between these extremes, there are infinite variations.

What makes the difference in the end is that female birds lay eggs, while female mammals give birth to live young that are very immature and need serious looking after—and, specifically, need looking after by their mothers. Only a few male mammals, such as wolves and many primates including ourselves, hang around to lend a hand. Most are off as soon as copulation is done—often within seconds—and are never seen again. But then, only the mother can gestate, and only she can feed the newborn infant (although male vampire bats, bizarrely, lactate up to a point). So, in most mammals, there is little useful for the male to do, and if he did hang around he would only compete for food. But most female birds lay their eggs within a few days of copulation, and from the moment the eggs are laid the males and females can share the child care—and so they often do. But sometimes, it seems, it suits them not to.

Whatever the details of their family lives, birds like all animals have to lay their eggs and raise their babies at the right time and in the right kind of place: sheltered as far as possible from the elements, safe as far as possible from predators and parasites, and within flying or swimming distance of a food supply that is as reliable as possible.

FIRST FIND A PLACE TO LAY THE EGGS

Birds, as everyone knows, usually lay their eggs in nests. The nest is in a sense an extension of the bird itself, like the web of a spider: nests evolve by natural selection, just as bodies and behavior evolve. As the Oxford zoologist Richard Dawkins put the matter, nests are an example of "the extended phenotype"—*phenotype* meaning the sum of all observable characteristics, physical and behavioral.

Some nests are very simple and some are immensely complicated—almost gratuitously so, it sometimes seems. In general we must assume that complicated nests must have evolved from simpler nests, but this does not mean that all simple nests must be primitive. Sometimes the simplest nests have probably evolved from more complicated ones—as if some ancestral bird, from a lineage of complicated nest builders, simply said to itself, "to hell with this," and started doing things more simply, and found that the easier version worked just as well. Every family has its black sheep, its backsliders, and sometimes the backsliders succeed. Over very long periods in any one lineage we might in theory find that nests evolve from simple to complicated, then back to simple again, and then become more complicated but in a different way, then simple again, and so on and so on.

Some birds make no nest at all: among those that do not are the guillemots and the Emperor Penguin. Guillemots somewhat perversely lay their single egg on a bare ledge on sheer cliffs in some of the world's wildest environments—a neatly shaped, conical egg that rolls around in a circle rather than off the edge. Emperor Penguins incubate their single egg while standing on sea ice, around Antarctica. Eggs laid on the ice would freeze within seconds, and there is nothing to build a nest out of, except more ice. So each parent balances the egg on his or her feet, protected by a fold of skin from the belly, like the pouch of a kangaroo, though the penguin's pouch is upside down. As described later, the male does the bulk of the incubating.

Most birds, though, do build some kind of a nest. The simplest kind is a "scrape"—a slight depression in the ground, usually scratched out with the feet, as in the Ostrich. Ostriches leave their scrapes bare. Other scrapers, like Emus and cassowaries, line them with vegetation. Many birds build platforms of vegetation, sometimes on the ground like many a duck, sometimes floating though lashed to the reeds, like grebes or Marsh Terns, sometimes up in trees like storks and rooks. Eiders build their nesting platforms along the beach in northern climes and line them with wonderfully soft down from their own bellies. This is good for the ducklings and good for the species as a whole because the Icelanders in particular have founded an industry on eiderdown. They don't harvest the down until after the ducks have left and so, unlike so many feather-based industries, it does the ducks no harm. Indeed, it does them good because the Icelanders take good care of them.

The archetypal bird's nest, as in most songbirds such as robins and thrushes, is cup-shaped, sometimes unlined, sometimes lined with mud, sometimes with soft vegetation, sometimes with feathers. Various Asian swiftlets build nests of saliva, stuck to the walls or the roofs of caves—the stuff of bird's nest soup. Swallows and martins build similar nests from mud, typically and often conspicuously stuck on walls, just beneath the eaves. In my favorite village in Italy, swallows, martins, and swifts share both the skies and the nesting space.

For some birds, the building of nests is their obsession, like a hobby that has taken over their lives. Weaver birds make beautiful, intricately woven, pendulous nests that hang in profusion from savannah trees like salamis in an Italian kitchen—yet, as we saw in the last chapter, the males commonly build several nests, one after the other, just to impress the females.

Among the heftiest of all birds' nests is that of the Hammerhead, better known as the Hammerkop. The nest, typically built in the fork of a tree over water (but otherwise on walls, banks, cliffs, or even on the ground) may be more than 1.5 meters (5 feet) across. It is no mere heap of sticks. It is domed, with a doorway framed with mud at the bottom, leading via a narrow passage to a hollow chamber within. Such a nest takes six weeks to build—and yet after a few months, the Hammerkop abandons it and starts again. Hammerkops may be found building their nests in any month of the year. Other creatures, mostly birds and reptiles, take over the abandoned nests. Hammerkops don't use their nest only for raising eggs, however. Typically, they stand on the top of it to copulate.

None, however, builds more obsessively than the ovenbirds of the passerine family Furnariidae. The 217 species extend among them from

The nests of the Baya Weaver are among the great architectural achievements of all animals.

Central America through all of South America out to the Falklands, but the whole family is named after just one of them that lives in Argentina and Brazil: the Rufous Hornero, *Funarius rufus*. To look at, the hornero is nothing special—just a small brown bird like a thrush—but it builds a remarkable, more or less spherical nest out of mud and dung shaped like an old-style baker's oven, which in Spanish is *horno* (while *Funarius* comes from the Latin for "furnace"; and *horno* and *furnace* are etymologically related, too). In Brazil the hornero is called "João-de-Barro," meaning "John-of-Mud." Inside the rough sphere are two chambers, protecting the three or four chicks from predators and wind (the entrance typically faces away from the prevailing wind). When building, timing is all—timing and luck. The hornero needs rain to work the mud, but if there is too much rain, it can't build at all. But when the nest is built and the mud dries, it is pure pottery and may last for several seasons—although the builders themselves use it for only one year, after which birds of many other kinds, presumably lacking the necessary energy and skills, are free to make use of them. The nests are prominent—no need to hide such fortresses away. Horneros build them in trees or on telegraph posts, electricity pylons, or indeed wherever they will.

Many ovenbirds build nests of thorns. Some spinetails (those in the genus *Synallaxis*) go one step further and adorn their thorny nests with the pellets of owls or the feces of dogs and cats. All this is presumably to repel intruders, which I imagine it would. Cordilleran Canasteros build vertical cylinders of thorny twigs, and if they can't find any, they build into cacti. Common Thornbirds are content with a great mass of thorny twigs that they first divide into two chambers, and add to in subsequent seasons. Eventually the mass of contiguous nests may be several meters across with a great many chambers, and although there is usually only one breeding pair of Thornbirds in residence at any one time, children from previous broods may occupy the spare rooms, like some European dynasty in their ancestral *Schloss*. Or then again, ovenbirds of different species, or birds from totally different families, may move into the spare rooms. The Firewood-gatherer decorates its big thorny nest with bones, metal, col- ored rags, or anything else it can find; birds of many kinds, from magpies to bowerbirds, have an eye for debris. Firewood-gatherers, like bower- birds, may be aesthetes: they may line the tunnel to the nest with bark or snakeskin, or the shells of snails and crabs.

A surprising number of birds, from kiwis to puffins to kingfishers to owls and many a songbird, nest in holes: in trees, in cacti, in banks, in cliffs, or in the ground (when holes may become burrows or tunnels). The nonbirder doesn't normally think of birds in holes—they seem entirely unsuited to such spaces—but that's the way it is. Some birds make their own holes, including most of the world's 185 species of woodpeckers. A lot of work goes into a woodpecker hole: up to 10,000 chips have been counted beneath the holes of Black Woodpeckers, requiring, one may imagine, tens of thousands of pecks. Woodpecker holes, with both sexes at work, can take ten to twenty-eight days to excavate. But many cavity nesters need their holes ready-made, and many of them rely heavily on the efforts of woodpeckers, for eventually woodpeckers abandon their precious holes and move on (though sometimes not for a decade). Sometimes, however, small woodpeckers, such as the Green Woodpecker, get pushed out long before they are ready to quit by jackdaws or starlings.

Among the hole nesters are almost a third of all the world's 165 or so species of ducks—bizarrely, it seems, since we might expect ducks to nest by the water, which indeed many of them do: Mallards by the pond edge, eiders on the seashore, and so on. Yet there are cavity nesters among the whistling ducks, shelducks, several groups of surface-feeding ducks, and some sea ducks. All of them need their holes ready-made. For some, crevices will do, or holes made by erosion or rot or termites. But most use holes made by other animals—and some are able to breed only in places where woodpeckers live and have prepared the site. Such ducks far prefer woodpecker holes because woodpeckers are astute and choose their sites well, as safe from predators as possible. The entrance to a woodpecker hole is small but the inside is big—roomy enough for a growing family of young woodpeckers, and indeed for ducklings, who leave the nest soon after hatching. The ducklings are content to jump from great heights because they are so light that they land softly enough. In similar vein, the great British biologist J. B. S. Haldane pointed out that a mouse could jump down a mineshaft and come to no harm while a man would break every bone.

So it was that in the 1830s the great Franco-American birder and artist James Audubon noted that a North American Wood Duck, alias "Carolina Duck," nested for three successive years in the abandoned hole of an Ivory-billed Woodpecker (the magnificent species that alas is probably but not certainly extinct). But the Wood Duck also nested in holes dug out by the similar and more common Pileated Woodpecker—and still does; nowadays there are no Wood Ducks in the wild except where there are Pileated Woodpeckers to provide them with somewhere to breed, although some have been conditioned to breed in nest boxes. In Japan, the Mandarin Duck, nearest relative of America's Wood Duck, breeds in the holes of Black Woodpeckers, and indeed probably relies on them. Goosanders, those sharp-billed fishing ducks, also nest in the holes of Black Woodpeckers in Japan, though they are also prepared to nest among boulders, or in a hole in the ground or in a bank.

Many sea ducks rely on woodpecker holes. In North America, almost all Buffleheads nest in the holes of the woodpeckers known as flickers. There and elsewhere, Smew, Hooded Mergansers, and Common Gold-eneyes favor the nests of large woodpeckers. In South America, Speckled Teal nest in the burrows of ground-nesting woodpeckers. Speckled Teal also nest in abandoned chambers within the communal nests of Monk Parakeets—huge, enclosed structures of sticks, built up to 20 meters (66 feet) high in trees and power lines. Because Monk Parakeets nest in colonies, the Speckled Teal finish up nesting in colonies, too, which is most unusual among ducks. Wood Ducks and Mandarin Ducks obviously could not nest in colonies because they can nest only where the wood-peckers nest—and the woodpeckers that they favor are solitary nesters. Ruddy-headed Geese and Falkland Steamerducks sometimes nest in the burrows of penguins, while the Torrent Duck of New Zealand, perpetu-ally running the gauntlet of mountain cascades in its search for small invertebrates, sometimes nests in the abandoned burrows of Sacred Kingfishers. Here is a mountain idyll: rushing water, beautiful blue Tor-rent Ducks, flashing kingfishers.

In Australia and Madagascar, where there are no woodpeckers, ducks such as the Radjah Shelduck nest in holes made by rotting fungi, ants, and termites. Chestnut Teal, Grey Teal, Sunda (Indonesian) Teal, Andaman (Sunda) Teal, and Madagascar Teal—all closely related one to another—nest in holes in the mangrove trees of the kind that fringe the oceans throughout the tropics and subtropics: holes made by termites and by fungi when branches fall off. But as the rot spreads, the tree dies, so the teal need a constant turnover of trees that are on their last legs ("postma-ture," as the foresters say) but are not yet ready to keel over. In short, they need the mangroves to be left alone, neither felled nor tidied up. The White-winged Duck nests in the forests of Southeast Asia and is more and more endangered as the forests are felled. Bizarrely, it hangs on in southeast Sumatra, where there seem to be very few trees left at all. But in those parts the foresters leave the Rengas tree—not because they like Rengas trees but because the trees produce a sap that irritates the skin and they prefer not to touch it. So the White-winged Duck continues to nest in Rengas holes.

Some ducks nest in holes made by mammals. The Cape Shelduck of southern Africa likes the big holes dug by aardvarks. The late Janet Tear, one of the world's most distinguished wildfowl specialists, suggested that shelducks would be far less common in Britain were it not for the Normans (though it might have been the Romans), who brought the rabbit to Britain, providing an abundance of burrows for shelducks to nest in.

Ducks are only one family among many, but their nesting habits illustrate some key principles: first, that nature is as nature is—we can't simply guess how it ought to be—and although it seems odd for ducks to nest in woodpecker holes and for big sea ducks to nest in rabbit burrows, this, in fact, is what they do. Indeed, who would have thought that ducks of any kind, with their great flat swimming feet, would nest in trees at all? Second, we also see the interrelationships in nature—which, again, are far from obvious. Who would have suspected, from the depths of his armchair, that Wood Ducks need woodpeckers?

Why do some birds favor small nests (or no nests at all) while others seem to dedicate their lives to their construction? In this, as in all things, it's swings and roundabouts. Big nests in general offer more protection, but they are also more visible to predators. If you are big enough, and nest high enough, then perhaps you have nothing to fear. Golden Eagles surely have little to worry about. But conspicuous nests neatly rested on thick forked branches are easy meat for cats, civets, ratels, monkeys, snakes, and hawks. One reason for laying white eggs rather than camouflaged eggs, which seem less conspicuous, it has been surmised, is that white eggs transmit more light, and so are less visible from below—but only if the nest that bears them is also flimsy.

Swing sets and highway intersections determine, too, whether pairs of birds or single birds opt to nest and raise their families as far as possible from the madding crowd or in colonies—colonies that in many species include many thousands of birds, sometimes tens of thousands, and occasionally millions.

Nesting in isolation, in your own territory, has obvious advantages. The birds stake out their own feeding ground and—once they have secured the borders—they can forage without competition. Predators have to seek out each nest individually; they may be lucky, and they may not. Many birds favor this approach, from Golden Eagles in their eyries, high on the cliffs, to skylarks on the ground.

Yet many birds, from gannets to vultures to starlings, benefit from feeding in flocks—which does not mean they would benefit from nesting in flocks, but it does mean they may have no immediate reason not to do so. For many birds, too, isolation is not an option. Most seabirds are obliged to nest in colonies because in the wild and wasteful oceans that are their habitat, occasional rocks or islands or promontories are their best or only option. Nesting seabirds are often packed in cheek by jowl—just within pecking distance. On Cape Bird in Antarctica, there may be 60,000 Adélie Penguins in due season, packed not into one continuous mass as gannets tend to be but arranged in numerous, discrete sub-colonies—though why they arrange themselves this way is uncertain. It is certain, though, that although they are packed in close, they do not get on well together. Each has to walk to the sea for food and is pecked by its neighbors all the way there and all the way back. As the New Zealand penguin biologist Lloyd Spencer Davis comments, penguins are "unsociably gregarious. They go to great lengths to seek each other out and then behave like selfish little brats."

But there are advantages in group nesting, notably protection against predators. To be sure, colonies of birds are far more conspicuous than isolated nests, and colonial seabirds that nest on northern continents are plundered by a host of mammals from weasels to bears. Their eggs and chicks are also plagued from above by Herring Gulls, skuas, and miscellaneous crows, including ravens, while the adults may be picked off by Peregrine Falcons. Bizarrely, on islands around Scotland, red deer introduced from the mainland ate the heads of tern chicks in their search for calcium, while introduced hedgehogs made short work of their eggs. On islands and island continents of the south, including New Zealand and Australia, terrestrial mammals used to be far less of a problem (there are no indigenous land mammals in New Zealand) but except on the smallest, wildest islands and rocks, all are now beleaguered by foxes, feral cats, stoats, rats, and others introduced by meddling Europeans. There have always been predatory birds aplenty in the south. The threat from above again comes from skuas; from the peculiar sheathbills, shaped like rotund gulls with chicken-like faces but seriously tough; and Giant Petrels, like uncouth and ugly albatrosses.

But there is safety in numbers. Even skuas find it hard to take chicks from the middle of a colony; they are far more inclined to pick them up from around the edges. A whole colony collectively can be more alert than a single bird, or an isolated pair. Together, too, nesting birds may mob the invader; by mobbing, small birds may give even some of the biggest birds a very rough time. I have often felt sorry for kestrels and buteos, wondering how on earth they make a living, with all the harassment around them. Terns are outstanding mobbers: beautiful elegant creatures that they

The Greater Flamingo, like all flamingos, breeds in communes, laying its eggs on towers of mud built in a swamp.

are—slim, swallow-tailed, hovering like kestrels, plunging like kingfishers—some kinds see off invaders by defecating over them, copiously and in convoy. Fieldfares, handsome relatives of thrushes, are great defecators, too. Kestrels have been known to die from such attacks, starving before they are able to unstick their feathers and take to the air again.

For added safety, seabirds often nest in mixed colonies: different birds bring different qualities to bear. On some islands of the Southern Ocean, Rockhopper Penguins nest among Erect-crested Penguins. On others, Gentoo, Chinstrap, and Adélie Penguins nest together. Terns commonly nest alongside other species of tern, or with gulls, skimmers, boobies, various auks, albatrosses, cormorants, or ducks.

Birds that breed in colonies—like mammals such as the gnus of Africa that breed in herds—commonly produce all their young more or less together. The theory is that predators are then overfaced: they can eat only a small(-ish) proportion of the eggs or chicks when there is a huge number all at once. Some birds are so committed to group breeding that they will not mate unless others—plenty of others—are doing the same. Thus, some zoos in recent years have fitted their flamingo enclosures with mirrors, to give the impression of group sex. This same tendency may have hastened the collapse of the Passenger Pigeon in North America, which numbered an estimated 3 billion in the late eighteenth century, still numbered hundreds of millions in the 1870s, yet was all but gone by 1900 and was totally extinct by 1914 (the last one died in the Cincinnati Zoo). It has been suggested that the last few breeding colonies, though perhaps with hundreds or thousands of individuals, were not big enough to stimulate breeding.

How birds look after the chicks, once born, is just as various as everything else they do by way of reproduction.

HOW TO RUN A FAMILY

The megapodes of Australia, New Guinea, parts of the Philippines, Indonesia, and other southern islands solve the problem of child care by not doing any.

Megapodes are fowl, like chickens and pheasants; some of them are known as brush-turkeys. They fly, and in the past must have flown from island to island: juveniles of the Orange-footed Megapode, which is the most widespread kind, have been known to land on ships by night (as birds of many kinds often do). Their name means "big feet," but their feet are not what they are famous for. The truly outstanding feature of the megapodes is their manner of reproduction. Like other fowl, megapodes are prolific layers: twelve to twenty eggs per season. But unlike other fowl, or any other kind of bird, they do not sit on their eggs to incubate them.

Instead, like turtles or crocodiles, they rely on various forms of external heat.

Some megapodes dig a hole in a warm sandy beach (not too deep), lay an egg, and fill the hole up again—similar to a marine turtle but not quite the same, because turtles lay up to eighty eggs per hole. Warmth in this case comes from the sun. Other megapodes use geothermal heat. So it is that 53,000 megapodes flock to a single group of volcanos in New Britain, laying their eggs not in one-time-use holes but in permanent burrows. The birds have to travel a long way to get to their breeding site. But the most famous of all the megapodes, and especially the Malleefowl of central Australia, lays its eggs in mounds of vegetation—compost heaps, custom-built for incubation.

Most megapodes are monogamous, and the pairs dig their holes and make their heaps together; then they both abandon the young before the eggs are hatched. From the outset, the babies have to fend for themselves. It all seems most peremptory. Yet no one could accuse the megapodes of idleness. Malleefowl incubation heaps are among the biggest structures that any animal of any kind builds when working alone or in pairs. (Beavers, termites, and especially corals may make bigger structures, but they work in gangs or, like corals, en masse.) The heaps are built carefully, too, from moist leaf litter and soil in a well-chosen spot; once it is built, the Malleefowl mother keeps it in good repair with daily maintenance. By some means or other she can gauge the temperature of the heap's deep interior, and maintain it to within a few degrees: 32 to 35°C (90 to 95°F) is optimum. Neither does she stint on the eggs. They are large (up to 230 grams—half a pound—each) and extremely yolky—one-half to two-thirds of the total weight of the egg—which means that the chick is well nourished and well advanced at the time of hatching. Inevitably, though, the heap is steamy and stuffy—very humid, low in oxygen, and high in carbon dioxide. Accordingly, the eggs have relatively thin shells with big pores to maximize the exchange of gas and water. Thus, the mother has evolved the very considerable gardening skills required to create a heap that does not either freeze the chicks to death or cook them alive (as an untended compost heap might do), while the chicks in parallel have evolved the respiratory wherewithal to survive within it.

But wherever they lay their eggs, on beaches or in hot earth or in compost heaps, the parents clear off before the babies hatch. The chicks must extricate themselves unaided. With might and purpose, they force the thin-shelled egg apart with legs, back, and head; the fluid then drains from their lungs, which dilate with all speed, and they must dig themselves out of whatever kind of hole their mother has left them in. Some are 30 centimeters (12 inches) underground or under compost and some are as much as 120 centimeters (4 feet) down. It may take them hours or days to get to the surface. Baby turtles have to dig themselves out, too—their mothers are long since back at sea by the time they hatch. Mother crocodiles lay their eggs in compost heaps like Malleefowl, but then they hang around, standing guard, and listen out for the plaintive peeps from inside the shell; at hatching time, they lend a helping hand—or at least a helping snout. Then the mother croc looks after her babies (and a mother croc is a good beast for a baby croc to have on its side). But, from the outset, the megapode chicks are on their own. A lot die in the first few months.

As always, there are variations. Some of the compost-heap builders—at least five species of them—may also lay their eggs in holes on beaches, and so make use of solar power. Others lay their eggs in mounds built by other megapodes. This means that they are opportunists, although they cannot truly be called "parasites," like cuckoos, since they don't expect their hosts actually to bring up their children. While most seem to be monogamous, others seem to be polygynous, male brush-turkeys build mounds that they vigorously defend while inviting females in to mate and then lay their eggs. But, then again, some females practice serial monogamy, mating with one male, laying an egg in his heap, then moving on to the next.

Is the megapode strategy merely primitive? Does it merely perpetuate some ancient reptilian practice? Surely not. Although the megapode breeding strategy resembles that of turtles and crocodiles, there are deep biological contrasts. Since megapodes are birds, the sex of the offspring depends entirely (or almost entirely) on its chromosomes: those with ZW chromosomes become female, and those with ZZ chromosomes become male. But the sex of turtles, crocs, and all other surviving reptiles depends on the temperature of incubation. If the eggs of turtles and tortoises are incubated at less than around 30°C (86°F), they turn out male. If they are kept at more than around 30°C, they become female. With lizards and alligators, it's the other way around: the cooler ones (less than about 30°C) are female and the warmer ones come out male. With Nile crocodiles, it seems, those incubated at less than 31.7°C (89°F) are female; from 31.7 to 34.5°C (94°F) degrees produces males; and if the eggs are hotter than

34.5°C, they turn out female again. In short, the biological functions of the megapodes and reptile modes of incubation seem quite different.

Megapodes, too, like all birds, are presumed to be descendants of *Archaeopteryx*, and although no one knows whether *Archaeopteryx* sat on its eggs to incubate them, there has never been reason to doubt this. Besides, it's clear from their fossil nests—which sometimes include fossil parents— that many dinosaurs also sat on their eggs, and birds are dinosaurs, too (or so it is widely accepted). Finally, as if to emphasize the point, there are three species of megapode that do incubate their eggs like normal birds. I would guess (although I confess I do not know) that these three are the primitive types. I would surmise that the geothermal and compost approaches evolved just once, from one particular megapode ancestor, and that the basic technique then evolved in various directions.

Presumably, in the past, the tactic served them well, or natural selection would have put a stop to it. But it clearly has its limitations. Megapodes have not spread far into Asia—indeed, their range seems to end where the habitats of cats, civets, foxes, and pigs begin. Now that human beings have obligingly brought cats, foxes, and pigs to the megapodes' own doorsteps—and, for good measure, pinch their eggs for themselves—megapode fortunes are diminishing. Twenty-two species are left to us. But another thirty-three species are known to have gone extinct over the past few thousand years, almost certainly human-assisted, and nine of the remaining twenty-two are thought to be Endangered or Vulnerable. Yet again, conservation is urgent.

All other birds have some kind of family life. Most are monogamous and some remain faithful throughout their lives, including some species of pigeon, most terns most of the time, and at least some geese like the Greylag and some swans like the Bewick's (or Tundra) Swan. Others are serially monogamous: one mate at a time, but with a new mate for each season, or indeed for every new attempt at breeding—which may mean several mates in a season. Many seem monogamous, but are routinely unfaithful, so that any one nest is liable to contain eggs or chicks that belong only to one of the ostensible parents, and sometimes to neither—but one or both of the parents will also have other offspring, who are being raised in other nests. As we saw in the last chapter, Cliff Swallows offer an extreme example: they are monogamous economically (both partners help to run the household) but not genetically (either partner may have offspring here, there, and everywhere).

Male Pied Flycatchers (which are Old World flycatchers) have their own approach. Some individuals are monogamous, but some might be called serial bigamists. Unlike most flycatchers, which build small cup-shaped nests in the forks of branches of trees, the Pied Flycatcher and a few others nests in tree holes. The male arrives at the breeding territory early in the season, bags a suitable nest hole, then seeks to attract a female—all of which, so far, is "normal" enough. But once the female is established in the home, then some males—in one study, around 15 percent of them—fly off and try to set up a second home. They don't fly far: anything from a few hundred yards to around two miles. A few may even set up a third home. None of the "wives" apparently knows of the others' existence.

With two or even more separate nests established, different bigamous males seem to pursue different strategies. Some males simply abandon the second female as soon as her clutch is laid and return to help the first mate to raise their family. The secondary female must cope alone—but has little chance of raising more than a few of her babies. Some males, however, help both or all three of their mates to raise their families—and at least in good years some manage to bring three entire clutches to fruition.

The male seems to do well out of this arrangement since he produces more offspring, but his prodigality has a downside. While he is away building a second family, his first partner may be impregnated by other males; when he does return to her to help her raise the chicks, he may find that some of them are not his. On the other hand, strictly monogamous males can be deceived, too.

At least at first sight, female Pied Flycatchers do not seem to benefit from their partners' excessive energy. The second and third "wives" often do badly. It may be that the males who manage to establish second and third families are far superior to the strictly monogamous ones, and it may be that a female who manages a part-share in such a supermale is better off than one who has the full-time attentions of a lesser bird. But perhaps the Pied Flycatcher is simply offering us a true example of a battle between the sexes—in which the male really does deceive his females and comes out with net benefit. We certainly cannot assume that life always works out for the best.

To be fair to the male Pied Flycatcher, he does help at least one of his "wives" to raise the babies, and the exemplary types that help both or all three of them sometimes finish up producing more than twenty viable offspring in a season—all inheriting a proportion of his apparently outstanding genes. So we can see how natural selection would favor such a strategy.

The males of a close cousin of the Pied Flycatcher, the Collared Flycatcher of eastern Europe, visits other birds' nests in the breeding season, evidently to check out how well they fare, for future reference. In one field experiment, scientists, being scientists, added more eggs to Collared Flycatcher nests in one area, and in a neighboring area they reduced the number of eggs in the various nests; the next year, most males turned up in the territory that had been given extra eggs. Once more we see that birds can be very aware of their environment; they know what they are doing, and they build their knowledge season by season throughout their lives. They can be deceived by meddlesome scientists only because they (the birds) are clever. Deception does not work on creatures that don't know what's going on in the first place.

Such awareness is even more evident—undeniable, indeed—in various birds that live habitually in regular groups, including the Black-capped Chickadees of eastern Ontario. Black-capped Chickadees live monogamously through the summer, each pair in its territory. They excavate the nesting cavity together, defend their territory together, and forage together. Yet there is some asymmetry in their relationship, for the male feeds the female and prevents other males from coming too close to her. In winter the couples break up and individuals join together to live in flocks of around ten birds, each flock with two separate dominance hierarchies—the males in one and the females in the other. Through all this time they get to know each other—personally, as individuals. Come the spring and new pairs form, the top-ranking male with the top-ranking female, and so on all the way down. Since the dominance hierarchy may stay much the same for several seasons at a stretch, so too may the pairings.

All does not run perfectly smooth, however. Females can and do mate with males other than their partners, and sometimes they declare themselves "divorced," bringing the relationship to a perfunctory end. The females do this mainly because they are appalling social climbers. Invariably (insofar as anything in nature is "invariable"), they trade in their partner for another higher up the hierarchy. In one study, scientists removed the highest-ranking females from the neighborhood—and immediately, all the females of lower rank (which was all of them) switched their allegiance to males of higher status. When the original, high-ranking females were returned, all the upstarts were kicked downstairs again. *Faute de mieux,* low-ranking females are content to mate extramaritally with higher-ranking males, thus acquiring a share of their hypothetically superior genes. When the scientists removed the lower-ranking females, their low-ranking mates remained mateless. Nobody wanted to know them.

In Black-capped Chickadees we see that wild creatures are not always the single-minded solipsists that we may suppose. They are immensely socially aware—as people are. They live all their lives in a social context, with all that that implies: competitiveness, cooperation, and above all awareness. Perhaps we see this most clearly in Arabian Babblers, from the family Timaliidae, which live in extended family groups of up to twenty-two throughout the Middle East. Typically, there is a dominant pair that does most of the breeding, as in wolves, with subordinates helping out at the nest. Neighboring groups of Arabian Babblers seem to regard each other with suspicion, and if necessary with hostility, like rival villages in some unpoliced frontierland. As is also the case with some mammals—meerkats are the most famous example—individual Arabian Babblers take up duties as sentinels, standing in conspicuous places and keeping a lookout for predators, while the rest of the flock searches for insects and other unwary invertebrates on the ground. When predators appear, Arabian Babblers are fearless mobbers—with the dominant birds leading from the front, like medieval kings in battle (or some medieval kings, at least). All in all, there seems to be a lot of selflessness in Arabian Babbler behavior—true communality.

Traditionally, these past few decades, apparent selflessness in nonhuman animals (and indeed in humans, too) has been explained in various ways (both of which suggest that the behavior is selfish after all, which for some reason makes it scientifically more respectable). The main explanation is that of "kin selection": animals will risk their own lives for their own relatives because they contain the same genes as themselves. The same applies to mobbing or to sentinel duty. These duties may seem risky, but the brave soul who dies in the execution of his or her duty (in Arabian Babblers both sexes take turns as sentinels, though the males generally do more) is thereby helping to save his or her own relatives who contain the same genes, so from a genetic point of view this behavior is still selfish.

Kin selection presumably applies also to the Arabian Babblers—the members of the flock are likely to be related to the sentinel. But there is another kind of explanation. In Chapter 7, we encountered the handicap

One individual—typically the dominant one—stands sentinel in groups of Arabian Babblers, while the others enjoy their feeding.

principle, first enunciated by Amotz Zahavi. It applies to Arabian Babblers, too, he said. For duty as a sentinel is not a form of drudgery or of penance, as it often is among soldiers who are put on sentry duty. The birds are keen to be sentinels—and it's the dominant birds that do it most. In truth they are saying to their potential rivals and mates, "Look what a fine fellow I am! Look what I can do!" The risk itself is the handicap. Others have disputed this interpretation, but for what my opinion is worth, it seems to me to be highly plausible, and it suggests once more that true sociality in animals pays; it entails some self-sacrifice and requires a high degree of awareness of who's who and what's what.

In hornbills and Emperor Penguins, very different birds in very different habitats, we see extreme cooperativeness between two monogamous partners with absolute division of labor and interdependence. Hornbills, as befits relatives of kingfishers, nest in natural holes in trees or sometimes in cliffs and banks. In almost all of the fifty-four species, when the pair has found a suitable hole, the female starts to close the entrance, building out from each side with mud. Then, when the entrance is still just big enough to pass through, she goes inside and closes the hole further, using her own feces and sometimes bits of vegetation. In some species, the male brings extra materials from outside, and in some he actually helps with the building, but in most he leaves the work to her. Soon she is walled in except for a slit, like a "loop" in the wall of a medieval castle for shooting arrows through. She can thrust her beak through the slit, and that's all. Once she is immured, she and her chicks depend entirely on her mate to bring them food.

It may seem very claustrophobic, but the vertical slit causes an upward draft and so provides good ventilation, and the female throws out all excrements to keep the inside clean and tidy (unlike some hole-nesting seabirds that build up a tremendous stink, which perhaps helps to deter predators). In most species the female takes the opportunity to molt all her flight and tail feathers while safe in the hole, and grow new ones—a convenient time to do this, but adding still more to her nutritional load. In some species, the females break out when the eggs hatch, and then may help to feed the youngsters. In others, she stays inside until the young are ready to leave. In all of nature, there are few examples of females that put their own lives (as opposed to those of their offspring) so decisively in the hands of their mates.

In many families of birds, we find at least some species that practice cooperative breeding: individuals other than the parents help to feed the young, and sometimes help to incubate. But hornbills as a family are the greatest cooperators of all: in about one-third of all species, scattered among all of the several genera, the parents are helped out by juveniles (usually from previous clutches) and sometimes by other adults (usually relatives).

Nature can be wonderfully cooperative. Female hornbills, like this Great Indian Hornbill, wall themselves into the nest hole and must then rely upon the male to forage.

The two ground-hornbills of Africa are different from the rest in their breeding as well as their anatomy, parasites, and feeding. Thus, although they nest in holes, female ground-hornbills do not wall themselves in. But the Southern Ground-Hornbill, from Tanzania and Kenya and thereabouts, does practice cooperative breeding—groups of up to fifteen helping to raise the young of the single, dominant breeding pair. In short, the wolf-like behavior that ground-hornbills bring to their hunting also applies to their breeding.

The Acorn Woodpeckers of the United States, which we met in Chapter 5, take family cooperativeness even further. Sometimes they live simply enough in pairs, but often they team up in somewhat extended families of fifteen or even more, with all who are up to the task sharing the incubation of eggs and the feeding of chicks. The adult males may mate with more than one female, the females with more than one male. There may often be two breeding females who may well be sisters and share the nest—which, since they are indeed woodpeckers, is a hole in a tree. All is not quite sweetness and light, for if either female is left on her own, she is apt to smash one or more eggs of the other, and a third of all the eggs may be lost in this way. Overall, though, the system works—if it didn't, it wouldn't happen, illustrating once more that in nature there are many, many ways to crack each particular nut, and whatever is possible, some creature or other, somewhere, is liable to try.

Emperor Penguins breed on the sea ice around the Antarctic up to 100 kilometers (60 miles) from the sea, and since their entire breeding cycle—courting, laying, incubating, guarding and feeding the young, and then recovering—takes virtually the whole year, they cannot avoid the Antarctic winter. So they do not try. They bite the bullet. The female lays her single egg as the southern winter reaches its height, in May. But incubation is left entirely to the male. She transfers it gingerly to him virtually as it is laid—if it rolled away for more than a few seconds it would surely perish—and from then on, for the next sixty-four days, two whole months, he rests it on his feet, cocooned in folds of skin from his belly. His legs, as well as the rest of him, are snug within layers of tight feathers. Once the egg is safely installed, the female clears off, back to sea to feed.

All the time he is incubating, the male cannot feed. He could not take the egg to sea, and if he left it behind on the ice, it would perish. Besides, the sea is far away—much too far for the female to ferry food to him. So from May to July—right through the teeth of winter—the males stand and starve. The average temperature throughout this time is $-25°C$ ($-13°F$). The average windspeed is 25 kilometers per hour (15 miles per hour), rising sometimes to 75 kilometers per hour (46 miles per hour). Wind chill is formidable. Penguins in general are beautifully adapted to the cold—fat, thick feathers and a wonderful system of blood vessels arranged to ensure that body heat is not carried to the surface—and Emperors are better equipped than most. Yet to survive such an onslaught of cold they also need to behave appropriately, and in practice they form "huddles" of up to 5,000 birds standing as close together as possible—about ten per square meter (10 square feet). Huddling reduces the loss of heat per bird by 25 to 50 percent. The ones on the windward side peel away and walk down the sides of the huddle to the rear. This movement is constant, so that no bird spends much time feeling the brunt of the wind;

each spends most of his time within the huddle, getting closer and closer to the front as more birds pile in behind, and the whole group shifts slowly downwind, hour by hour. The attitude of Emperor Penguins is well adapted, too. They are wonderfully nonaggressive. If they were fractious, as many birds are—if they were anything like gannets, for example—the whole system would collapse. It is an exercise in absolute cooperativeness.

When the Emperors' eggs hatch, still in midwinter, the males feed the chicks on "crop milk," as described in Chapter 5; this is a huge physiological strain on an animal that has fasted for weeks in one of the harshest environments on earth (matched only by some of the wilds of Siberia). Then, in July, the females return. Their absence is not feckless, but strategic. Then the male is free to go and feed and for the following five months—right through to the end of the calendar year—both parents help in feeding the young. Even so, the chicks grow slowly. When they fledge, in late spring (about October), they are still about 40 percent short of adult size. Even though they have only to cope with the summer, many die. The adults are able to breed again by the following May.

King Penguins are similar to the Emperors but somewhat smaller, and they live somewhat farther from the pole. They have two breeding seasons, the main one from November to December, which in those southern parts is midsummer, and another from February to March, or late summer to autumn. But their entire breeding cycle, in contrast to the Emperors, takes more than one year. So they can breed only twice every three years, and at any one time in any one colony, you will see chicks and juveniles of all ages and adults in various stages of molt. Just like a human village, really.

One last twist. At exactly the same time as Emperor Penguins are laying their eggs in Antarctica in the depths of the southern winter, various kinds of geese are laying their eggs in the Arctic summer. This on the face of things seems a great deal more sensible, but it has its downside, too. The Arctic summer is agreeable and the sun never sets—but it is short. Geese have to time their arrival precisely, raise their chicks to fledging within a few months (in stark contrast to the King and Emperor Penguins), and then make a sharp getaway, together with their newly fledged chicks. In short, toughing it out like a penguin is one way to cope with the realities of the world, and of the seasons; migration is another.

All we have looked at so far, however, are variations on a theme of monogamy. Many birds have a quite different approach to family life, but one that again demonstrates the division of labor between the parents and—at least sometimes—their cooperativeness and mutual dependency.

BOHEMIANS

At first sight, the family arrangements of Wattled Jacanas on the Chagres River in Panama seem fairly conventional, at least by mammalian standards. The jacanas divide the waters with their floating lily leaves into territories, and the boundaries of each are patrolled and protected by the avian equivalent of a doorman—big, dressed in black, with a red face and a long, yellow probing beak. Within the territory, far more peaceable, is a group of smaller jacanas, looking after the eggs and the children that hatch from them. Here, you might suppose, is the classic model of the warrior chief with "wives": he provides the protection, and they take care of the domestic arrangements.

But in Wattled Jacanas, the protectors are the females, and the smaller, long-suffering toilers in the middle are the males. Furthermore, whereas the eggs and chicks that each male takes such care of all belong to the boss female, DNA tests reveal that they do not all belong to himself. Male Wattled Jacanas often find themselves taking care of other males' children.

The family arrangements of the Wattled Jacana were first described formally in the *Proceedings of the Royal Society of London* in 1998, and one of the scientists later commented: "It's about as bad as it can be for these guys." Overall, as Joan Roughgarden points out in *Evolution's Rainbow*, "The investigators, themselves male, were outraged." But she takes a feminist view and, she says, had the arrangement been the other way around—a protective male with a harem of compliant females—this "probably wouldn't have provoked such outrage."

In truth, too, says Professor Roughgarden, the male Wattled Jacanas do not do so badly out of the arrangement as it might seem. The males who look after the children of other males are looking after their own children at the same time, while enjoying the protection of the big, bullying female. By the same token, some children of any one male are being looked after by other males, so in the end it is even-steven. The arrangement may look odd through conventional human eyes, but it works; and when you analyze it, you can see why it should.

In such cases, biologists are wont to speak of "role reversal." Such a term is sociologically laden: it implies that if the female is protecting, while the male is the primary carer, then some rule of nature is being flouted. But the males are the principal child carers in many kinds of bird, from several quite unrelated orders, so this is natural, too. We cannot reasonably say that roles are being reversed unless the roles are standardized—which in truth they are not.

Jacanas belong to the great order of the shorebirds and waders (the Charadriiformes), and in other shorebirds, too, the conventional parental roles are reversed. Among them are the phalaropes. In the Red-necked Phalarope, the females are 10 percent larger than the males. In the Grey (or Red), the difference is 20 percent. The females of Wilson's Phalarope are 35 percent larger than the males. The females use their bulk like female jacanas, to defend the breeding territory. The females have the brighter plumage, too, and display their finery as they compete for mates. The males are more cryptically colored because, like the females of many other birds from many ducks to many songbirds, they are the stay-at-homes who do all of the incubating.

Grey Phalaropes spend their winters on the open oceans, but breed on temporary ponds inland, in the high north. The breeding season is extremely short. The females arrive at the breeding site in June and find a mate within a few days—although some arrive already paired. In general, the female decides where she will lay her eggs, but the male makes the nest, a scrape beneath a tussock near to water. She lays, and from then on all the care is left to him. When the clutch is complete (four eggs), the females of Red-necked and Grey Phalaropes look round for any other males who may still be spare and mate with them. Even though the female does no incubating, she does remain at the colony, keeping a weather eye, and will lay new eggs if any of her clutches are destroyed. For his part, the male rarely leaves the nest at all. When the chicks hatch, he broods them for a few days (meaning that he keeps them warm and secure beneath his body) and takes them to water like a mother duck. But at fourteen to twenty days, when they are fledging or almost so, he abandons them. The whole exercise, from the time of mating to the complete independence of the chicks, takes forty-eight days. This is in startling contrast to the year and a bit that frigatebirds or King Penguins need to raise their young. After breeding, phalaropes gather in great flocks before migrating south.

Emus are, of course, quite unrelated to phalaropes, but their breeding is in many ways similar. In Emus, as in phalaropes, the females are bigger. In contrast to rheas, where only the male booms, the female Emu is the one with the voice; both sexes of Emu grunt and hiss, but the female Emu is the one with the boom. The air sac in the neck of the female Emu, leading via an aperture to the windpipe, acts as a resonating chamber. In Emus, too, the males are the main child carers.

Emus start to pair up in the height of the Australian summer—December to January. Each couple then defends a territory of up to 20 square kilometers (8 square miles). In April, May, or June (autumn and winter), the female lays a clutch of between nine and twenty eggs on a platform of leaves, grass, or sticks, on the ground, preferably under a bush or tree. Then the males do all the incubating, and while they do, many at least of the females move on, sometimes to pair with other males and to lay new clutches. Sometimes a female will hang around as the male incubates, helping to defend him and her babies with her bulk and her huge booming voice, but when the eggs hatch, the male becomes very aggressive and ungratefully drives his faithful companion away. If people venture close, they, too, are attacked. No doubt after eight long weeks of sitting still, he is feeling a little irritable.

Emus on the whole are nomadic: as we saw in Chapter 5, they go where the vegetation is. Only when the male is incubating do Emus need to stay in one place, and since he neither eats, drinks, nor defecates during his long vigil, he remains independent of local supplies. Once the chicks are hatched, the male looks after them. He does not exactly lead them, like a mother duck. They seem to do their thing and he trails sheepishly behind. After five or seven months he has had enough. The chicks go their way and he goes his. But then he is ready to find another mate to begin the whole cycle again.

The Ostrich provides yet another variation on a theme of cooperative breeding. In contrast to the Emu, the male Ostrich is bigger than the female, at up to 2.4 meters (8 feet), and 115 kilograms (250 pounds). In the breeding season—typically the dry season in East Africa—the male stakes out his territory, anything up to 20 square kilometers (8 square miles); patrolling, flicking his snow-tipped wings, inflating his bright-hued neck (red, pink, or blue), shattering the landscape with his powerful, booming roar. The display that tells other males to keep their distance also brings in the females, and when a suitable one comes by, he squats down low, spreading and waving his wings alternately—wings that are small compared to his own great bulk but big in absolute terms—and she, suitably impressed, stands bowed with trembling wings. It is, it seems, the conventional scenario of macho male and submissive, quivering female.

But there convention ends. Within his territory the male Ostrich scratches out a number of shallow scrapes. Many females may be drawn into his domain, but only one becomes his partner; she, known as the "major hen," forms a pair bond with him, at least loosely. She very probably will already have a home range of her own; like a female European Robin, she is an independent person. She selects one of the scrapes that is on offer and lays up to a dozen eggs in it—not one per day, as is the most common pattern among birds, but every alternate day, so the whole laying takes about three weeks. The eggs are huge—the biggest laid by any living bird—and have shells up to 2 millimeters (a twelfth of an inch) thick. To lay a dozen of them over about three weeks is a huge physiological feat—but on the other hand, compared to the Ostrich hen's own body weight, the eggs are among the smallest laid by any bird, so perhaps it is not quite so extraordinary. Incubation does not begin straightaway, so while the later eggs are still being laid the earlier ones must be guarded.

But there is a significant twist to this story. For in addition to the major hen, a number of "minor" hens come by. Ever obliging, the male fertilizes them, too, and they, too, lay at least some eggs in the chosen nest. After each minor hen has laid her egg or eggs, she clears off and takes no further part in the proceedings—although she may also lay eggs in other nests in the area, belonging to other males. By the time the clutch is complete, it may contain up to eighty eggs. Only a dozen or so belong both to the resident male and to the major hen. Some of the rest belong to the male, but not to the major hen. Some belong to neither of the resident parents, since some at least of the minor hens are already inseminated when they arrive. Incubation begins when the clutch is deemed complete, and the chore is shared; the resident, major hen incubates by day and the male takes the longer watch, by night. The chicks hatch in about six weeks. The chicks are precocial—able to leave the nest straightaway. The male and female accompany and guard them against all comers until they are big enough to fend for themselves. The chicks from several nests usually gather in one big flock. They will be guarded by adults who are probably parents to at least some of them, but not necessarily. Despite all this care, however, only about 15 percent of the hatchlings survive their first year of life. Of those that do survive, the females are sexually mature at two years and the males may breed at three to four, and in the wild an Ostrich may live for forty years—so each Ostrich may have thirty or more opportunities to breed and the low survival rate in any one clutch is not the disaster it would be if they were short-lived.

How could such a peculiar system have evolved? Why do the major hens tolerate their mate's infidelity? Why does she help to look after eggs and chicks that are not her own? Why is the male content to look after eggs and chicks laid by minor hens who may have been impregnated elsewhere? What's in it for each of the players?

The main issue is vulnerability. Ostriches live in savannah or semidesert. The eggs, laid on the ground, are prey to all kinds of predators, including at ground-level hyenas, for whom the thick eggshells are no deterrent at all; and from above, Egyptian Vultures, which drop stones on the eggs to break them. Only about 10 percent of the Ostrich's eggs finally hatch. The major hen makes sure that her own eggs are at the middle of the clutch, so they are protected from hyenas and other terrestrial predators by the ones at the periphery, which are much more likely to be plundered. Indeed, the outside ones are sacrificed from the start because an incubating Ostrich can cover only about twenty eggs—just a quarter of those in the largest clutches. So although the major hen does finish up helping to protect at least some babies that are not her own, that is a small price to pay for the protection afforded to her own eggs. How the major hen knows which eggs are her own, so that she can keep them in the middle, goodness knows. Scientists know whose eggs are whose because they can mark them with felt-tip pens.

The male is happy with his family arrangements because he spreads his genetic options wonderfully: some of his babies are beautifully looked after by the major hen, as well as by himself, and the ones he sires by the minor hens at least have some chance of survival. The minor hens seem to get a rough deal—but if they did not mate with a dominant mate who had his own territory, they would not mate at all, or if they did, then their offspring, without territories to call their own, would be bound to perish. To be sure, the offspring of a minor hen have a very low chance of survival in any one nest, but the minor hens spread their options by laying eggs in more than one nest. They still have a fairly rough deal, but it is better than no deal at all. Most interesting, though, is the way the psychology of the birds has adapted to the harsh realities of desert life. The resident females in most essentially monogamous marriages are very jealous of usurpers.

But all creatures must know what side their bread is buttered if they are to survive; and life in the midst of hyenas requires a little give and take.

The family life of South America's Greater Rhea seems to combine elements both from the Ostrich way of doing things and from the Emu way. In spring and summer, before the breeding season gets properly under way, each dominant male Greater Rhea establishes a territory—displaying, fighting off other males, and booming his great booming call. The female is voiceless. The male is polygynous: once he has laid out his stall, the dominant male sets out to attract not just one but a harem of females, anywhere from two to a dozen. Once copulation starts, he makes a nest—a scrape with a rim of vegetation.

The hens cruise for mates, in groups. Each female may mate with several males. With each new mate she stays around long enough to lay several eggs in or near his nest; and the male uses his beak to roll them into the nest if they have not been placed there already. When a female has had enough of any one male, she drifts off to spend time with another, laying her new partner's eggs. She repeats this until she decides she has had enough for one year, whereupon she clears off to return to the serious business of feeding, taking no further part in family life.

For his part, the dominant male becomes more and more aggressive as his nest fills up until in the end—when he has anything from ten to twenty-five eggs, though sometimes as many as seventy—he drives any remaining females away with as much vigor as he would apply to other males or predators. Then he incubates for thirty-five to forty days. When the chicks hatch, they first rest up in the nest for a few hours and then the whole surviving brood sets off with their father as their guardian. At least, he will be father to most of them, but probably not to all. In their defense he is fearless. Guardian males have been known to attack airplanes. The gauchos, herding their cattle out on the pampas, commonly ride with dogs to keep the male rheas at bay. Gauchos are not afraid of rheas, for they are not afraid of anything. But they don't want their horses spooked.

But there are complications in the family lives of Greater Rheas. Not all males in any one year can be dominant males. It seems, indeed, that in any one year only about 5 percent of males get to breed. But some of the rest form partnerships with the dominant male and help him out—and those that do form partnerships generally manage a few copulations of their own. Some of these subordinates take over the incubation duties, leaving the dominant male to try to attract a new harem elsewhere. Generally the dominant male incubates this second clutch himself. Subordinates incubate just as successfully as the dominant males, but they are not so good at guarding them once they are hatched. In this, there are shades of the relationship we saw between male Long-tailed Manakins in their efforts to attract mates: one male acts as the servant to the other—but in the hope and expectation of some reward that at least is better than nothing. Just to round out the complications in the lives of the Greater Rhea, some nondominant males manage some copulations but do no incubating; and some take no part in breeding at all.

Overall, the family arrangements of Greater Rheas seem to favor the females. Although in any one year only about 5 percent of male Greater Rheas manage to raise broods, about 30 percent of females produce at least some offspring. Like Ostriches, rheas are long-lived—up to twenty years—so failures or subordinate roles in one year may be made good by successes in other years. Nature has no sense of justice—it feels no obligation to be *fair*—and some individuals in any population are bound to do better than others and some fail altogether. But it is not too difficult to see how, even among Greater Rheas, there is method in the apparent madness. Dominant males do better by eliciting the help of subordinates; subordinates do better by pandering to the dominant males than they would if they simply went it alone; those who fail to breed keep their powder dry for another year; and the females get the pick, and do nothing except lay eggs.

But also, among birds, there are ways of cooperating that have shaded by degrees into frank exploitation—indeed, into parasitism.

EGG DUMPERS AND BROOD PARASITES

One way for a male to improve his reproductive options is to impregnate more than one female—which increases the number of his offspring, their genetic variety, and the variety of environments in which his offspring are brought up. A triple bonus, in short.

Females, though, face greater problems—another example of nature's asymmetry. Like the males, they can, of course, increase the genetic variety of their offspring by judicious adultery. But adultery does not increase the total number of a female's offspring, since any one male would be perfectly able to fertilize all the eggs that she can produce. She does not need more than one male for this. Besides, the mere ability to lay eggs is not what counts. Most birds, and perhaps all, could lay more eggs than they actually do—and some can lay many times more. What limits output is the number of chicks they can successfully raise, whether they do this alone or with a partner. Child rearing takes a greater toll in energy than mere laying—just as lactation in mammals takes a greater long-term toll than gestation, despite appearances. Fowl, like chickens and pheasants, increase the number of chicks they can raise by encouraging them to for-age for themselves—the mother provides protection but the offspring find their own seeds and worms (while the male struts his stuff some-where else or feeds himself up for the next bout of male rivalry). So it is that fowl can lay up to twenty eggs with some hope of raising a reasonable proportion of them, while most birds are ill-advised to lay more than about four, and many birds of prey that lay a meager two expect the second one to die and finish up raising only one. Jacanas have solved all the problems up to a point—upping the total output of offspring, their genetic variety, and the variety of upbringing—by recruiting several "husbands." Essentially, female jacanas have taken on the conventional traditional "gender role" of the male. But, still, a female jacana cannot produce as many offspring as, in theory, a polygynous male bird might.

Yet there is another, more accessible way for females both to increase their output of offspring and to increase the variety of environments in which those offspring are brought up. Lay an egg, or several, or lots, in somebody else's nest, and let the new host, or hosts, bring up the babies. The act of laying eggs in the nests of others is called "egg dumping." The portentous term for the general infiltration of offspring into the families of others is "brood parasitism."

A surprising number of female birds from many different orders dump eggs on their "conspecifics"—other members of their own species. Since eggs may be dumped in just a few seconds, and there are twenty-four hours in a day, it is hard to catch birds in the act. But with DNA fingerprinting to match the eggs in a nest against their carers, dumping seems to be emerging as a routine tactic. Indeed, DNA tests often reveal that some of the eggs in any one nest belong to neither of the ostensible parents—and egg dumping is the only explanation. Several species of ducks and geese have long been known to dump eggs on conspecifics. So, too, do moorhens, House Wrens, and House Sparrows. As we have al-ready seen, some female Cliff Swallows may race in to a neighbor's nest to lay an egg—sometimes before the very eyes of an incumbent male—or else dump eggs or chicks that they have prepared earlier. A third or even a half the nests of starlings may contain eggs dumped by others, and the same goes for Purple Martins.

It would be easy, for those who like to draw moral lessons from nature, to write all this down to fecklessness, chicanery, or at least absent-mindedness. Since in some cases (as sometimes in Cliff Swallows) the dumper may throw out one of the foster parents' own eggs to keep the numbers even, it can seem positively villainous. But most of the birds that lay eggs away from home also raise their own family of chicks in their own nests—and they are liable to be dumped upon in their turn. If this happens, then the arrangement emerges as an exercise in cooperativeness—far from reprehensible and indeed commendable. Any one nest is liable to be trashed by predators or by storms. Any one brood is liable to be lost. But if a female (and her partner) can salt away a few eggs among their conspecifics elsewhere, then their genetic legacy may still be passed on. Indeed, the phenomenon is common enough in human societies. To be sure, the dumpers in traditional human societies generally asked permission first. On the other hand (as in *Jane Eyre*), the fostered children often seem to be treated less well than the foster parents' own offspring. Among birds, so far as is reported, there is no such discrimination. Ostriches clearly do know whose eggs are whose within their vast communal nests and are happy to sacrifice those with lesser credentials. But Ostriches are professional communists, with highly evolved behavior that is refined to the task. For most birds, dumping remains a sideline, albeit an important one, and most foster parents are mere amateurs and are not so discriminating.

Some birds, though, lay some or all of their eggs in the nests of others of different species. Here the expression "brood parasitism" seems to be more apt. Cuckoos are the best-known practitioners: "cuckoo in the nest" is common parlance. But several other groups of birds from various orders are adepts too, including the honeyguides of Africa and Asia, the widow-birds of Africa, and various species of cowbird from the New World.

There are many kinds of cuckoo—140 species in the family Cuculi-dae—of whom fifty-seven are full-time brood parasites. (Some writers say "only fifty-seven are brood parasites" and others say "as many as fifty-seven are brood parasites." I leave you to judge.) The nonparasitic cuck-oos mostly live unexceptional family lives. Some of the parasitic types resemble—that is, they mimic—the birds they intend to parasitize, which seems very sensible: presumably it enables them to infiltrate. Others, however, resemble hawks. The so-called European Cuckoo (*Cuculus canorus*), mimics the Eurasian Sparrowhawk, while the closely related Indian Hawk-Cuckoo (*C. varius*) resembles the Shikra Sparrowhawk, and Large Hawk-Cuckoo (*C. sparveroides*), also from India, mimics the Besra. Such mimicry may seem strange, since cuckoos who look like hawks are mobbed by small birds, including the ones whose nests they seek to commandeer. They would surely find life easier, one might suppose, if they looked more innocuous. On the other hand, cuckoos decked out as hawks might well create panic as they fly around the woods and reedbeds, and so may cause their intended hosts to leave their nests unguarded. I know of no research to test this (which doesn't mean there isn't any), but something like this must surely be the case or natural selection could not have favored such meticulously hawkish garb in so many species.

Whatever the truth of this, cuckoos are very good at what they do. The female European Cuckoo first stakes out a territory, and then keeps close watch on the local songbirds. European Cuckoos between them commonly parasitize Dunnocks, European Robins, reed-warblers, Meadow Pipits, Tree Pipits, redstarts, and Pied Wagtails. Over their whole range—which spreads east into India—European Cuckoos are known to parasitize more than 100 different hosts.

But herein lies an obvious problem: unless the cuckoo's egg bears a reasonable resemblance to those of its host, the host will throw it out. The intruded egg is bound to be bigger than the host's because European Cuckoos are sparrowhawk size and their hosts in general are robin size, but it should be very similar in color and pattern. But different host species lay different colored eggs, and an egg that will fool one host will not necessarily fool another. So female European Cuckoos have to specialize—different individuals homing in on different hosts. Some specialize in Reed Warblers, and lay Reed Warbler–like eggs, while redstart specialists lay redstart-like eggs, and so on.

Yet this leads to more problems. First, different cuckoos must lay eggs of many different colors and patterns. In principle this is not too difficult. It is just another exercise in polymorphism, in which different individuals of animals from the same species may look or behave in one of several different ways. As we have seen, there are many examples among birds—the plumage of male Ruffs is highly various, for example. But also, each

Reed-warblers are among many small perching birds that seem happy to lavish their care on the monstrous young of cuckoos.

mother cuckoo has to make sure she lays her own particular eggs in a nest in which it would pass muster—and this is conceptually more difficult. Yet this is clearly not insuperable, either; it requires, simply, that each female cuckoo gear her reproductive strategy very specifically to one particular species of host, without too much deviation. In general, female cuckoos simply lay their eggs in the nest of the species that brought them up, so the tradition is passed from generation to generation.

But this raises yet another problem. If a female cuckoo focuses exclusively on Reed Warblers, and her daughters also focus only on Reed Warblers while another lineage of cuckoos focuses exclusively on Pied Wagtails, with no crossing of boundaries, we should finish up with two different species of cuckoo—or indeed, with as many species of cuckoo as there are host species to parasitize. How come this doesn't happen? The answer lies with the male cuckoos. The males are promiscuous, and as far as they are concerned a female cuckoo is a cuckoo is a cuckoo, whether a Reed Warbler specialist or a Pied Wagtail specialist; so the daughters of the cuckoo that specializes in Reed Warblers may well be the half-sisters of the daughters of the cuckoo that specializes in Pied Wagtails. This means that the populations remain functionally separate and yet, genetically, are all interlinked, and therefore all remain as one species. So that solves that.

Except, of course, that if the daughters of the Reed Warbler specialist share genes with the daughters of the Pied Wagtail specialist, then they should finish up laying the same color eggs. But there is a way around that, too. The genes that determine egg color in cuckoos are sex-linked: they are inherited only through the maternal line. So a daughter will lay the same color eggs as her mother, whoever the father is. A clever piece of research in the 1960s showed that the gene that determines the egg color of cuckoo eggs is to be found on the W chromosome. Only the females have a W chromosome—the male has two Zs. So all the theoretical problems are overcome. But it must have taken an awful lot of evolution to reach this point.

Anyway, the ever-watchful female cuckoo sooner or later sees her chance, preferably a nest that already contains a few eggs and is unguarded. She flies down, warily. If no one is about, she removes one or two eggs from the host's clutch, and lays one of her own. The whole operation takes only about fifteen seconds. Then, to compensate for all her hard work and nervous tension, the female of some species of cuckoo eats the egg (or eggs) that she has removed. Some female cuckoos lay up to twenty-five eggs in a season, each in a different nest—even more than a pheasant does. The average is around nine.

The eggs of brood parasites hatch extremely rapidly—around twelve days in European Cuckoos. This is a key adaptation. The European Cuckoo babies hatch before all or most of the host brood, even if they are laid later. Then begins one of the most extraordinary, and most grisly, demonstrations of behavioral adaptation in all of nature. Even before its eyes are open, the baby cuckoo starts to shove all the other eggs in the nest over the side. If some have already hatched, it shoves the chicks out. If a meddlesome scientist places some other object in the nest, that too is thrown out—everything, until only the cuckoo chick is left. The chick does all this by bending its head forward; spreading its stubby, naked wings, resting the host egg or chick against its back; and pushing down with its legs. One by one, out they all go. This behavior was first reported in 1788 by the English physician Edward Jenner, who, when he wasn't watching birds, also invented vaccination. In some species of cuckoo the chicks do not throw out the host's children. The Great Spotted Cuckoo, the Channel-billed Cuckoo, and the Asian Koel are among those that leave the host babies (often crows) in situ. But the chicks of these species snaffle most of the food, so the host chicks die anyway, or else the growing cuckoo tramples them to death.

So whether or not the cuckoo chick throws out the host offspring, it receives all the food that would otherwise have fed the entire brood. It grows rapidly and may soon dwarf the parents, who continue so conscientiously to feed it. Birds can seem remarkably aware and alert, able to take an overview of life—as indeed the female European Cuckoo does as she sizes up the potential hosts in her territory. But at times—quite often, actually—birds seem to behave like robots, living their lives according to reflexes, or hard-wired responses, that do not vary whatever the circumstances. So it is that gannets are hard-wired to repair their nests meticulously, tenderly replacing every twig that falls from it, yet the same gannet will watch one of its eggs roll away and be eaten before its eyes by Herring Gulls, without apparently registering the loss of the very thing that is the point of its whole breeding endeavor. Similarly, a male frigatebird at the nest might stand complaisantly by while some other frigatebird male slaughters his own chick, again before his very eyes and within pecking distance. In the same kind of way, or so it seems, Reed Warblers and Dunnocks are hard-wired to thrust caterpillars into the open beaks (the "gape") of their chicks. The baby cuckoo presents them with a beak that is even more gaping than that of their own babies, and instead of perceiving it to be a monster, they feed it more vigorously still. After the young cuckoo has fledged and left the nest, it continues to beg—and such is the allure of the ever-open beak that even passing strangers, including small birds from species other than the host, who had nothing to do with the cuckoo chick's upbringing, will stop to offer it the tidbits that they intended for their own brood. The chicks of the Great Spotted Cuckoo also mimic the begging call of their hosts. But it's the gape that really does the trick.

We have already seen that the seventeen species of honeyguides from the family Indicatoridae have a wonderful line in commensal feeding and a rare predilection for beeswax. They are also accomplished brood parasites. Favored hosts are barbets—which, like the honeyguides themselves, are related to woodpeckers. Since some barbets are highly social and gang upon intruders, it is risky to take liberties with them. But the honeyguides take those risks. The honeyguide's nestlings, when they hatch, have temporary hooks at the tips of their beaks expressly to puncture and smash the host's own eggs or kill host chicks. As they grow, their calls become louder and more insistent until a fledging honeyguide sounds like an entire nestful of host chicks calling together. The foster parents look after their uninvited guest for a short time after fledging, but at least in some honeyguide species the fledgling is soon recognized as an imposter and mercilessly chased. Then the youngster follows any passing honeyguides in their search for bees' nests. Bizarrely, immature honeyguides are dominant to the adults; and the immature Greater Honeyguide, which has a distinctive plumage, dominates all other honeyguides of whatever age or species. Human observers should not moralize, but this really does seem like a nightmare. Honeyguides may offer some small compensation to their hosts, however. Some at least offer a protection racket—defending pairs or groups of hosts that they themselves have already parasitized against further honeyguide attacks.

Host birds might draw at least some comfort from such "protection," but on the whole brood parasites seem to be seriously bad news. Sometimes the parent interloper and sometimes the parasite chicks kill the host's own eggs or chicks. If any host chicks do survive, they tend to be seriously underweight at the time of fledging, which must reduce their chances of survival (since birds in general are known to fare badly if they are underweight at fledging). Indeed, some brood parasites, including some cowbirds (of which more later), have driven entire populations of hosts to extinction.

Yet nature is never simple. As brood parasites, America's cowbirds are as bad as any. But in an extraordinary eleven-year study—truly a classic in the annals of ornithology—Neal Smith, of the Smithsonian Tropical Research Institute in Panama, has shown that sometimes, weird and counterintuitive though it may seem, brood parasitism by the Giant Cowbird can actually be of benefit to its hosts. In some circumstances, the hosts even seem to welcome the Giant Cowbird into their fold.

Among the various cowbirds that practice brood parasitism are the Brown-headed, the Screaming, and the Shiny Cowbirds, all from the genus *Molothrus*, and between them cowbirds have been known to impose their eggs on more than 220 other species. But Giant Cowbirds—*Scaphidura oryzivora*—are from a different genus. The ones that Neal Smith studied between 1964 and 1975 specialized in just two host species, both icterids: the Chestnut-headed Oropendola, *Zarhynchus wagleri*, and the Yellow-rumped Cacique, *Cacicus cela*.

Both the Chestnut-headed Oropendolas and the Yellow-rumped Caciques nest in colonies up to a hundred strong in sites that they return to year after year for decades and perhaps (for all anyone knows) for century after century. Often several species of oropendolas and caciques nest together. Despite their name, the Giant Cowbirds that parasitize them are far from huge—the males weigh around 130 grams (4½ ounces) and the females are only around 75 grams (less than 3 ounces), which means that the females are quite a lot smaller than the female oropendolas (at 112 grams, or 4 ounces) and not much bigger than the female caciques (at 65 grams, or a little more than 2 ounces). In short, you don't need to be a big swaggering cuckoo dressed up like a hawk and frightening everyone to death to be a successful brood parasite.

Oropendolas and caciques build nests like hanging baskets, similar in principle to those of weavers; and they hang them from the thinnest, outermost branches of umbrella-shaped trees where they are as safe as possible (which does not mean perfectly safe) from snakes, opossums, and marauding toucans.

But also, crucially, the oropendolas and caciques have one further protective strategy. They like to build their nests near the nests of bees of the genus *Trigona* and of various species of wasp. The wasps sting, of course, while *Trigona* bees do not sting but do bite and should not be provoked. The birds seek out the insect nests—the insects do not seek out the birds—and the birds that are lucky enough to find bee or wasps' nests are happy to delay their own breeding until mid- to late February, when the insects, which are highly seasonal, become active. The birds that wait for the bees or wasps to come online lose a couple of months from the breeding season and can raise only one brood a year. Those Chestnut-headed Oropendolas and Yellow-rumped Caciques that fail to find insect nests to nest alongside begin breeding around Christmas and can raise two broods a year, or even three. Yet the birds still prefer to nest alongside insect nests. Why?

You might suppose that the answer is obvious: the bees and wasps provide yet more protection against opossums and toucans. But it turns out that for these oropendolas and caciques, big vertebrate predators are not the main problem. Far worse are botflies, which lay their eggs or deposit their young larvae on the nestling chicks. The fly larvae burrow right through the chick, feeding as they go, and a small chick with more than ten botfly larvae is generally doomed. But those oropendolas and caciques that nest close to wasps and biting bees suffer far less predation from botflies. What it is about wasps and bees that botflies don't like—why exactly they keep away—is not known.

What of the Giant Cowbirds? They range from Mexico down to Argentina and specialize almost exclusively on oropendolas and caciques—although of various species, not just the two that Neal Smith studied in Panama. Oddly, they are generally seen only during the breeding season of their hosts. Where they go in the off-season is again not clear. But as soon as the hosts start breeding, the female cowbirds are out in force (though the male cowbirds are rarely seen in the colonies). Giant Cowbirds mate promiscuously, in large flocks, on flat areas of short grass. Presumably any one female might produce eggs from several fathers, spreading her genetic options.

Now to the point of the story. Neal Smith found, first, that oropendola and cacique colonies that did *not* have bees or wasps in their midst had many more cowbird chicks in their nests than those that did have their own bees and wasps. He also showed that this was no accident. Colonies of host birds that did have their own resident bees and wasps actively repelled cowbirds—and both the males and the females did the chasing. In colonies that did not have resident bees and wasps, only the females bothered to chase marauding cowbirds—and they did not chase the interlopers any more often than they chased other birds of their own species. In short, the birds in colonies that lacked bees or wasps gave the cowbirds an easy ride.

Host birds in the bee-less and wasp-less colonies went on giving the cowbirds an easy ride even after the invaders had laid their eggs. For oropendolas and caciques in general seem less tolerant of cowbird eggs than, say, Reed Warblers are of European Cuckoo eggs. Although cowbirds do try to mimic the eggs of their hosts, as cuckoos do, it is possible to see which is which—and the oropendolas and caciques certainly seem able to tell the difference. If they don't like the look of any egg in their nest, they stick their lower mandible into it, and then flick it out of the nest with their tongues. Eggs that are thus "spiked" have a characteristic "spike hole." Neal Smith gathered egg shells from around the host birds' trees—since the grass was short he was able to find all of them—and found a great many more cowbird eggs with spike holes under the trees that also had bees and wasp nests than around the trees with no insect nests. In other words, the hosts that had bees or wasps nests discriminated far more carefully against cowbird eggs than the birds who had no such nests. To prove the point, Smith added eggs to the hosts' nests that were clearly different from the hosts' eggs. Birds that had wasp or bee nests threw out these obvious fakes. Birds that did not have wasp or bee nests were more inclined to put up with them.

So what's going on? Well, unlikely though it may seem, the cowbird chicks, which generally hatch earlier than the host chicks as parasites are wont to do, groom their nestmates. For them, the botfly larvae that are preparing to burrow through the host chicks' skin, are tasty morsels. For oropendolas and caciques that lack the protection provided by bees and wasps, cowbird chicks are a reasonable substitute. Oropendola and cacique chicks that had no bees or wasps to protect them, and no cowbird chick to groom them, were heavily infected with botflies. But in host nests that lacked bees and wasps, but did have their own resident cowbird chick, infestation was down to 8 percent.

Of course, in general, cowbirds are bad news for oropendolas and caciques. But botflies are even worse news; and for Chestnut-headed Oropendolas and Yellow-rumped Caciques that can find no wasp or bee nests to shelter beside, it is worth paying the toll that the cowbirds exact. So they tolerate their resident cowbird chick in the same way that a householder might tolerate a giant and ill-tempered guard dog in a country overrun by bandits. Doubtless they would rather not have the dog, either, but it is better than the alternative.

Neal Smith's study—like all good science—raises yet more questions. The oropendolas and caciques without their own attendant bees and wasps behaved differently toward cowbirds than host birds that did have their own bees and wasps. They also treated the cowbird eggs far more leniently. So—are the bee-less and wasp-less oropendolas and caciques genetically different from the ones that had bees and wasps? Are we looking at two discrete populations? Or do the birds adjust their attitude toward cowbirds according to circumstance—whether or not there are bees or wasps around?

Smith tested this, too. With the aid of a U.S. Air Force "high-lift" truck with an arm that can stretch up to 22 meters (73 feet), he transferred nestlings from bee-less and wasp-less colonies into nests in colonies that had bees and/or wasps, and vice versa. Then he observed how these nestlings behaved when they grew up and laid eggs of their own, two years later. Answer: whichever way the transfer was made, the transferred birds behaved in the manner of their adopted site. Birds transferred to colonies with bees or wasps chased cowbirds and threw out cowbird eggs; those that were transferred to wasp-less or bee-less colonies were more tolerant.

How do we interpret this? Whichever way you look at it, the host birds seem to come out rather well. Conceivably, they might understand the score perfectly, and say to themselves: "Ah! We lack bees or wasps! Therefore we must go easy on the cowbirds!" More plausibly, so we might suppose, the rule is simply to do what the majority does. But even such follow-the-trend behavior is not trivial: it requires the birds to appreciate what the others are doing, and then to do the same thing. Chasing an invading cowbird is not conceptually simple. The birds observing this behavior have to know what they are looking at, and then do likewise. So it is clever.

How did brood parasitism evolve in the first place? It is impossible to answer such questions definitively, but we can at least speculate. Dumping of eggs in other birds' nests of the same species is commonplace and we can explain this in the first instance as a kind of mistake. Female birds are presumably stimulated to lay their eggs by the sight of a nest of a familiar kind, whether they have built it themselves or not. Since such a "mistake" carries some advantage—it may increase the number of offspring successfully raised—natural selection is liable to favor it, up to a point.

From laying extra eggs in other birds' nests of the same species to laying in nests of other species seems a fairly big step, but in Yellow-billed and Black-billed Cuckoos, we find what looks like an intriguing halfway house. In most years, these birds incubate their own eggs and raise the young, just like most other birds. But in years when food is particularly plentiful, the females lay extra eggs in other birds' nests—either of their own or of other species. But they also raise their own broods, just as a wren or a duck will do after a spot of dumping. We can imagine that European Cuckoos began this way and became so good at it that natural selection began to favor the ones that did it full time, and did not waste time and effort fending for their own children.

Once the parasitism of one species on another becomes habitual, an arms race would develop between parasite and host. As the centuries pass the parasites mimic the eggs of their hosts more and more accurately, while the hosts become more and more adept at detecting the minute differences so that the parasites need to be more and more accurate.

Eventually we reach the fine-tuned versatility of the European Cuckoo—to which its hosts have yet to mount a convincing defense. But the relationship that Neal Smith revealed—in which the hosts can apparently discriminate foreign eggs pretty well, but sometimes choose not to—adds yet another layer of subtlety.

More broadly, Neal Smith's study shows once more that life in the wild—the stuff of ecology—is immeasurably complicated. The relationships at work are far from obvious. It took Neal Smith eleven years of remarkably ingenious and conscientious study to work out just this one relationship—among a parasitic bird, a parasitic fly, one kind of bee and three kinds of wasps, and two small perching birds—and he points out that each of these creatures has a lot more going on in its life than he did not include in his study, for if he had, then the research would have got out of hand. You can't do science at all unless you simplify: pick out something you can actually study from among the melee. But in the world as a whole, there are probably between 5 and 8 million species, and the subtropical Americas where the Giant Cowbirds and their host birds live probably contain about a million of them; the relationships among any several of them are liable to be at least as intricate, and each of those relationships, directly or indirectly, impinges on all the others. On a practical note, a well-meaning conservationist who wanted to protect oropendolas and caciques might well begin by shooting all the cowbirds—yet, in some circumstances, he would be signing their death warrant. Expressions involving angels fearing to tread come to mind.

Konrad Lorenz's pet cockatoo flies off with Frau Lorenz's knitting wool.
Do parrots have a sense of humor?

9

THE MIND OF BIRDS

"THE MIND OF BIRDS"—NOW THERE'S A CONTENTIOUS title. It raises three huge questions (each of them in truth a cluster of questions) plus a whole range of subsidiary issues, each of which has occupied philosophers for as long as there have been philosophers and scientists for as long as there has been science—plus, of course, some of the world's most imaginative writers of fiction.

The first huge question is whether there is such a thing as "mind" at all. Is it really a discrete "thing," or a "force," like electromagnetism? Or is it, as some modern philosophers and neurophysiologists like to claim, just a side effect—something that we think or feel is happening inside our own heads but really is just the noise of firing neurons?

The second huge question is whether the things that go on in the things that we call minds are really any different from what goes on in a computer. Never mind birds—are human beings mere robots, driven by our own on-board computers (which in turn are programmed by our genes)? Of course, we can do things that computers cannot (and they can do things that we cannot), but is the difference just a matter of scale? If we made a computer big enough and complicated enough, could it chat with us, make jokes, and write plays that we could never have dreamed of (not just echoing ideas that had been fed into it)? Or is there, in the end, a qualitative difference between minds and computers, reflecting the fact that minds are alive and computers are not? Do life and mind go together? What is life, then?

The third huge question is whether there is any fundamental difference between what goes on in our own heads and what goes on in the minds of birds (if indeed they have any). Is it that we, human beings, partake of some special quality called "mind" that could not be replicated by a computer, while birds (and all other nonhuman creatures) are merely robots? René Descartes in the seventeenth century suggested that all nonhuman animals were simply automata—clockwork mannequins like the ones that were so popular in his own day. Thus, they would hardly have qualified even as robots, since robots in general are more versatile. Mannequins could be made to *look* as if they knew what they were doing—as if they had intentions, goals, feelings, and could work things out—but we know that clockwork mannequins can do no such things, that their apparent cognizance and angst and joy are simply an illusion. The same applies to nonhuman animals, according to Descartes. He proved it, with impeccable Gallic logic. Thoughts are composed of words, he declared. Animals may communicate up to a point, but they do not have verbal language. Therefore, or indeed *ergo*, they do not think. Q.E.D.

Then there are the subsidiary questions. Whether or not we decide that "mind" is a useful concept, how in practice does the brain work?

And what exactly are emotions? Do other animals have emotions—and if so, are they the same as ours? A bird (or a mouse or even a spider) may look frightened, but does it really feel fear? Does it feel anything at all, or does it just look as if it does—like a mannequin?

Then there are the practical issues. If we did make an ultra-complicated computer that could talk to us sensibly—not just "Good morning. I am your friendly computer. Please switch me on"—but a real, to-and-fro conversation full of allusions and puns and jokes and off-the-wall reminiscences, could we really say that it was, in the end, the same as us, or would this still be an illusion? Alan Turing, who in effect invented the modern computer, thought that if we got to the point where we really could not tell the difference between a chatty computer and an amiable dinner companion, then we might as well conclude that there isn't any. A computer that truly seems mindful *is* mindful. Turing was a genius, and I hate to argue with geniuses, but this seems to me a cop-out, of the "If it looks like a duck and quacks like a duck then it is a duck" variety. The American philosopher John Searle provided some kind of riposte in the 1980s when he showed that it is possible, at least in theory, to create a machine that converses with us, and yet in reality does not "understand" a word that is being said. Actually I think Searle is wrong, too, because a machine that really could converse convincingly would have to respond nimbly to jokes, and to see the sense in ungrammatical sentences of the kind that we all speak all the time; it is very hard to see how a machine could do this unless it really did "understand." (This, of course, seems to be in line with what Turing said, proving that if you think too much about these things, you end up going around in circles.)

Then we come to the ultimate teaser: what goes on in the heads of nonhuman animals? They see the world about them, of course, but what do they perceive? What do they make of the world? Do they think? Do they work things out from first principles? How do they regard their fellows? In a famous essay of 1974, the Yugoslav-American philosopher Thomas Nagel asked, "What is it like to be a bat?" Well—what is it like to be a bird? And, of course—how could we possibly know, since we cannot even know for certain how our fellow human beings think, or feel, or generally perceive the world?

And then—an issue that interests me mightily—do the minds of birds work in the same way as ours do? Or are there qualitative differences between bird brains and human brains, or bird brains and mammalian brains in general, that are as profound as the contrasts between human minds and computers? Why should it not be so? Birds and mammals shared a common ancestor that lived more than 300 million years ago— and we know that that ancestor had a very small brain indeed, and unless the laws of physiology were different in the deep past from how they are now, we can be sure that that ancestor was no intellectual. It was, in fact, dim. So insofar as mammals are bright (as we are, and chimps, and pigs, and elephants and whales and dogs and squirrels), and birds are bright (or so it often seems), we must have evolved our brightness independently. We have seen throughout this book that although birds do most of the things that mammals do, and a few do more besides, they do almost everything differently. Time and time again the two groups have reached similar endpoints—but almost always they have done so by different routes. Might the same be true of our brains? Either way, the answer would be interesting. We learn by making comparisons. If bird brains are different from ours, then they would provide us with a wonderful model by which to reassess our own minds. As Rudyard Kipling put the matter in a slightly different context, "And what can they know of England who only England know?" If it turned out that birds think in the same way as we do, and feel in the same way as we do, then this would suggest that there is a serious in-built limit to the number of ways in which it is possible for a living creature to think and feel. This would raise the very interesting question—why should this be so?

So it goes on. The questions multiply. Some are the kind that only philosophers could ask, but most are the kind that six-year-olds ask—and six-year-olds generally manage to hit the nail on the head. Adults do not answer the questions they thought about as six-year-olds. They merely stop asking them, to focus on marriage and mortgages. Education and maturation, in reality, are largely an exercise in shutting down, in channeling wild thoughts along practical and lucrative lines, a systematic beating out of the imagination.

But some philosophers and scientists and writers have gone on thinking about these questions, and the answer to all of them is—nobody knows! On every single issue listed above, and a great many more that I might have listed, there is dispute. In the end, it seems to me, the battle lines are drawn up just as they were in ancient Greece, or as they became drawn in the eighteenth and early nineteenth centuries in Europe, as the Age of Enlightenment gave way to Romanticism. Thus, on the one hand we have people (including many scientists and philosophers) who veer toward a "classical" or "positivist" view of life, wanting to base their worldview on careful assessment of the "known facts," which in practice means whatever can be seen, felt, and measured; on the other, we have people (including many scientists and philosophers) who might be called "Romantics" or sometimes even "transcendentalists," who feel that there is always likely to be something else going on besides what can be directly seen, touched, and measured, and that that "something else" is important. That "something else" includes wild, always elusive concepts such as "mind" and "feelings" and "consciousness."

Yet the position is not quite hopeless. Claims, which various philosophers have made of late in very fat books, that we now understand "mind" and "consciousness" are of course premature—and always will be, for it is empirically and logically impossible to provide definitive answers (or at least to provide definitive answers that add up to more than a hill of beans). But psychologists have definitely improved vastly on Descartes' confident assertion that nonhuman animals are mere automata.

THE ROAD TO BEHAVIORISM

Biology truly began to emerge as a science—or a series of sciences—in the course of the nineteenth century. Darwin was pivotal, changing the whole worldview of biology. So, too, was the birth of biochemistry at the start of the nineteenth century—blurring what had been an absolute distinction between the processes of life, on the one hand, and chemistry, on the other. Physiology came into its own not least through the Frenchman Claude Bernard, whom we briefly encountered in Chapter 1; as physiology merged with biochemistry, as it was bound to do, it began to seem as if all of life's processes could be explained in terms of chemistry. The science of electricity had been growing apace since the eighteenth century, and its role in biological systems—not least in nervous impulse—became more and more apparent—and so emerged the discipline of electrophysiology.

Louis Pasteur's ventures into microbiology (he looked at everything from the ailments of silk moths to rabies) added to the general feeling that all of life's phenomena, including the most mysterious diseases that had often seemed merely to be demonic, could be understood in materialist terms. The philosophy of positivism came on board at the end of the nineteenth century, leading to the idea that materialist notions are the ones that deserve to be taken the most seriously, since they are based on direct observation of repeatable, physical phenomena that can be measured and experimented upon. Science partook both of the materialism and of the positivism: the materialist notion that the world can be understood in terms of what can be directly observed and measured; and the positivist notion that the ideas rooted in measurement are the ones that should be taken most seriously.

Clearly, not everything we may choose to take seriously can be rammed into this hard-nosed version of science. God obviously cannot. But materialist-positivists could reconcile science with religion either by schizophrenia—allowing God to surface only on Sundays—or by espousing atheism and declaring that God is nonsense. Scientists qua materialist-positivist-atheists are still with us, and indeed write bestselling books.

Human values and emotions could not be rammed into the materialist-positivist mold either—beauty, love, aesthetics, honor, honesty. Again, they could be dealt with by schizophrenia—round off a hard day in the laboratory with an evening of Schubert—or by simple denial. So we find some modern scientists who speak of all human values simply as evolved devices for survival, or as various kinds of illusion. The materialist-positivist approach can be seen as extreme rationalism, and hence can reasonably be viewed as an extreme manifestation of classicism, even though it may seem far from the Ionic columns of Greece that we generally associate with the "classical" age.

But the study of psychology seemed to fall into the cracks between the different ways of thinking about the world—and still does. It embraces a series of seemingly essential and very high-sounding concepts that do not obviously belong in biology at all—or in any conventional science: mind, consciousness, awareness, understanding sensibility, thought, intelligence, emotion, subjective experience, worldview. But all these phenomena seem to be rooted within our own heads and are influenced by our bodies (adrenaline from glands around the kidneys produces the sensation of fear, for example). So they ought (should they not?) to be containable within the grand broad sweep of materialist biology—physiology, biochemistry, and electrophysiology. But how can we measure mind? How can we even begin to get a handle on subjective experiences—apart from our own experiences, which are, by definition, subjective? If psychology is about what goes on in the brain, and we cannot measure the things that seem to go on in the brain, then how can psychology ever be a science?

Scientists or would-be scientists in late nineteenth century came up with the whole range of different and often contrasting approaches. The nonmaterialist approach was best represented by George Romanes (1848–94). To some extent he became a protege of Darwin—they became great friends. Romanes, like Darwin, was primarily a naturalist and he sought to root psychology in knowledge of what animals actually *do*, and particularly of what they do in a state of nature, when they are left to themselves. To this end he set out—in a very Darwin-like fashion—to assemble all the facts. These facts, of course, included anecdotes because, after all, it is hard to observe animals in the field and wild animals do what they do when they choose to do it, and it can be very difficult to see them doing anything interesting even once, let alone repeatedly or to order. Indeed, as Romanes wrote in *Animal Intelligence* in 1881, "I have fished the seas of popular literature as well as the rivers of scientific writing." In much the same way, in the build-up to *Origin of Species*, Darwin had chatted to breeders of pigeons and fancy goldfish, and to all manner of gardeners.

Yet anecdotes can take us only so far. Scientists feel that truly to be rigorous they must at least quantify, and anecdotes—accounts of one-time incidents—do not lend themselves to quantification. Without quantification there can be no mathematical analysis, and hence (so scientists tend to feel) there can be no truly critical appraisal. Anecdotes by themselves are what a judge in a court of law would call "inadmissible evidence." Romanes is an extremely important figure in the history of animal psychology and is now coming back into fashion, but because he seemed to deal in Kipling-esque "Just-So" stories, he was pushed aside in his day and remained on the sidelines for more than a century.

So what aspects of animal psychology can be measured? Mind, consciousness, happiness, fear—these are far too elusive and subjective. The answer was, and is—behavior. Hence, a whole generation of scientists arose who sought from various angles to get a handle on the psychology of animals by observing what animals do in situations where their behavior can be reliably observed and measured. They were the behaviorists, and they worked mainly or exclusively in the laboratory, where all can be controlled.

THE BEHAVIORISTS

A key figure in the history of behaviorism was a Russian who expressly denied that he was a behaviorist: Ivan Pavlov (1849–1936). Pavlov took the idea of the reflex—which had been well demonstrated since at least the eighteenth century—and from it developed the notion of the "conditioned reflex." His key experiment (albeit somewhat grisly) has become the stuff of mythology. First, he showed that when dogs are shown meat, they salivate. The meat is the "stimulus" and the salivation is the "response," and the two together demonstrate a "simple reflex." Then, whenever he produced meat, he also sounded a bell or flashed a light (it didn't really matter which). After a time the dogs would salivate when they saw the light or heard the bell, even without the meat. They had learned to associate the sight or the sound of the neutral "secondary" stimulus with the real stimulus (meat). This was the "conditioned reflex." Once the conditioning took hold, it was strong: dogs took a long time to stop salivating in response to lights or bells long after they stopped getting meat at the same time. But after a time, if light or bell were repeatedly offered with no meat to back them up, the reflex was "extinguished" and the dogs stopped salivating in response. But if meat was again given with light or bell, they soon got back into their original routine. A memory of their original training remained with them.

Then came several key Americans. John Broadus (J. B.) Watson (1878–1958) was the first "true" behaviorist and indeed he coined the term "behaviorism." He argued that entire routines of behavior could be seen as strings of reflexes. The response that concluded the first reflex in the chain acted as the stimulus for the next reflex in the chain—and so on and so on. Learning could occur, as Pavlov had demonstrated, by association: new stimuli could be substituted for existing ones, and so the routine would be altered. But still the behavioral routine was nothing more than a string of reflexes, joined end to end. Watson left academe in 1920 and went into advertising, where he was immensely successful. Many a modern ad owes its origins to J. B. Watson and his mechanistic view of human psychology. Ads work; to an extent, therefore, he was right. We are all Pavlov's dogs up to a point.

But the giant of behaviorism, who dominated psychology for about half a century, was Burrhus Frederic (B. F.) Skinner (1904–90). Skinner talked of "operant conditioning." He showed that animals would learn to carry out particular responses to particular stimuli if they were rewarded for doing so, and would stop making particular responses if they were punished for doing so. In this way, he said, behavior could be "shaped." For instance (his basic experiment), when animals are placed in cages, the first thing they do is to explore—pressing this, pecking that, depending on what is available to do and what kinds of animals they are. They do everything there is to be done in that limited environment more or less randomly. But if a rat in a cage presses a conveniently placed lever, or a pigeon pecks at a disk that just happens to be there, and is rewarded with a pellet of food, then it soon learns to go on pressing or pecking. If, on the other hand, such behavior results in a nasty electric shock, the rat or pigeon stops doing it. By simple reward and punishment, Skinner showed that animals could quickly be taught to do the most remarkable things. Pigeons could be taught to play bowls. He invented the "Skinner box," in which animals could carry out behavioral routines for days and days without anybody being present: their responses to different stimuli were recorded automatically. Thus, he accumulated huge amounts of data that could be analyzed statistically—the stuff of hard science.

Skinner also applied his findings to humans. He and his friend and mentor Robert Yerkes (1876–1956), most famous for his work on primates, spoke of "social engineering"—shaping the behavior of us all by rewards and punishments. To be fair, Skinner showed that rewards work much better than punishments, so in practice, people (and animals) learn much more quickly if they are treated kindly than if they are given a hard time. He applied his findings to prisoners. They were rewarded for toeing the line and for general niceness, but instead of being punished outright for unsocial behavior, they were simply told to go and sit on a bench until they mended their ways (a technique now commonly used on children). This worked—and one cannot help feeling it is a lot more humane than much of the treatment that is commonly handed out in many or most of the world's prisons. Even so, "social engineering" has a sinister ring to it; and since those were also the days of eugenics and enforced sterilization of people who were deemed to be substandard (in the United States, Sweden, and various other countries), it seems small wonder that the golden age of positivist biology was also the golden age of totalitarianism. Two of the most chilling novels of the twentieth century parodied such ways of thinking: Aldous Huxley's *Brave New World* and George Orwell's *1984*.

I may seem to digress, but in truth not much. Behaviorism ruled animal psychology until late into the twentieth century. It was the main game in town when I was at university in the early 1960s. It was combined with a mortal fear of anthropomorphism. In a professional journal for teachers I came across a letter from an earnest school marm, urging her colleagues not to allow their hapless pupils to suggest, for example, that cuckoos lay eggs. What cuckoos really do, she fumed, is "exhibit egg-depositing behavior." No wonder children fled from science in their droves, bewildered and disappointed (and still do). But behaviorism seemed very powerful. With enough ingenuity, it seemed possible to explain everything that animals do, or indeed that humans do, in terms of Watson's end-to-end strings of reflexes and Skinner's operant conditioning. Skinner even tried to explain the human facility for language in behaviorist terms. As small children, he argued, we hear lots of words spoken, and we imitate them. When, by chance, we apply the right word to an object or action or color, our mothers say, "Well done, Billy!" or Susie or whatever we are called. If we get it wrong, then we are greeted at best with silence. We don't get bananas if we say "pears." By such means we learn the language of Shakespeare, or Sophocles, or the Upanishads, depending on where we are brought up. Or so the behaviorist enthusiasts suggested.

Behaviorism was primarily about how the brain works, and in particular how it learns. The agenda was to find general principles that would apply to *all* brains, whether pigeons or rats or human beings. To an extent it succeeded. Some of the behaviorist notions about learning—including operant conditioning, and reward and punishment—clearly do apply. We can indeed be "shaped" to some extent by Skinner's rewards or by Watson's ads.

Pavlov's earliest, pre-behaviorist experiments seemed to suggest that the brain is a blank slate, a *tabula rasa*. Dogs salivate in response to meat—that is clearly a built-in reflex. Indeed, it seemed at first that any kind of stimulus at all could be substituted for the meat and the dogs would still learn to salivate in response to it: infinite flexibility. By the same token, by Skinner's process of "shaping," any animal should be able to learn to do more or less anything that was within its physical capability. But these assumptions turned out *not* to be true. Pigeons did learn to play bowls, but it was well-nigh impossible to teach rats to stand on their hind quarters in response to reward or punishment, even though they do this naturally all the time. In other words, the *tabula rasa* is not as *rasa* as all that. There are some things the brain of any particular animal will do and some things it will not, even though the things it will not do seem just as easy as the things it will. If we compare the brain to a computer, then we can say that it runs on a particular operating system.

In general, behaviorism has some great philosophical strengths (which is why it had such wide appeal). One strength is that it really is a good idea, if we want to get a firm handle on things—even things as elusive as animal psychology—to measure all that can be measured, which is what the behaviorists set out to do. Behaviorism also partook of some very basic principles of what is considered to be good science. Scientists always seek the simplest explanations—certainly if there is a straight choice between simplicity and complexity, it seems best to opt for simplicity.

But behaviorism (like all science!) also contains some deep philosophical traps. One was enunciated by Albert Einstein, who was a great philosopher of science as well as an extraordinary physicist. Explanations should indeed be as simple as possible, he said, *"but no simpler!"* It began to turn out that the workings of brains, including bird brains, cannot be explained by the very simple mechanisms proposed by Watson and Skinner. More broadly, just because the behaviorists decided for the purposes of experimentation to exclude the concept of consciousness—and awareness, and feelings, and mind—that does not mean that they do not exist.

There is another trap that, again, all science is prone to. Scientists like to get clean results. For this they need to observe repeatable phenomena that they can quantify and analyze mathematically, which often means statistically. But animal behavior seems messy. If you just watch animals doing their thing, it is usually very hard to make head or tail of it. Watson and Skinner and their associates got the clean and quantifiable results they did only by getting their animals to perform very simple tasks that had been carefully tailored to elicit particular responses. The experimental animals ran through mazes, with the option of going right or left. They pressed levers. They pecked at disks. In short, in this simple, circumscribed environment they could do only the things they were allowed to do—which were simple and circumscribed things. But how does this circumscribed, repeatable, measurable, and scientifically respectable behavior relate to what animals do in the wild? If Shakespeare had been put in a Skinner box, he presumably would have learned to press a lever in return for candied medlars or pickled lampreys or whatever it was he ate. But could we infer from this that he was also capable of writing *Hamlet*? Would we even get a hint of his wider abilities? And what, actually, is more important?

So it was clear from the outset, at least to some people, that the behaviorist agenda was limited. To begin with, there was obviously a need to study animals when they were given freedom of action, when they could decide for themselves what they wanted to do and how to do it.

Then again, the behaviorist agenda was mainly about learning: how animals learn to modify their behavior in the light of experience. But it seems clear that animals (and people) do a great deal that is not directly learned. In some species of migrating birds, including swallows, the youngsters set off for the south after their parents have left, with no one to lead them. They may learn to read the night sky before they go and learn a great deal about the details of the journey as they travel, but they certainly do not learn to migrate in the first place, or indeed have any opportunity to learn where Africa is, or Argentina, or wherever they intend to go. Here is another demonstration that the brains of animals, including birds and people, do not come into this world as a *tabula rasa*. In computer terms, animals are born or hatched with brains that, to a significant extent, are not merely running on a particular operating system, but they come with programs preloaded. The sum total of all the inbuilt mental "programs" is what we mean by "instinct." Instinct is yet another of those qualities that we know as a matter of folklore must exist, in animals in general and in ourselves, and it needs to be studied if we are to make sense of fellow creatures and, indeed, of ourselves.

So it was that in the mid-twentieth century some biologists set out to study "natural" behavior: what animals did when they were making their own decisions. But to make things easier for themselves, and to give more reliable and quantifiable results, these biologists did not simply observe animals in the wild, like traditional naturalists. They aimed to give the animals freedom of choice, but to do so in circumstances where the observer could indeed observe and quantify what could be quantified. A key part of this agenda was to study and make sense of the instincts that shape the general behavior of all animals, no matter what else those animals might learn as they go through life. Instincts are characteristic of each species. Add flexibility to those basic instincts (flexibility acquired not least by learning), and we have done much to encapsulate the overall personality.

The formal study of natural behavior, and particularly of the instincts that underpin it, is called "ethology." The two great ethologists of the mid-twentieth century were the Dutchman Nikolaas (Niko) Tinbergen (1907–88) and an Austrian from Vienna, Konrad Lorenz (1903–89).

THE ETHOLOGISTS: TINBERGEN AND LORENZ

Tinbergen and Lorenz were great friends (albeit temporarily divided by the Second World War, since they were on opposite sides), but they were very different. Tinbergen was the field naturalist *par excellence*, but he followed up his field observations with artfully designed trials, in the field and in the laboratory, to make sense of what he had seen. The ability to design simple trials or experiments that are truly instructive is perhaps the most important quality of a scientist. Lorenz by contrast has been compared to a farmer. He lived with many kinds of animals, including many birds, in his grand semi-rustic *Schloss*—a real-life Dr. Doolittle. Only by such close contact over many years leading to true empathy could anyone really understand animals, he said, although he greatly admired Tinbergen's skill in designing experiments that were both simple and penetrating.

To paraphrase, Tinbergen envisaged the behavior of animals largely as a collection of instinctual packages—little, embedded routines. Each routine was evoked by some "releaser"—a particular stimulus. The way that Herring Gull chicks solicit food from their parents is one famous example. The adult Herring Gull arrives at the nest with a bellyful of food. In there, it is of no use to the chick. But the chick pecks at the adult's beak—and at this signal, the parent opens its beak and disgorges the food, already partly digested, helped by the chick, which reaches deep into its throat.

But Tinbergen perceived that the chicks do not peck at the adult's beak haphazardly. They target the bright red spot, near the tip of the beak, that is characteristic of Herring Gulls (and related species such as the Great and the Lesser Black-backed Gulls). Most traditional birders would make such an observation and leave it at that—an anecdote. But Tinbergen, being a true scientist, wanted to see if the chicks were indeed pecking at the red spot, and if so, what it was about the spot that invoked their pecking. Tinbergen, in fact, as no one had ever done before (or at least so convincingly), turned a field observation into a testable hypothesis and then tested it—thus providing the kind of solid foundation that natural history generally lacks. First, he showed that baby Herring Gulls would peck perfectly happily at the beaks of stuffed Herring Gulls, or indeed of

*One of the first formal demonstrations of animal instinct in action:
baby Herring Gulls peck at a red spot on the mother's beak, which prompts the parent to
disgorge food from its crop.*

model Herring Gulls. But they would not peck at the beaks if the red spot was painted over. Yet they would peck with vigor even at crude, cardboard cut-out models of Herring Gull heads provided the red spot was present. They would even peck at pencils with a suitable spot. Evidently, the red spot "released" the pecking instinct of the chick, and the pecking in turn "released" the disgorgement. So the dialogue was established, mechanically. There are echoes in this of Watson's idea that stimuli provoke responses, which in turn act as stimuli for further responses and so on.

Tinbergen went on to reveal the important and widespread phenomenon of super-stimulation. A crude cut-out head of a Herring Gull with an extra-thick bill and big bright spot elicited a stronger response from the chick than a real head of a real Herring Gull. He went on to show the same phenomenon in other contexts—for example, that Herring Gulls prefer to sit on artificial eggs that are bigger than their own than to incubate the real thing. Indeed, they will contrive to incubate monstrous creations that they can hardly sit on at all.

So it is that cuckoos are able to palm off their eggs on much smaller birds. The host birds have an in-built urge to sit on eggs—and the bigger the egg, the more enthusiastically they may sit. They stuff the beaks of the cuckoo chicks even more conscientiously than those of their own children because feeding is stimulated by the gape, and a cuckoo's beak gapes wider than a pipit's or finch's. Similarly, birds-of-paradise with extra feathers stuck to their tails, or cockatoos with artificially enhanced crests, attract more mates. Super-stimulation is very widespread in nature and in many ways important—and the insight begins with Tinbergen's genius for good observation and simple experiment. Perhaps we should not extrapolate too readily from animals to people, but the padded shoulders of men's jackets or the high-heeled shoes of women that exaggerate the length of leg, and lipstick and mascara and all the rest, suggest that the principle of super-stimulation works for people, too. Tinbergen was a pivotal figure. Everyone should at least read his *The Study of Instinct*, of 1951, and *The Herring Gull's World*, of 1953.

But I find Lorenz even more extraordinary. Lorenz seems to me to be a hybrid. He was born into an age and in a place, Vienna, that was the home of logical positivism, which around the time of the First World War grew out of nineteenth-century positivism, and hence was the philosophical center of hardline science—positivist, materialist, rigorous, scrubbed down to a skeleton. His own masterwork, *King Solomon's Ring*, was published in 1952, and all the way through he warns us not to be carried away, not to be anthropomorphic: not to suppose that animals feel the things that we do, or partake of consciousness and foresight, or any other such abstractions, just because it sometimes looks as if they do. Their behavior, he stresses, is merely "instinctive." He seems to want to suggest, indeed— to put the matter in modern parlance—that animals are programmed robots, just a step or two ahead of Descartes's mannequins.

But Lorenz's own instincts seem to rebel. He identifies with his animals. Empathy, he says, is an essential prerequisite to understanding. In his preface to *King Solomon's Ring* he reveals his admiration for Rudyard Kipling, who, in his *Just-So Stories* and *The Jungle Book*, describes the wiliness of snakes and the wisdom of wolves and the detached arrogance of the panther precisely as if they were human; and for the Swedish author Selma Lagerlöf, who wrote in similar vein in the children's book *Nils Holgersson*. Lorenz says of them both: "They may daringly let the animal speak like a human being, they may even ascribe human motives to its actions and yet . . . they convey a true impression of what a wild animal is like." Only scientists who are very confident, as Lorenz had good reason to be by 1950, dare to risk the scorn of their harder-nosed colleagues with such sentiments.

In his own book, Lorenz tells us that the Greylag Geese that wandered freely in and out of his house, and who left their indelible dark green marks on the carpets, are "delightfully affectionate" and "long for human society." They "fall in love" and mate for life, and the ganders throughout show "courageous chivalry." For years Lorenz maintained a colony of jackdaws on the roof. He found them a delight, but suggests that if the rest of us find jackdaws too much to handle, then we might like to try a starling— "a poor man's dog." Like a dog, a starling "cannot be bought ready-made. It is seldom that a dog, bought as an adult, becomes really your dog. . . . It is the personal contact that counts." But if even a starling seems too demanding, "let me recommend you a siskin. . . . Of all the small birds I know it is the only one which, even when captured in maturity, not only becomes tame but also really affectionate." Contrast the siskin with the gardener's friend, the robin. A robin can be trained "in a surprisingly short time, to approach its keeper voluntarily"—but only if duly rewarded. The siskin will approach its keeper only after several months but, says Lorenz: "Once it has taken this step, it approaches his keeper for his company's sake and not in the expectancy of food. "Such a 'companionable tameness' is much more endearing to our human mentality than the highly material cupboard-love of the robin" (given, after all, that robins like us because they think we are pigs).

In similar vein, Lorenz describes the antics of his Sulphur-crested Cockatoo, Koka: "The cockatoo paid court to my mother in a very charming way, dancing round her in the most grotesque fashion, folding and unfolding his beautiful crest and following her wherever she went." And:

> One of the nicest cockatoo-tricks which, in fanciful inventiveness, equaled the experiments of monkey or human children, arose from the ardent love of the bird for my mother who, so long as she stayed in the garden in summertime, knitted without stopping. The cockatoo seemed to understand exactly how the soft skeins worked and what the wool was for. He always seized the free end of the wool with his beak and then flew lustily into the air, unraveling the ball behind him. . . . Like a paper kite with a long tail he climbed high and then flew in regular circles round the great lime tree which stood in front of our house. Once, when nobody was there to stop him he encircled the tree, right up to its summit, with brightly coloured woollen strands which it was impossible to disentangle from the wide-spreading foliage.

Or then again, one day Lorenz was alarmed to see his old father shuffling away from the verandah gripping his middle, as if in great pain, and "swearing like a trooper." The cause was not colic, but Koka, who had nicked the buttons from the old man's suit as he slept and laid them out neatly, "following the whole outline of the professor: here the arms, there the waistcoat, and here, unmistakably, the buttons off his trousers."

Two points in particular emerge from these passages. First, Lorenz speaks anthropomorphically, even though he warns his readers not to be anthropomorphic. Second, and more broadly, he is intimating, even though he is apparently reluctant to do so, that the buttoned-down, quantified, stripped-to-the-bone approach to animal psychology—the approach encapsulated in behaviorism—just would not do. The American philosopher of science Thomas Kuhn in the 1960s declared that, at any one time, scientists in any one field mostly tend to conform to the same "paradigm," which in this context means "worldview." Most of the time, he said, they go along with the prevailing paradigm and try to shove all their new observations and ideas into it. But after a time the anomalies accumulate. Then we get a "paradigm shift." The kinds of observations that Lorenz made, and the ways in which he described them, seem to me to constitute just such a shift. They did not spell the end of behaviorism, for behaviorism—in its place—is still useful. But they did show, very clearly, that behaviorism could not achieve what its most zealous proponents had supposed it could. It could not adequately describe or throw convincing light on the actual behavior of animals when they are doing their own thing. After several decades of behaviorism, the field of animal psychology was still wide open.

Many of Lorenz's most revealing observations were made on birds. But it seems to me that the buttoned-down, positivist, behaviorist paradigm was finally shifted, once and for all, from the 1960s onward, by the work of Jane Goodall on wild chimps in the Gombe Stream Reserve in what was then Tanganyíka. She showed that chimps do many things that no travelers' tales or armchair philosophizing had ever come near to guessing: that wild chimps not only use tools but also make them, although tool making up to then had been taken as one of the defining features of humankind. She also showed that they were formidable hunters—far from the amiable vegetarians that were commonly portrayed. Most profoundly of all, she showed that different chimps differ enormously in their talents, their social skills, and their emotional responses. Each chimp is a personality. The fate of particular chimp communities depends not simply on how many there are in the troop, or the sex ratio, or the available resources, and all these other quantifiable demographic and ecological details, but on *who* is in the troop—whether the troop is led by intelligent, sensible individuals or by some erratic tyrant. Biologists who seek truly to understand the lives of wild chimps cannot operate simply as ecologists. They must borrow the tools of the historian, record who did what to whom, and trace the social ramifications. Although chimps are not human, and it is a mistake to treat them as such, it is impossible to get to grips with them *without* being anthropomorphic. In the 1980s, in the one conversation I was privileged to have with her, Jane Goodall told me that she was never able to express what she really felt about the chimps until she started writing popular books, and shook off the bared-to-the-bone assumptions and vocabulary of formal science.

Today there is a new generation of field biologists who at one level can properly be seen as naturalists: they watch what animals do, without too many preconceptions about what can be going on in their heads and what cannot. But unlike their amateur antecedents, they bring the rigor of science to bear whenever this is possible. They study particular groups for hour after hour, year after year, as Jane Goodall studied her chimps, and quantify whatever can be quantified. When animal behavior is studied closely without pre-judgment, in all its richness, it becomes very hard to describe what is going on in terms that seem halfway plausible without allowing concepts to creep back in that, in an earlier age, were rejected out of hand: consciousness, mind, happiness. We are witnessing a paradigm shift.

The shift was already evident in the late 1980s, when the Royal Society in London held a conference on "How Animals Think." That title is wonderfully provocative—for few respectable scientists in earlier years would have dared to suggest that animals "think" at all. At that conference I spoke to Professor Herb Terrace of Columbia University, New York, and he said: "Descartes argued that conscious thought requires verbal language of a human kind, and so he argued that animals cannot think because they do not have human language. Now we can see that we have to explain how animals think even though they don't have human language." We may note in passing that most of our own thinking—even conscious thinking—does not depend on words, either. Ideas come to us "in a flash," but we may then need 2,000 words to express them (or a lifetime, as Einstein said of his own thoughts on relativity). If you want to work out some really difficult problem, don't think about it—sleep on it. Soon after that meeting I spoke to Professor Patrick Bateson of Cambridge University, and he said: "Anthropomorphism, properly applied, can be heuristic." In other words, it can be at least as useful to compare other animals to humans as a way of trying to understand them as it is to compare them to mannequins or robots. At about that time, too, the Dutch primatologist Frans de Waal wrote: "After decades of systematic underestimation of animal mental capacities, a period with some risk of overestimation could not hurt." After all, he said, "The object of science is not to produce the safest theories, but to produce the best and most stimulating ones."[1]

So we should be open-minded. To be sure, we should not throw caution to the winds. We will never advance our understanding if we simply sit around and make up stories as Kipling did. But anyone who truly wants to understand the minds of animals—or, indeed, our own minds—must be in search of a new paradigm because, at present, we don't have one.

We should not simply decree, *a priori*, that animals are clockwork toys, or robots, unless proved otherwise. But neither should we abandon our critical faculties—we should not assume that a crow or a parrot that does apparently brilliant things should be offered a university chair. Neither should we suppose that our own ways of doing things are necessarily superior, or that our world is the best; a crow is not necessarily inferior to a professor just because crows are not versed in professorial matters. If animals sometimes fail to fit in with our own way of looking at the world, we should not assume that they are stupid. We should at least ask ourselves how well we would do if we were thrust into their world.

With such notions and their accompanying caveats in mind, what should we make of the following?

VERY CLEVER BIRDS

Birds do much that seems to be of a mental kind that is way beyond our compass. Few of us could navigate from Texas to Argentina or from Colchester to Cape Town even with a map—and certainly could not have done so when we were infants, and had no one to lead us, as baby swallows do. No one that I know could hide 30,000 pine seeds and expect to find most of them again months later, under snow, as the Clark's Nutcrackers of North America routinely do. Very few of us could remember the hundreds of strophes that some songbirds string together as they seek to impress their rivals and potential mates. All this, and much more, seems immensely "clever."

Yet there is a huge caveat. All of us are aware of the *savants* syndrome—epitomized by Dustin Hoffman in *Rain Man*: a nice but sad individual who could remember the details of every plane crash there ever was, and remember the sequence of countless playing cards, but alas could not perform the basic tasks required to get through the day. In the 1980s, the American psychologists Leda Cosmides and John Tooby provided a general explanation for the syndrome. The standard view until then was that human brains are endowed with "general intelligence." Brains were conceived as all-purpose thinking machines, which generally speaking is how the behaviorists saw the brain: if you are good at one thing, you are likely to be good at others, too. To some extent this is true. The brightest kids at school often do seem to excel in just about everything. But to a significant extent, said Cosmides and Tooby, the accepted model was not true. To a significant extent our brains—and the brains of all animals—are more like Swiss Army Knives—those astonishing penknives that may contain thirty or more specialist tools, for cutting fruit, boring holes, and, famously, getting stones out of horses' hooves.[2]

In the Swiss Army Knife model, the brain is "modular": different modules for different functions. In people whom we feel are seriously bright, information and skills seem to flow freely from one module to another, and the result manifests as general intelligence. But in most of us, much of the time, the flow between modules is not quite so free, so that a person (or an animal) may be extremely adept in some fields and fairly useless in others, even though in principle one field might learn from another. Dustin Hoffman's character was an extreme example. Pocket calculators are *savants* in mechanical form—better at mental arithmetic even than Alan Turing, but absolutely lacking his astonishing ability to speculate. The calculator performs its mighty arithmetical feats with the aid of a few simple "algorithms"—built-in tricks that are guaranteed to solve particular kinds of problems. We use algorithms when we tie our shoelaces (make a loop, cross it over, tuck it under, etc.), and those of us old enough to remember the days before calculators also had to learn a tedious but surefire algorithmic approach to long division. Presumably, Clark's Nutcrackers employ just such an algorithm to solve their nut-finding problem. It is wonderful but only (perhaps) in the same way that the pocket calculator is wonderful. You would not trust either to manage your worldly affairs. I wonder—whimsically, because I see no immediate prospect of an answer—whether the brains of birds are *more* modular than those of mammals, and whether, if this is so, this is because of their respective ancestry. Birds and mammals belong to quite different lineages: birds are diapsids, while mammals are synapsids. Are diapsid brains innately more modular than synapsid brains? The question seems well worth asking.

Yet birds do other things that seem hard to explain in terms of algorithms, or of dedicated modules.

Take, for example, the Carrion Crows on one Japanese university campus, which since the early 1990s have developed a marvelous technique for cracking the walnuts that they pick from the trees that grow by the roadside. They put the nuts in front of the traffic. They place them when the lights are red and the cars are stopped. When the lights turn

Greater Black-backed Gulls are among the several birds that break the shells of armored animals—such as clams—by dropping them from a height. Cormorants, meanwhile, stick to hunting underwater—but then must hang their wings out to dry.

green and the cars start up, they fly to the top of the traffic lights and keep watch. When the lights are red again, they return to collect their booty, conveniently crushed. If the cars miss, they sometimes move the nuts to one side. If the cars go on missing, they sometimes take them up to the telephone wires and drop them in front of the moving traffic. Carrion Crows and various gulls commonly do much the same thing with clams, as we have seen, dropping them onto stones to break in the way that Lammergeiers drop bones. But walnuts do not break so easily as clams because, when fresh, they have a soft but tough outer casing. Crows in California are now beginning to learn the same trick as their Japanese counterparts, or so it is reported. I have seen the same technique in China—employed not by crows but by farmers, who spread maize over the road for a bit of free dehusking. I thought the farmers were very ingenious and I do not see in principle why we should not give similar credit to the crows.

Even smarter, it seems—and now widely regarded as the intellectual stars of the bird world—are New Caledonian Crows. New Caledonian Crows not only use tools, which many birds do, they *make* their own tools. Until Jane Goodall showed that chimps can make their own tools, too, this was thought to be the sole prerogative of human beings—and indeed was often presented as our defining characteristic.

During a three-year field study in New Caledonia, a New Zealand biologist, Dr. Gavin Hunt, showed the crows doing this in the wild. They use their beaks as snippers to make hooks out of twigs with which they would pull grubs out of crevices in trees (yet another of the birds that do what a woodpecker does, but without woodpeckers' ability to hack their way into wood and probe with a snake-like tongue). They also make rakes from leathery leaves.

At the University of Oxford, biologists led by Alex Kacelnik are still plumbing the extent of the New Caledonian Crows' abilities. In one early experiment they presented two crows with small buckets of meat at the bottom of long tubes and gave them two bits of wire to fish the buckets out with—one hooked and one straight. Question: Would the birds select the straight wire or the hooked one? The male, Abel, immediately saw the point and grabbed the hooked wire. He retrieved the bucket after a few tries and then he flew off with the wire, leaving the female, Betty, to make do as best she could—yet another victim of male chauvinism. But no problem. As first observed by Alex Kacelnik's colleague, Alex Weir, Betty the crow simply bent the wire to make a hook for herself. She was no slavish tool user. She was a bona fide tool maker. Presented with the same problem again (scientists are not content with a one-time event), she did it again. She can also make hooks from other materials. She is not fixated on wire. Oxford has twenty or so New Caledonian Crows. I have watched them in the company of Professor Kacelnik, and it seems, from his commentary, that nothing they do in their aviary is merely random, as casual observers might assume. Every glance and interaction seems to have some point to it. Now Oxford scientists have fitted small cameras to New Caledonian Crows in the wild. It will be very interesting, indeed, to look at the world from their point of view and see from closest hand what they get up to.

Parrots are the other acknowledged prodigies of the bird world. One of the favorite tricks of New Zealand's Keas is to drop stones on the roofs of houses (corrugated-metal or roofs of bungalows are much favored in those parts) until the owners emerge, and then fly off with much psittacine chortling (or so it seems). At about the age of eight, I and my hooligan chums used to do much the same.

But the most accomplished of all parrots recorded so far—up there it seems with Betty the New Caledonian Crow—is Alex the African Grey, famed not for his apparent sense of humor, like Keas or like Lorenz's cockatoo, but for his apparent ability to speak.

Alex was bought in a pet store when he was one year old, in 1977. But since he was lucky enough or unlucky enough to be bought by a psychologist, Irene Pepperberg (then a research associate at Perdue University), Alex became perhaps the most intensively studied and educated parrot in all of history. Indeed, his name is short for Avian Learning Experiment. Alex learned human speech, as parrots commonly do—but not, claims Dr. Pepperberg, in mindless mimicry. He seemed to speak with meaning and made up his own sentences.

Alex seemed to have a wide range of concepts—objects, verbs, adjectives, prepositions, and general politenesses. By 1999, when Alex was into his twenties, Dr. Pepperberg reported that he could identify fifty different objects and distinguish among seven different colors and shapes, and understood concepts such as bigger, smaller, same, and different. Shown an

Geniuses of the bird world: New Caledonian Crows are among the few nonhuman animals not only to use tools but also to make the tools.

object, he could describe its shape, color, and the material it was made from. If he was asked to describe the differences between two objects, he would say, for example, that one was smaller and the other was bigger, and if there was no difference he said "none." He was also learning the concepts of "over" and "under." Overall he had a vocabulary of around 150 words—and seemed to understand what they meant. He also had a sense of number, and could differentiate among quantities up to six. When he grew tired of the testing, he said, "I'm gonna go away," but if the experimenter looked annoyed, he would say, "I'm sorry." If he wanted a banana—"wanna banana"—and was given a nut, he either just stared and asked again, or else took the nut and threw it at the researcher. Alas, he died, unexpectedly, on September 6, 2007, at Brandeis University in Massachusetts, where Dr. Pepperberg was then working—still young for an African Grey Parrot, at thirty-one years old. There are two other African Greys at Brandeis, but we must see if with further training they become as smart as Alex. We cannot know what else he might have done.

But although we should be prepared to err on the side of generosity when interpreting studies such as Dr. Pepperberg's, we have to be critical, too. If a human child did the things that Alex did, we would indeed say that he or she understood the language. Such understanding, we would say, is not just a question of attaching particular sounds—words—to particular objects, or actions. Human language is open-ended; we can use the words we know in an infinity of combinations, in any order, to produce original sentences of potentially infinite variety to express a potential infinity of meanings. Noam Chomsky, whose name I took in vain in Chapter 6, in the discussion of birdsong, pointed out in the 1960s that the key to this is the human grasp of syntax—the rules of grammar that imbue the strings of words with "meaning." He suggested that human beings have a specific language "module" (the first and prime example of Tooby and Cosmides's more general notion of the modular brain). Nonhuman animals simply do not. In the 1970s, various American primatologists tried to teach chimpanzees and bonobos (previously known as "pygmy chimps") to use human sign language. The chimps, like Alex, performed many impressive feats that suggested that they were indeed using the language creatively. But critics—notably Herb Terrace—said they were doing no such thing. They had simply learned, in the way described by B. F. Skinner, that if they made particular signs they were rewarded. There was no Chomskian syntax in there at all. Professor Terrace has raised the same objection vis-à-vis Alex. I do not presume to comment. I am merely relating the facts, as they now stand.

I suggest, though, that if we really want to get a handle on what animals can do, then we should look at their social lives.

BRAINS AND SOCIALITY

Brains are expensive organs. Even in human beings, they account for only a small proportion of body weight and yet they take a very large share of the body's energy budget. Most animals get by with fairly small brains, and many creatures—including trees—manage to live complex lives with no brains at all. Intuitively we can see the advantage of big brains, once they have evolved. Clever creatures like us call the shots. We can decide whether trees live or die, and they have no such control over us. But it is still very hard to explain how big brains evolved in the first place. Christian theologians traditionally told us that God simply wanted to create a creature "in His image" with whom he could have a satisfactory relationship. Evolutionary biologists have suggested that the very first human beings with their proto-brains had at least some ability to make better tools, and so they were more likely to survive than their dimmer counterparts, and the bigger the brains, the better the tools they could make.

In more recent times, however, biologists have begun to suggest that animals are brainy (or some of them are) for social reasons. Geoffrey Miller, now at Stanford University, suggests that clever creatures can attract more mates because they are more inventive and creative than the average. The British biologist Nick Humphrey (and, later, Robin Dunbar) has a broader thesis—that animals do well by living in big cooperative groups, and complex groups can cohere only if the individual members are socially aware. They must know who's who, and remember everyone's personal history, and how they can and should behave toward their fellows. They must know who they should defer to; who they should submit to; who they should avoid; who is worth mating with; and who they can reasonably hope to mate with, without being beaten up for their pains, either by the potential partner or by some superior.

We have already glimpsed the social subtleties of Black-capped Chickadees and Arabian Babblers. Neither can easily be explained if we assume that those birds are simple automata, operating by rote. North America's

Western Scrub-Jay, *Aphelocoma californica*—yet another crow—again displays extraordinary social awareness, in a quite different context.

It would be good for my general argument if the Western Scrub-Jay had a highly complex social life, but in truth it does not—or at least, not by the standards of its closest relatives in the genus *Aphelocoma*. For Western Scrub-Jay parents only rarely employ helpers at the nest, while the Florida Scrub-Jay, *A. coerulescens*, usually employs one or several helpers, and the Mexican Jay, *A. ultramarina*, is so cooperative in its breeding that its nests commonly contain chicks from several different sets of parents. So while then the Western Scrub-Jay may not itself have a very complex social life, it certainly belongs to a lineage that does.

But the Western Scrub-Jay's claim to intellectual fame comes mainly from its very refined approach to stashing away food for later, or caching, as revealed in particular by Nicola Clayton and her colleagues at the University of Cambridge. It seems that the common ancestor of all the corvids must have been a cacher, at least of modest skill. Then (it seems) several lineages of corvids independently lost the ability to cache, while at least two lineages independently developed caching into one of nature's most miraculous skills. One of those lineages now culminates in Clark's Nutcracker, and the other reaches its apotheosis with the Pinyon Jay, both in North America (a continent especially rich in acorns, which are very good for caching). By comparison with these two megastars, the Western Scrub-Jay is only a moderate cacher. But to the simple crafts of storage and spatial memory it has added various layers of behavioral flexibility. Indeed, some feel that Western Scrub-Jays may be demonstrating mental skills that traditionally were thought to belong to us and us alone.

One ingenious set of experiments at Cambridge has revealed the Western Scrub-Jay's special ability to relate the past to the future. For the duration of the trial the birds lived in a three-roomed mini-"motel." They spent their days in the big central room, where they fed on ground pine nuts. This meant that they were well-fed—not hungry—but they had no opportunity for caching. Crushed bits of nut are not good for this. They then spent the night in one of two bedrooms—lights out, so they went to sleep. In the first room, when they woke up the next morning, there were nuts and worms to eat: excellent hotel service. But when the jays were (randomly) assigned to the second room, they found when they woke next morning that there was nothing to eat.

When they had had time to learn which room was which—three-star hotel or traveler's nightmare—the birds were given whole nuts during their days spent in the dayroom. These they were able to cache—in trays of sand specially laid on in the side rooms.

Without any prompting, the jays invariably cached far more nuts in the room where they knew they were likely to be hungry than in the room where breakfast was regularly served. Note that when they did the caching (during the day, obviously) they were not hungry. They were not simply projecting their present state into the future. They knew that in the future they were likely to be hungry and were taking proper precaution. Neither, obviously, were they simply caching surplus food in a routine way. They knew where stores were most likely to be needed.

Western Scrub-Jays also adjust their caching to social circumstance. In this they are not unique, for it has long been known that various crows in the wild go to some lengths to protect their food stores from pilferers. Ravens prefer to cache their spare nuts behind obstacles, where other ravens cannot see, and they are very reluctant to cache anything if other ravens are watching them. They wait until the observers are gone—or else they pretend to cache when others are watching, but do not, to throw the would-be thief off the scent. Northwestern Crows have also been shown to do this. Some crows, too, come back and move their caches—perhaps, and presumably, because they suspect that other crows were watching them when they cached them the first time. On the other side of the coin, Pinyon Jays and Mexican Jays are known to be extremely accomplished thieves. They watch others do the caching and then remember exactly where all the good things are. The card game Concentration, also known as "pairs," in which the players try to pick up two red sevens or two pictures of Sylvester or Tweetie-Pie, depending on the age of the opposition, shows how difficult this can be. If Pinyon Jays played Concentration, they could set themselves up as the most tremendous hustlers.

In the laboratory, Nicola Clayton, with her husband Nathan Emery and various colleagues, have shown extra subtleties in Western Scrub-Jays. Like ravens, they are reluctant to cache if others are watching them. At least, if they know that they are being watched as they cache, then they come back later and move the stores—typically to new sites that the observer had no reason to know about. But there is more to it than that. For Western Scrub-Jays are good pilferers besides being reasonable cachers. It does not seem to occur to them to move their caches when they are observed unless they themselves have previously done some thieving. Then they seem to know what it is like to be a thief—and then they apparently realize that others are not to be trusted either.

Various kinds of caching crows seem to know that some lookers-on are more likely to steal their caches than others. So it is that Steller's Jays steal caches from Gray Jays, but Clark's Nutcrackers do not. In the laboratory, Gray Jays were happy to cache when they were observed by Clark's Nutcrackers—but they did not cache when they were watched by Steller's Jays. But Gray Jays did cache when they were watched by other Gray Jays. Does this mean that Gray Jays are not thieves, or that Gray Jays for some reason are happy to let other Gray Jays steal from them? (Which seems somewhat less likely.) Such observations are still somewhat scattered, however. There is much research yet to be done.

Western Scrub-Jays "cache" food to use later—but if they notice other birds spying on them, they come back later and move it.

The feats of Western Scrub-Jays raise two issues of profound scientific and philosophic significance. First, their preference for caching seeds in places where previously they have had bad experiences suggests that they may have "episodic memory." The point is not simply that they have learned from the past, as any creature may do (as even a plant may "learn" up to a point), but that they remember particular places and circumstances, and how they felt when they were there; they can project this memory into the future, and take steps to ensure that they do not suffer the same fate again. Until now, such ability to remember the past in detail and transfer the memory to the future has been thought to be strictly human.

Second, the way that several kinds of crow watch out for observers when caching and move their caches if they are observed (but only if they themselves have been thieves and know what thieving is, in the case of the Western Scrub-Jay!) suggests that they may have "theory of mind." That is, they may realize that other creatures also have minds, that other creatures may know some things that they do not know, and may not know things that they do know. Theory of mind is beautifully illustrated by the old music hall routine—"Little did he know that she knows that they knew that he had forgotten that they knew . . ." More prosaically, once an animal has "theory of mind," it can guess what others may do and can develop the arts of deception—fooling the rival into false belief.

In practice, however, "theory of mind" is proving—like "consciousness," "mind," and even "feeling"—to be a remarkably elusive concept. Naturalists and animal lovers in general have often assumed that other animals have it just because they seem to. Thus, Mark Twain's character Jim Baker in *A Tramp Abroad* (1880) strongly implies that Blue Jays are possessed of it: "A jay hasn't got any more principles than a Congressman. A jay will lie, a jay will steal, a jay will deceive, a jay will betray; and, four times out of five, a jay will go back on his solemnest promise. The sacredness of an obligation is a thing which you can't cram into no blue jay's head." But critical experiments are hard to devise and even harder to carry out. Some do suggest at least controversially that some other animals, including chimps and elephants, have some intimation of theory of mind. But so far humans are the only animals that can be shown beyond doubt to know that other individuals have minds, too, and that others may be thinking different things. If Western Scrub-Jays are also possessed of such subtlety (or, indeed, Blue Jays), then human claims to uniqueness are chiseled away once more.

Still, though, we should not be carried away. We can envisage ways in which the Western Scrub-Jays might achieve all that they do just by following simple rules of thumb, as in: "If you see another crow looking at you when caching, then re-cache." It may not be necessary for the crow that does the re-caching to infer that the observing crow actually "knows," or "realizes," or "understands" anything. The caching crow re-caches when it sees another crow almost in the way that a pigeon in a Skinner box may learn to peck a particular shape if it knows it will be rewarded. Again, we just have to keep an open mind.

But I want to leave the final anecdote to Konrad Lorenz, who, together with Jane Goodall, has surely done the most to change the way modern scientists look at animals—and to increase our respect for them. The tale again comes from *King Solomon's Ring* and it again concerns his beloved jackdaws, which lived semi-wild on his roof. Jackdaws, too, establish a pecking order within each flock—but jackdaws, Lorenz tells us, contrive to be as peaceful as possible. "After some few disputes, which need not necessarily lead to blows, each bird knows which of the others she has to fear and which must show respect to her." Furthermore, the high-rankers peck only the ones who are next in line in the hierarchy, which they see as a potential threat; the ones way down the order are beneath their attention, and the high-rankers tolerate them or ignore them altogether. This is in stark contrast to farmyard hens, which peck everyone in their flock that is inferior to them (so that if the birds are too crowded and the one at the bottom of the pile has nowhere to run, it is liable to be pecked to death, and goes on being pecked even after death).

What, though, determines the pecking order in the jackdaw colony? "Not only physical strength," says Lorenz, "but also personal courage, energy, and even the self-assurance of every individual bird are decisive. This order of rank is extremely conservative. An animal proved inferior, if only morally, in a dispute, will not venture lightly to cross the path of its conqueror."

In one of Lorenz's jackdaw flocks the boss male—prosaically named "Double-aluminium" after the bands that were fitted to his legs—had "fallen in love" with "a little jackdaw lady, who hitherto had been maltreated by 80 percent of the colony." Accordingly, she became "the wife of the president"—and the news of her promotion spread immediately.

From the very day that Double-aluminium took up with this erstwhile nobody, she "may no longer receive so much as a black look from any other jackdaw. But more curious still—the promoted bird knows of its promotion!" Indeed, she began to throw her weight around in a way that was most unbecoming to a jackdaw lady of her status. Yet as Lorenz points out, "It is no credit to an animal to be shy and anxious after a bad experience but to understand that a hitherto existent danger is now removed and to face the fact with an adequate supply of courage requires more sense." Just like the Black-capped Chickadees, the individuals in the jackdaw flock not only knew who was who, they knew—apparently from minute to minute—what was what.

Even here, we should not get carried away. Scientists in artificial intelligence (AI) these past few decades have shown how very simple robots with very simple programs can do the most remarkable things. The nests of termites, for example, are most extraordinary. The bit that sticks above the ground is only the half of it; underground is a labyrinth of tunnels that provide air-conditioning and keep the mound at agreeable temperatures even though they may stand in open desert on the Equator. The whole edifice is built out of particles of mud, each installed separately. It seems impossible to see how the termites could build such a thing unless each of them carried some master plan in its head and knew how each particle of mud would contribute to the whole. Yet robots with two or three simple rules embossed in their chips can do the same thing (or at least in principle). By analogy, we can see how, in principle birds, too, might build the most extraordinarily complex societies just by following a relatively few, simple, built-in rules. That remains possible; and if we are interested in getting to the truth, then we should bear such possibilities in mind.

So what answers can we give to the opening salvo of questions? How in particular can we respond to Thomas Nagel's question (in its modified form): "What is it like to be a bird?" Well, "nothing is definite" is the short if disappointing answer. But then, on matters such as this, definitive answers are not possible. Yet there has been progress. At the very least, it is no longer enough to think of other animals as clockwork toys. Shelley may not have been correct in detail—skylarks in search of mates and outfacing rivals may not literally feel "blithe." But it seems nonsense to suggest, as for so long was taken for granted, that they feel nothing at all. We can be as sure as we can be about anything that to be like a bird is to be like *something*. It is not like being a stone or a washing machine.

We still do not know what the minds of birds are like, or what is in them. But at the start of the twenty-first century, biologists are at least acknowledging that it makes sense to ask the question. To insist, as many did insist in the early twentieth century, that birds do not have minds at all seems merely perverse. Of course, they do.

IV

Birds and Us

Peregrines are taking to city life worldwide. For them, the towers are cliffs—just right for nesting; and there are pigeons aplenty to feed on.

10

LIVING WITH BIRDS AND LEARNING FROM BIRDS

I

T IS A PRIVILEGE TO LIVE IN THIS WORLD, TO BE SUR-
rounded by so many goodly creatures and to be aware of them. Yet
we, collective humanity, are not behaving as if that were so. Conventional economists treat the earth—or the bits they find useful—as a
"resource." The goal of all the world's most powerful governments is "economic growth." Other creatures hardly get a look in.

So it was that in 2000, in *Threatened Birds of the World*, Birdlife International estimated that one in nine of all birds—about 1,100 species in all—
are at significant risk of extinction within this current century. In some
groups the position is far worse: about one in four of all pheasants are
threatened; well over half of the albatrosses; perhaps two-thirds of the
parrots. Yet Birdlife International made that estimate before the full impact of global warming was appreciated. If the seas rise as much as the
worst-case scenarios predict—or even half as much as the worst predict—
then all bets are off. Every group of living creatures, including our own,
could be in a tailspin.

Some shrug their shoulders. Extinction happens, they say. It's just the
way of the world. All species go extinct eventually, just as all individuals
die. Since the time of *Archaeopteryx*, so it has been conservatively estimated,
there must have been at least 150,000 different species of birds. Yet only
10,500 are with us now. That implies a total extinction rate of 87 percent.
Today's projected rate—a mere 11 percent or so—seems small beer.

But the 140,000 extinctions that birds have already suffered took
place over 140 million years. One species of bird has been lost, on average,

every 1,000 years. In contrast, modern records show that the world has lost at least eighty species of birds in the past 400 years—one every five years, or 200 times what might be called the "background rate." The 11 percent that are now endangered include about a thousand species. If we lose them all in the next century, which seems increasingly plausible, then this will be a rate of ten per year—10,000 times the background rate. Besides, many of the extinct bird species that we know about left descendants—new species that include the birds we see around us now. The ones that will be wiped out in the next century will leave no descendants.

Of species that are known to have gone extinct in recent centuries, most were wiped out by human beings—although not quite all. Tidal waves and typhoons and volcanos can and do wipe out entire habitats, and sometimes those habitats harbor entire species. Most of the birds that were wiped out by human beings suffered at the hands of Europeans—although again, not quite all. The fifteen or so species of moas who for millions of years were the top herbivores in New Zealand were finished off by the Maoris before the Europeans arrived. The Maoris arrived from the South Seas around AD 1000, and it seems that the moas were all gone by the end of the fifteenth century. Maoris ostensibly have a strong conservation ethic—traditionally, for example, if they want to cut down a tree, they first hold a ceremony and ask its permission (or so I was told by a Maori lawyer). Yet they developed a "moa culture," eating the birds and their eggs, wearing their feathers, and fashioning tools and ornaments from their bones. The problem, so some have speculated, was that on their native islands they were maritime people, used to catching fish from small boats. When people fish only from small boats, and do nothing otherwise to harm the marine environment, the yield is infinitely renewable. Perhaps, then, they assumed that the supply of moas would be infinite, too. But alas it is not so. Island creatures are particularly vulnerable because their numbers are inevitably small, and they have nowhere to run. When the moas went, the great Haast's Eagle, which preyed upon them, went, too.

But Europeans have caused most of the recent extinctions and, again, most of the classic examples are among island species. So it was with the Dodo, which lived on the Indian Ocean island of Mauritius. The first description, by the Dutch explorer Jacob Cornelius-zoon van Neck, dates from 1601, and the bird was last described in the wild by the English sailor Benjamin Harry, chief mate on the Berkley Castle, in around 1680. From first mention to obliteration took just eight decades. Traditional accounts suggest that the Dodo was as its name suggests—very fat and very stupid; all too easily put into the cooking pot, like many an island bird (and this, too, is how the boobies got their name, although boobies are still with us). In various reconstructions, not least the one in Oxford's Natural History museum, the Dodo does indeed look like an irredeemably amiable, feathery football; this is the version made famous by Sir John Tenniel's illustrations in Lewis Carroll's *Alice in Wonderland*. Some sailors reported that the Dodo was tasty, while others said that it was foul. Perhaps it depended on season—succulent after the wet season, stringy after the lean times of the dry season. Possibly—even probably—it may not have been as easygoing as it looked. It might indeed have put up quite a strop. In fact, so many moderns think that it wasn't the sailors who finished it off, but the creatures they brought with them, including rats, who stole its eggs, which it laid on the ground, and perhaps killed its chicks. This is a common story. Many an island species has been wiped out by exotic creatures introduced by human beings. Indeed, extraneous introductions are probably the second-greatest single cause of extinctions among all land animals, after loss of habitat.

About 160 years later, the Great Auk met a similar fate. The Great Auk was even more like a penguin than the existing guillemots and murres; 75 centimeters (30 inches) in length, upright in stance, black and white, and flightless. It was also highly successful. Once it lived on islands throughout the North Atlantic, from Newfoundland to Norway and as far south and west and east as Florida and the Channel Islands. But on land it was easy to catch; and in the wild seas where it lived, it was good food for passing sailors, who also hunted it for its plumage—boiling the birds to loosen the feathers over fires fueled by the fat from previous killings. By 1800 very few Great Auks were left. In 1844, a collector commissioned some sailors to catch some of the few survivors, and on June 3, 1844, on Eldey Island off the southwest of Iceland, the sailors duly caught and strangled the last pair that were ever recorded.

The tale of the Stephen Island Wren is just as sad but also comic, in a macabre kind of way. The Stephen Island Wren was one of the tiny coterie of New Zealand wrens that seemed to be the sister group of all the other passerines—a distinction, indeed. Stephen Island, its only known habitat, is a rock in the Cook Strait, between the two main New Zealand islands. It is one square mile in area and so this tiny bird, just 10 centime-

Dodos—thanks to early taxonomists—were seen and depicted as amiable and rotund. But at least at some times of year they may have been lean and mean.

ters (4 inches) long, also had one of the tiniest of all known avian habitats. It was strikingly decked out in yellow and black check. It apparently lived in burrows and may well have been nocturnal. It was rarely seen at all and was never seen in flight—and since it had short round wings it was, perhaps, the world's only flightless passerine.

The first Stephen Island Wren to be known to science was discovered not by an ornithologist or by any human being but by the cat that belonged to the lighthouse keeper, who brought it proudly to its master in the way cats do, one day in 1894. About six months and several captures later it brought home the last of the species ever to be seen. Thus, this tiny but distinguished bird made one of the briefest of all appearances on the ornithological scene and is the only species ever to be discovered, and then finished off, by a domestic cat. Libelous suggestions surround its passing—or at least they would be libelous if the people involved were still alive. Collectors raced to obtain specimens, including the banker-turned-naturalist Walter Rothschild, who had nine in his museum at Tring, Hertfordshire. And it has been suggested (doubtless scurrilously) that the lighthouse keeper himself—a Mr. Lyall, after whom the wren is named—helped his cat to polish off some of the last few, in exchange for the no doubt generous fees. In his excellent *Extinct Birds*, Errol Fuller quotes *The Canterbury Press* from 1905: "It would be as well if the Marine Department, in sending keepers to isolated islands . . . were to see that they are not allowed to take any cats with them, even if mouse-traps have to be furnished at the cost of the state." Now there's a revolutionary thought.

All the examples so far are of island birds—which are indeed the most vulnerable. There are many more examples—scores and scores, in fact. Yet one of the most spectacular extinctions in all of history was of a species that seemed until very recent times to dominate an entire continent: the Passenger Pigeon.

The earliest known mention of the Passenger Pigeon dates from 1534. In 1637, one T. Morton wrote from New England of "Millions of Turtle-doves on the greene boughes, which sate pecking of the full, ripe pleasant grapes." A Peter Kalm described flocks in the spring of 1749 that were a mile across and several miles long, "and they flew so closely together that the sky and sun were obscured by them. . . . When they alighted . . . less firmly rooted trees broke down completely under the load." In a monograph of 1955 published by the University of Wisconsin, A. W. Schorger suggested that, at its peak, the Passenger Pigeon accounted for at least one quarter or perhaps as many as 40 percent of all the land birds in the United States. In due season millions upon millions of them lived, as the Wild Turkeys did, on the American chestnut trees. They marauded from place to place in search of food, like queleas or Emus or some quails, turning up in any one place every few years, flying, it is said, at 60 miles per hour. They were clearly built for both speed and stamina, and such a feat is eminently plausible.

But the pigeons were shot, and shot, and shot again, decade after decade. In one hunting competition in the nineteenth century, 30,000 corpses were needed to claim a prize. Perhaps—or so it has been suggested—they would not breed at all except in big flocks, like flamingos only more so—so perhaps, as their numbers fell, their breeding declined even faster. Whatever the fact of the matter, by 1914 only one was left—an old bird called Martha, who breathed her last in the Cincinnati Zoo.

Today, most people are aware of "conservation" and even politicians speak routinely of "the environment"—although it isn't obvious that most of them know what "the environment" actually is. To most presidents and prime ministers, "environment" means "real estate" or, more specifically, "golf course." It's green, isn't it? And as Ronald Reagan famously asked, "How many trees do you want?"

Indeed, we have new and far more efficient ways of eliminating species than in the past. Our technologies are more powerful and the world's economy is ever more tightly focused on the creation of wealth. Messy old communities of hunter-gatherers and traditional farmers that often got along well enough with wild creatures are being tidied away with all possible haste, in the names of economic growth and "development," which are seen as progress, which is seen self-evidently to be a good thing. The wave of extinctions is shifting from rare and vulnerable islanders to the mainstream. Most species of all living creatures live in tropical forest, and tropical forest worldwide is being felled at a horrible rate. An area of forest the size of Wales (or New Jersey) is removed from Amazonia every year, or so it is conservatively estimated. Thousands of species of all kinds go with them—probably even more than you might expect because many creatures in tropical forests (including many of the trees themselves) have remarkably limited distribution, and when their particular patch is gone they too must go, just as surely as any Dodo on its oceanic island. Many creatures must be disappearing—including many birds—before humanity has even become aware of them. I say "humanity" and not "scientists" because we cannot take it for granted that all of them were ever known even to local tribespeople, brilliant naturalists though they often were and are. Neither does the fault lie only with the tropics—far from it. The boreal forest of North America, both in the United States and in Canada, has often been felled of late as rapidly and extensively as anywhere. The fate of the Spotted Owl, as various factions competed to use land for different purposes, remains a *cause célèbre*.

To be sure there are laws to encourage sustainable forestry, and I have been in parts of the Amazon where enlightened foresters conscientiously follow the rules. It is a definite improvement. But the laws are intended primarily to maintain the flow of timber. It is far from obvious that they are always good for animals. In general there are many conflicts between conservationists and foresters who want to grow usable timber, even when those foresters are content to maintain native trees. Most obviously, traditional foresters are wont to remove "post-mature" trees—the ones that in effect have stopped growing and are gently rotting on their feet. But these are often the ones that other creatures like best. Red-tailed Black-Cockatoos were disappearing from western Australia when I was there a few years ago because the ancient eucalypts were being cleaned out—but the old trees were the ones with the holes, where the black-cockatoos nested. Here, too, lay another problem: black-cockatoos are long-lived, and so only the local naturalists and professional scientists noticed the decline. The birds were still there, but they were not breeding. But it should be possible to provide birds with places to breed. Many other environmental problems are not so solvable. In Indonesia, home to one of the world's greatest tropical forests, up to 90 percent of the logging is said to be outside the law.

We have many new ways of killing wild creatures, too, without really trying. Industrial chemistry is an obvious culprit. In the 1960s, in many places the world over, the magnificent Peregrine Falcon was brought to the brink of extinction by organochlorine pesticides, such as DDT and particularly Dieldrin. Scientists had "proved" that although these noxious agents kill insects stone-dead, they are "harmless" to vertebrates in the recommended doses. But organochlorines accumulate in the body fat. Creatures at the top of the food chain pick them up from the fat of the creatures they feed upon, and so accumulate vast concentrations in their own body fat. This caused Peregrines and other birds of prey to lay eggs with thin and fragile shells, which broke as they were incubated. No scientist could reasonably have anticipated this—but this is precisely the point. Nature is always several steps ahead of us. The point is not simply that much remains unknown. It is that it is logically impossible to know what it is we don't know. So we cannot always anticipate what it is we should have been investigating. In practice we simply make mistakes and then try to pick up the pieces.

Surely, though, we learn from these mistakes, and so get steadily better? Would it were so. In India, large birds of prey are traditionally among the most conspicuous birds of all. When I was first there in the late 1980s, Delhi had the greatest concentration and the greatest variety of birds of prey of anywhere in the world—eagles, vultures, kites. The underlying reason for this was not good: there was plenty of rubbish, including many a corpse. But they did no one any harm; they were just part of the scenery. In India and south Asia in the early 1990s, the Long-billed, Slender-billed, and Oriental White-backed (now Indian White-rumped) Vultures each reached a population of around 40 million. Yet by 2004 their populations had fallen by 97 percent and the World Conservation Union declared them to be Critically Endangered. To be sure, there are still thousands of them. But—as the Passenger Pigeon showed us—it is not the absolute number that counts so much as the rate of decline and the failure to correct the underlying cause. The cause in the case of the vultures is diclofenac, used commonly in medicine as an anti-inflammatory (very good for acute attacks of gout), but also given to cattle in large quantities these days apparently to ease the strain of being a modern cow. Why modern farmers feel they need to keep the pharmaceutical industry in business when they have such troubles of their own is one of life's mysteries. Vultures in India eat dead cattle (in Europe and North America these days they have much less chance to do so, which deprives them of their food supply) and, it turns out, the diclofenac wrecks their kidneys. Who'd have thought it?

Many people, of course, don't like vultures. Their demise, some would doubtless conclude, is a thoroughly good thing. But even on a purely practical level, such a view is horribly misguided. The dead cattle fester. They harbor anthrax. Feral dogs come to feed upon them, or so it is reported— and feral dogs are far more dangerous than any wild wolf ever was because they are stupider than wolves and bolder, unafraid of humans. As the dogs spread, so does the threat of rabies.

In fact, modern people have a hundred different ways of making life impossible for our fellow creatures. The Great Auk was simply slaughtered—but the auks that remain to us are not exactly having a great time. Since they spend their lives at sea, they are extremely vulnerable to oil spills. Many are caught in fishing nets. Most suffer from competition as commercial fishermen pinch their fish, not necessarily to eat, but to feed to salmon on salmon farms. Fish farming does not benefit wild fish to the extent that we are sometimes led to believe. Again, such examples could be multiplied a hundredfold. At least, though, you might suppose, we no longer commit the sins of old-style collectors, who paid local people to round up the last of the species. Do we not? In truth, the international trade in exotic animals is worth between 10 billion and 20 billion dollars a year—third in magnitude only to the arms trade and the trade in narcotics. An estimated one-quarter of the exotic animal trade is illegal. If you want a Hyacinth Macaw, and have enough money, some desperate or cynical soul somewhere will get one for you. Or half a dozen. Even if they are the last ones left.

Yet there are some encouraging tales. The Bald Eagle, North America's magnificent fishing eagle, is one such. Benjamin Franklin held it in great contempt—"a bird of bad moral character"—but the newly emerging, democratic United States knew how to overrule even the most charismatic leaders and on June 20, 1782, it adopted the Bald Eagle as its national symbol. Thus elevated, the Bald Eagle should surely have flourished—but life is rarely so simple. The people who made space for the bird on their coins and dollar bills also built cities and factories along the coasts where it liked to live and cleared its various inland habitats. They also accused it of stealing salmon and placed bounties on its head, the last of which was repealed only in 1953. Then came what was very nearly the *coup de grâce* in the form of industrial toxins—DDT, which thinned the shells of their eggs as with the Peregrine and others, and PCBs, polychlorinated biphenyls, organic compounds with a host of industrial uses from electrical wiring to flame retardants and adhesives, that wrecked their livers. In Benjamin Franklin's day there must have been hundreds of thousands of Bald Eagles. By 1963 there were only 417 nesting pairs in all of the contiguous forty-eight of the United States combined. The eagles had disappeared from the southern states altogether. They did continue to thrive in Alaska.

But in 1962 Rachel Carson published *Silent Spring* and the tide began to turn. In 1967, the Bald Eagle was declared a Protected Species. In 1972, DDT was banned for most purposes, and in 1973, PCBs were banned from situations where they were liable to contaminate the environment at large. In 1973, the Bald Eagle was placed on the U.S. Endangered list, and America's Fish and Wildlife Service and many other groups, public and private, began to protect their habitats. The National Wildlife Refuge system provided habitats for wintering and nesting. Various centers bred the eagles in captivity and many were released into their former haunts. By 1981, the population in the contiguous forty-eight states had doubled. The census of 1993 showed 4,500 nesting pairs. In 1995, the threat was officially declared to be reduced—the eagle's status was downgraded from Endangered to Threatened.

Now there are more than 6,000 nesting pairs in the United States, with an estimated 50,000 to 70,000 (though some say 80,000 to 100,000) in Alaska. In June 2007, the Bald Eagle was removed from the endangered list altogether. Bald Eagles, like all birds of prey, are slow-breeding; two eggs each spring, of which usually only one survives. But the birds are long-lived—more than thirty years in the wild—so that a single pair may produce more than twenty youngsters in a lifetime. Thus, even with such an apparently slow-breeder, the rate of increase can be prodigious—once the threats are removed. Indeed, in the lower forty-eight states, it has multiplied 25-fold since the 1960s.

The Bald Eagle has recovered partly through direct intervention—banning of noxious chemicals, protection of habitats, some captive breeding—but mainly because attitudes changed. Other birds have learned to adapt to the new habitats created by human beings. Many, for example, thrive on farmland or pastureland as they never did in wild nature, from the Emus of Australia to the starlings and skylarks of Europe. Others have taken to city life. The suburbs have the potential to be particularly rich precisely because they provide so many and varied habitats, lovingly created (albeit mainly for other purposes) by hosts of assiduous gardeners. Some birds are flourishing as never before precisely because of the suburbs, like the European Robin that has been featured throughout this book, or the Northern Cardinal in North America, which can feed longer in the north because of all the birdfeeders. However, it seems that with the modern vogue for decking—the garden conceived not as a touch of nature but as an extension of the living room—suburban birds are suffering.

As always, nature springs surprises. Some of the birds that take to cities are not what you might expect. We have seen how many birds of prey there used to be in Delhi, and I have seen trees around Indian and Pakistani cities dripping with kites (which I took to be Brahminy Kites). Peregrines have taken wonderfully to big cities in North America and Europe—Chicago, Salt Lake City, New York, Prague, Berlin, London—picking off the feral pigeons and the odd rat. Both nightjars and nighthawks seem elusive and are therefore presumed to be shy, and yet—amazingly—several species worldwide have adapted to city life. North America's Common Nighthawk nests on flat roofs and hawks for insects around the streetlights, like bats. The Band-winged Nightjar has been a significant resident in Rio de Janeiro since 1955, and the Savanna Nightjar is a regular sight in Indonesian cities, including Jakarta and Surabaya. I

It's impossible to predict which birds will take to life in cities. Among the least probable but most successful are the Nacunda Nighthawks of Brazil.

was pleasantly surprised to find a large colony of herons, high in the trees, more or less in the heart of Istanbul. Herons nest in force in central London, too: I have seen people at London Zoo watching the herons that perched on the roof of the elephant house while ignoring the elephants (in the days when they were still kept in London).

But the human city-dwellers have to be receptive. Dr. David Goode, employed as London's first full-time ecologist, told me that young mothers had complained about herons—fearing they would stab their babies with their sharp beaks and carry them off. Nightly, when I used to work in central London, along with a million other Londoners and visitors, I was privileged to witness one of the great wildlife spectacles of the world: the roosting of the starlings, in their great swirling flocks. Yet people complained. They made a mess. They made noise. Goodness me! Sometimes you could hardly hear the traffic. But all these creatures are proving that people and birds can often live together—with mutual benefit, if people only care to see it in that light. We could, indeed, and should, contrive deliberately to make cities far more wildlife friendly than they are, adding nest boxes and sometimes providing holes for birds to enter, as pigeons were encouraged to enter medieval lofts. The starlings have largely gone from London now because the agriculture around London where they used to feed by day is no longer suitable. Here is another problem: the success of wild creatures in one region often depends absolutely on events in other places where the people of the first region may have no control. This principle applies on both the local scale and the global—and particularly to birds, for whom half the world may be their habitat.

All too often, as well, the impression of improvement is false. It is good to see more Northern Cardinals in the United States; but this primarily reflects the spread of suburbia. Southeast England in recent years has been brightened by flocks of Ring-necked Parakeets. But they did not find their way here naturally; they escaped from aviaries. Like most exotics, they are spreading largely at the expense of local birds. The skeins of Canada Geese that fly past my window in Oxford are a wonderful sight, but they again are escaped exotics and they again are flourishing at the expense of the natives. Even so, well-meaning "conservationists" of an amateur kind resist attempts to cull them. Recently the Little Egret took up residence spontaneously in Britain and soon, some say, the Cattle Egret will follow. They, too, are good to see because they will have come here spontaneously. They have not been officiously introduced.

But these apparent successes may well reflect the beginnings of serious global warming, which in net will be hugely destructive. The creatures most affected in times of climate change are the ones that already live at the extremes and whose habitats are liable to be wiped out altogether. If the planet dries, then the creatures of the wetlands must disappear. If it floods, then the terrestrial creatures are in trouble. As it warms, the creatures that are adapted to supreme cold find they have nowhere to run.

So it is, apparently, with the four penguins that breed on the continent of Antarctica itself: the Adélie, Chinstrap, Gentoo, and the magnificent Emperor. The Antarctic Peninsula is taking the brunt of global warming: it is warming five times faster than the global average and the sea ice has diminished by 40 percent since 1980. The Southern Ocean has warmed discernibly to a depth of 3,000 meters (10,000 feet)—and as the surface waters warm, the currents that drive nutrients up from the depths, to feed the plankton that feeds krill and the fish, are enfeebled. So the krill and fish on which the penguins depend are sadly depleted. Human beings are taking too many fish as well—and also krill, the chief food of the Chinstrap—not for human food, of course, but for fish farms. Some populations of Chinstrap and Gentoo have fallen off by a third or up to two-thirds. Some have halved over the past fifty years. The Emperors manfully incubating their eggs through the teeth of winters that are warmer but still fierce find themselves facing stronger and stronger winds, while standing on ice that is ever thinner. The ice breaks up too early in the summer, and many a chick is lost. They now face competition, too, from Gentoos and Chinstraps spreading southwards as the climate warms. Ironically, perhaps, the warmer air of the extreme south can now hold more moisture than before, which means there is more snow; and the Adélie Penguin is compromised because it prefers to raise its chicks on bare rock, which tends these days to be snow-covered. On the northwest of the Antarctic Peninsula, where warming is most dramatic, two-thirds of the Adélies have been lost in the past twenty-five years. They, too, will lose out to the Chinstraps and Gentoos.

On the other hand, there have been many heroic rescue attempts these past few years and decades, and some have been remarkably and encouragingly successful. Truly there has been a shift of attitude. In previous generations the keenest naturalists were often the most anxious to send the last of each species to the taxidermist.

Among the classic rescues was and is the Hawaiian Goose, or Nene—

soul survivor of what was once a suite of related species from Hawaii, some of them very large and some of them flightless. Once the Nene was common but it was largely seen off, as is so often and so drearily the case, by introduced predators—mongooses, pigs, and cats—and by hunting, and by loss of habitat, as Hawaii became more and more agricultural. The last ones lived in the uplands—but only for the same reason that, for example, the Golden Eagle in Britain now survives only in the Scottish Highlands and the lakes of northwest England: the uplands were all that was left to it. By 1952 only thirty Nenes were left in the wild. But Britain's Sir Peter Scott, pioneer conservationist, set up a breeding program at Slimbridge in Gloucestershire, now a prime site of the Wildfowl and Wetlands Trust, and the species has not only flourished in aviaries and gardens worldwide but has been returned to the wild, where the population now stands at around 500. I was introduced to a big sloppy Labrador at Slimbridge whose job it was to teach juvenile Hawaiian Geese, candidates for reintroduction, to be afraid of dogs. In the wild these days, fear is a necessary survival tactic.

Classic, too, has been the rescue of the Mauritius Kestrel. Terrestrial birds of prey on islands face a double whammy: their populations are bound to be smaller than those of herbivores simply because they are high on the food chain, and are even smaller on islands because all island populations are small (unless the creatures feed at sea). Even before Mauritius was taken over by the French, there were probably fewer than 350 breeding pairs. Then in the 1950s, the usual European appurtenances and hangers-on—DDT, mongooses, cats, and crab-eating macaques, big monkeys introduced from Asia for goodness knows what reason—ate or otherwise destroyed their eggs. By 1974 only four individuals were left. But in 1979, the Welsh ornithologist Carl Jones, backed by Gerald Durrell of Jersey Zoo, established a wildlife sanctuary on Ile aux Aigrettes, off the southeast coast of Mauritius. He climbed the trees that harbored the last remaining nests and removed the eggs. Meanwhile, the wild birds laid replacement eggs and so the population increased. Some of the chicks remained in captivity to build a reserve population and some were returned to the wild, using the "hacking" techniques developed by falconers of adding to the broods of wild birds. Today there are about 800 in the wild, with reserve populations in captivity. The Mauritius Kestrel is still officially classified as Vulnerable—but as Carl Jones says in his melodious

Welsh lilt, when the rescue first began it had been reduced to "a bag of bones." So here, beyond doubt, is success.

A classic tale that still provokes controversy is that of the California Condor. Condors truly flourished in the Pleistocene, when they ranged far and wide through the Americas, feeding on the many corpses of the vast array of gigantic mammals, elephants, rhinos, horses, ground sloths, and countless bison, which either simply died or were conveniently slaughtered by a comparable array of predators that includes lions, wolves, and a giant long-legged running bear. But the California Condor has long since been more or less confined to the U.S. West Coast, and the only other condor is the Andean. Be that as it may, the California Condor was hugely reduced in the nineteenth century as its habitat was destroyed and it was shot—and because it was poisoned by the lead bullets that were commonly embedded in the corpses it did find to eat. By 1987 only twenty-two were left, and they were removed en masse to begin breeding programs in the San Diego Wild Animal Park and Los Angeles Zoo. Certainly this saved them from extinction, yet the move was not universally popular. In particular the influential Sierra Club argued that the birds should be allowed to "die with dignity."

Like Carl Jones with the Mauritius Kestrel, the condor breeders accelerated reproduction by removing the first egg, so that the captive birds laid another. Then they hand-fed the chicks from the eggs that were hatched in incubators, using glove puppets in the shape of condor heads so that the young condors grew up thinking they were indeed condors, and not zoo keepers. Just one year after the last wild California Condor was captured, plans began for reintroduction. Again it was a pioneering effort—as all these endeavors must be, for each raises its own problems and all, at least in detail, are unprecedented. Biologists from the U.S. Fish and Wildlife Service first released Andean Condors (one sex only—females—since they did not wish to establish another colony of exotics) to see if the habitat was suitable for condors at all. It was, so the Andean Condors were rounded up and sent home to South America. California Condors were released into California in 1991 and 1992, and into Arizona, near the Grand Canyon, in 1996.

The newly released birds have met many a hazard: Golden Eagles; power lines, which they have been trained to avoid; and more lead poisoning from hunters' bullets. In a bill that is supposed to apply from January

2008, hunters will have to use nonlead bullets when they are in condor territory. In the same way, in the 1980s, anglers in Britain were encouraged to use nonlead weights on their fishing lines, since the lead ones were obviously killing off the Mute Swans. I remember at the time that a hard-nosed minority of fishermen protested to the last. Numbers of wild California Condors are still topped up with adolescents bred in captivity, but there are some signs at least that they are becoming re-established. Only time will tell.

The Whooping Crane, the only crane that lives exclusively in North America (and America's tallest bird), is posing one of the hardest challenges of all—and invoking some of the most ingenious techniques yet attempted. Once Whooping Cranes were fairly widespread throughout the American Midwest, but now they breed only in the taiga of Alberta in Canada and, recently, in a wildlife refuge in Wisconsin. In winter they migrate south to the coast of Texas and to inland Florida.

For one reason and another, by 1941 only forty-one Whooping Cranes were left in the wild. Attempts were made in the mid-1970s to raise the numbers by cross-fostering with Sandhill Cranes—putting Whooping Crane eggs in Sandhill nests. The Sandhills were good foster parents, but the fostered chicks, when they grew up, thought they were Sandhills themselves and refused to mate with their own kind. This attempt was abandoned in 1989. Then in the 1990s, a team of U.S. and Canadian conservationists tried to establish a population in Florida—a group that would breed there and stay there year-round, not bothering to migrate anywhere. By 2006 there were about fifty-three birds in Florida, but the death rate was high and they were not reproducing.

But also in the mid-1990s there began one of the boldest conservation projects ever attempted. Although Whooping Cranes can breed full time in the south, they clearly prefer to breed in the north and then migrate. But young Whooping Cranes, unlike young swallows, cannot find their own way on their first migration. But if there is no existing flock to lead them, how can they migrate at all? Enter Operation Migration, working in conjunction with the Whooping Crane Eastern Partnership (WCEP). The conservation scientists first raised Whooping Crane eggs artificially in Wisconsin in the now established manner, with Whooping Crane glove puppets so that they knew what species they were, and then got them to follow a microlight plane—using a technique pioneered by Bill Lishman and Joe Duff, who had first tried it out on Canada Geese. The plane then led the young birds from Wisconsin down to Florida for the winter, flying east of the Mississippi. They found their own way back.

Fantastical though it all seems, the technique seems to be working. Operation Migration began in 2001, and more and more birds have been added each year, except for a total wipe-out—a storm—in 2006. Fifty-two birds raised in Wisconsin now winter in Florida; and a few—though still young by Whooping Crane standards—have begun to breed in Wisconsin, having made their way back from Florida. Konrad Lorenz, who persuaded columns of ducklings to follow him around his garden, would surely have been proud of such an endeavor.

Worldwide, now, scores of captive breeding programs are in progress. They are highly organized and, for all kinds of reasons, they need to be. No population of any creature can be considered viable or reasonably safe from extinction unless it contains several hundred individuals, at least. If there are too few, genetic variation is lost, which reduces their resilience as the environment changes. Breeding must be controlled, at least up to a point, to avoid inbreeding. No one institution—zoo, wildlife park, aviary—can afford to keep hundreds of any one kind of animal, and that would not be a good idea in any case, for if they were all in one place they might all be wiped out by the same typhoon or tidal wave or epidemic. So the captive individuals of any one animal are raised in small populations in many different zoos, but treated as if they were one big breeding population, with suitably arranged marriages where necessary—marriages arranged not to produce "improvement," as in domestic livestock, but the complete opposite: to maintain as much genetic variation as possible, runts and all. When I was involved with the London Zoo in the early 1990s, the beautiful, white, plumed Bali Starling, with its pale blue beak, was high among the breeding priorities, while other aviaries around Britain had populations of their own—ready to be exchanged where appropriate with the London birds. Prominent among quite a few serious breeding populations at the Chester Zoo, where I also became briefly involved, was the Waldrapp, also known as the Bald Ibis, then reduced to a single wild population in Turkey (though there are attempts to reestablish it in North Africa). In many a zoo and aviary these days you will see notices on the enclosures to say that the incumbents are members of a breeding program organized at a national or continental level, and sometimes at an international level.

Some deride such efforts. In general, captive breeding is somewhat hit and miss. Some of the most-endangered species have bred well in captivity, but some simply refuse to do so, sometimes for reasons that remain unknown. Some, after breeding in captivity, fail to thrive or breed when put back into the wild, perhaps because their habitat has changed in some subtle way that only they appreciate, and sometimes perhaps because captive breeding has disturbed their psychology. Furthermore, there is no way of telling which will succeed and which will fail. Ring-necked Parakeets and Budgerigars have taken to the wild unasked and uninvited time and time again, after being reared in back gardens and people's front rooms. Other parrots high on the endangered list have failed to thrive in their native haunts even after the most careful rearing and reintroduction by dedicated professionals. After more than twenty years' endeavor it has not proved possible to re-establish the Arizona Thick-billed Parrot in the wild, or at least not convincingly, the Arizona Thick-billed being one of the very few parrots in recent historical times that has been native to the United States.

But the derision is misplaced. Captive breeding should not be seen as the end of the line. It is a stake in the future. If the California Condor had been allowed "to die with dignity," it would now be gone. We cannot claim that it is saved: the wild populations are still highly circumscribed and need assistance. But we can hope that, at some time in the future, conditions in North America will again be propitious and the remaining few will spread once more. One of the principal lessons from all biology is that we must think in the long term; in biological terms, "the long term" means the next 10,000 years, or the next million. A thousand years should be a standard unit of political time. If we continue along our present political and economic course, then we will be lucky to survive this century in a tolerable state. If we get our own affairs in order, then our descendants could still look forward to a very long future. If, now, we try as hard as possible to conserve as many of our fellow creatures as possible—I am inclined to say "by whatever means"—then our descendants might still enjoy the company of other creatures, just as we have been able to do, for thousands of years to come. If we let them disappear, then the future will be so much the bleaker.

So what are the prospects? The present world is dominated by the twin ideas of "development" and "progress," both of which, conventionally, are measured mainly in material terms. North America and Western Europe are still among the richest areas of the world, and for this reason alone they are also commonly deemed to be the most "advanced." So at least to some extent they seem to represent the future, if present trends continue. So what of birds? How are they faring in the "developed" world?

The matter is not simple. It is not easy to count birds, and in any case, as we have seen, security does not depend only on numbers. In general, big populations seem safer than small ones, but some birds that live in highly specialized niches have remained fairly constant even though, as extreme specialists, their numbers are always low; meanwhile, the Passenger Pigeon has shown us that even the most prodigious populations may crash within a few seasons. Total range matters: species that live only in one spot are more likely to be wiped out than those that live more widely. For birds that live only around the Atlantic coasts of the United States and the Gulf of Mexico, hurricanes are an increasing threat. Also significant is the rate of decline—which can only be assessed by measurement over time. Finally, conservationists must judge the current threats and predict future trends: which forests are being felled, which marshlands are drained, and whether wildlife-friendly pasture is liable to give way to hostile, intensive cereal growing (for example, for biofuels) as a country's economy changes. (I hesitate to say "develops," for much of the change is for the worse.)

But with enormous expertise and huge effort it is possible to reach sensible conclusions—and recent reports from both sides of the Atlantic are now showing mixed fortunes. Some species are indeed doing well, partly by adapting to the new habitats that human beings are creating (like America's suburban Blue Jays and Northern Cardinals), partly by escaping the competition that they encountered in the less compromised wilderness (like the House Sparrows that have adapted to cities worldwide), and partly because some of them are benefiting from the balmier climate (like Britain's Little Egrets). But the general trend is down.

The State of Europe's Common Birds, 2007 is the latest report by the European Bird Census Council, which worked with Birdlife International, Britain's Royal Society for the Protection of Birds, British Trust for Ornithology, and many national societies to provide accounts of 124 species from twenty countries between 1980 and 2005. Twenty-nine of the 124 have increased, and another twenty-seven seem to have remained stable. But fifty-six—nearly half—are in decline, and some of them dramatically so. (The position of the remaining twelve remains uncertain.)

Among those that have done moderately well are the Common Whitethroat, Rook, Common Buzzard, European Goldfinch, Common

Rosefinch, and the Great Spotted Woodpecker. The ten who have done best of all are the Collared Flycatcher, the Common Chiffchaff, the Black Woodpecker, and the Hawfinch, which live in forest; and the Common Raven, Common Buzzard, Blackcap, Eurasian Green Woodpecker, Eurasian Collared Dove, and Common Wood-Pigeon, which live in various other kinds of country—moorland, scrub, heath, and so on.

None of the birds that are doing well are recognized primarily as farmland birds. For, as farming over Europe intensifies, the farmland birds show a decline in their populations of 44 percent—and the eastern European countries that have now joined the European Union are beginning to go the way of their western neighbors. Of the ten bird species that are worst hit, five rely heavily on farmland: the Crested Lark, Grey Partridge, Northern Lapwing, European Turtle-Dove, and European Serin. Among forest birds, the Lesser Spotted Woodpecker and the Willow Tit are sadly declining; and among birds of moorland and scrub the Northern Wheatear, Eurasian Wryneck, and Common Nightingale are doing very badly. The Common Nightingale, Lesser Spotted Woodpecker, and Eurasian Wryneck are all reaching Endangered status in Britain; the Lesser Spotted Woodpecker and the Eurasian Wryneck no longer breed in Britain.

In North America, in 2007, North America's National Audubon Society and the American Bird Conservancy published their "Watch List." They listed fifty-nine birds as "Species of Highest National Concern," also known as the "Red Watch List." Among them are several who have turned up in this book: the Greater Prairie Chicken, the California Condor, the Whooping Crane, the Thick-Billed Parrot, certain species of the Spotted Owl, the entire species of the Seaside Sparrow, and the Ivory-billed Woodpecker (for whom hope springs eternal). Also included are Steller's Eider, the Spectacled Eider, and the Ivory Gull. The American ornithologists found that another 119 species were declining and rare and included them on the "Yellow Watch List." Among them are the Greater Sage-Grouse, Montezuma Quail, Bar-tailed Godwit, Roseate Tern, Black Skimmer, Emperor Goose, Trumpeter Swan, Clark's Grebe, Swainson's Hawk, Ross's Gull, Elf Owl, Antilean Nighthawk, Blue-throated Hummingbird, and the magnificent Elegant Trogon. Declining passerines included the Pinyon Jay and the Island Scrub-Jay, Chestnut-collared Longspur, a titmouse, a gnatcatcher, the Wrentit, various thrashers, some warblers and sparrows, a towhee, and so on.

On both sides of the Atlantic, global warming is blamed for much of the decline, plus changes in farming and woodland practice—which in general means intensification and industrialization: bigger fields, with fewer hedgerows, and big machines and industrial chemistry in place of human labor. The fortunes of the various birds have little or nothing to do with what kind of bird they are. In Europe, the Black Woodpecker and the Eurasian Green Woodpecker have been doing well while the Lesser Spotted Woodpecker has been suffering badly; in the United States, the Ivory-billed Woodpecker is already gone. Among pigeons, the European Turtle-Dove is in decline while the Eurasian Collared Dove and Common Wood-Pigeon have been on the ascendant. Among crows, Europe's Rooks are doing well while ravens are flourishing and so, too, in America, is the Blue Jay; but America's Pinyon Jay and Island Scrub-Jay are on the "Yellow Watch List." Of course, we find a similar pattern among many different groups worldwide. Some of the world's commonest birds are pigeons, although thirteen species of pigeon are currently considered to be Critically Endangered and eight have disappeared since 1600 (including, of course, the Dodo and the Passenger Pigeon). Overall, the bad news far outweighs the good news.

So what's to be done? Beyond doubt, we need good science. The misuse of science—for example, to replace traditional farming with the industrial kind—has made life much harder for birds and for animals at large. But without good conservation science—which nowadays includes molecular biology and computer science and climate prediction, besides the many, complex threads of traditional ecology—the cause is lost.

But we get the kind of science we pay for. The present-day world economy is geared to the creation of wealth. Economic growth is the principal goal and boast of modern governments and is typically achieved at the expense of the places where wild creatures live, their food supply and breeding grounds, as forests are felled to make way for beef and rivers are drained to irrigate commodity crops and cities are built on the most productive ground and so on and so on. Many heroic and brilliant conservation projects are taking place, and some of them are reasonably funded, but very few indeed can truly be said to be in the economic mainstream. All are simply eddies in the mainstream of economic "development," eminently fragile. Ecology is among the least well funded of the sciences.

We have the kind of economy we have, and we do the kind of science we do, for all kinds of reasons, including those of history. But the main reason in the end is that we just don't care enough. We do not get angry enough when we see animals, or indeed people, treated badly, or when forest is swept aside. Those who do protest are commonly derided as extremists. They are perceived to be in the way of economic progress, and so to be "unrealistic." Yet they are the defenders of reality—the *real* realities of landscapes and of living creatures. It's the present economy, which recognizes no limits to financial growth, that is unrealistic.

So it's attitude that matters most. We need to give a damn. Hope for the future—for ourselves and all other creatures—lies not with new technologies or even economics but in the possibility that human beings, collectively, might undergo a change of heart, of mindset, of the things we have come to take for granted and the way we look at the world. This possibility is discussed in the next and final chapter.

EPILOGUE:
A MATTER OF ATTITUDE

I WAS LUCKY ENOUGH TO BE BORN IN THE RIGHT PLACE AT the right time, even if it did not always seem like that at the time. I was able to go to a school where biology was taught excellently and then to a university for more of the same. My job since the 1960s has enabled me to converse with some of the world's greatest biologists. Yet I now feel more and more that much of what I was given to believe was wrong. Life isn't the way that modern biologists conventionally construe it to be.

Charles Darwin, with his *On the Origin of Species by Means of Natural Selection*, published in 1859, set the tone of modern biology. I believe that Darwin himself was a very nice man. He argued face-to-face with slave-owners in Brazil on their own turf at a time when slavery was the norm; his friends on both sides of the Atlantic remained loyal over decades; he was courteous to everyone; he acted as unofficial squire and adviser to the local villagers; and he was devoted to his wife, Emma, and to his many children. His great insights—first, that evolution really is the way of the world; and second, that lineages of living creatures become better adapted to their surroundings as the generations pass by means of natural selection—were and are wonderful. I was first exposed seriously to his ideas in 1959 when I was in the sixth form, at the brilliant exhibition at London's Natural History Museum, to celebrate the centenary of *Origin of Species*. I have remained a good Darwinian ever since, or so I like to think.

Yet I fear that *Origin of Species* suffers from a fatal flaw in emphasis—and that the world has suffered since as a result. Darwin's thesis as it stands, or as it is commonly construed to be, has reinforced and largely given rise to a worldview, a mindset, that is technically wrong and deeply pernicious. Darwin was obsessed by living creatures. He was the pivotal theorist of biology, and also one of the greatest field naturalists. Yet, with horrible irony, his ideas have been interpreted or misinterpreted in ways that are helping to destroy the very creatures that he devoted his life to: plants, beetles, birds, all of us.

For Darwin, like all of us, was a child of his time; and the early nineteenth century, when he was growing up, was a very tough time, indeed. It was the height of the Industrial Revolution that had given rise to what William Blake a few decades earlier had called "those dark Satanic mills." Industrial cities were growing faster than they could cope. Bubonic plague was no longer a serious threat in England, but cholera, typhoid, and tuberculosis definitely were. The British Empire was at its height, too, and commerce was taking it over. In the eighteenth century, the British in India had mostly treated the Indians as equal partners: they socialized and intermarried. By the early nineteenth century, the relationship was that of master and servant. When Darwin was born, Britain was fighting Napoleon, and it really was a life-and-death affair that the British simply could not afford to lose. It was not like one of the modern wars, fought in somebody else's country, where the invaders can blow the whistle when they have had enough and go home. In the early nineteenth century, too, for a whole variety of reasons, orthodox Christianity was losing its appeal. The decline of faith throughout the nineteenth century is one of the great themes of modern religious history and philosophy—not least by those who experienced it, from Dostoyevsky through Lord Tennyson through George Eliot and Thomas Hardy and, indeed, Darwin himself. Then, in the mid-century, came the philosophy that I feel might be construed as the *coup de grâce*: the positivism of Auguste Comte. In truth, Comte's ideas were complex and fundamentally humane (George Eliot was a great fan), but the notion that comes through is that nothing in the universe is real except what we can see, touch, and measure. In other words, Comte's positivism was both materialist and atheist.

In short, Darwin grew up and wrote *Origin of Species* in an age that was brutal, ruthless, competitive, predominantly commercial, and increasingly atheist and materialist. All of these attitudes were underpinned and justified by the physical harshness of the world, and a growing body of philosophy and science.

As has often been pointed out, Darwin's great book reflects all this. His thesis was largely inspired by the English economist/cleric Thomas Malthus, who argued that the human population must rise exponentially, which means that the growth becomes faster and faster each year, whereas the output of food could increase at best arithmetically—by just a fixed amount each year. Therefore, said Malthus, there will soon be more people than the world can feed and then the human species must collapse. Darwin extrapolates this idea to all of life. All lineages of all creatures must produce more offspring than their environment can support and so, from conception to the grave, like it or not, they are all inexorably locked in competition. They must cope with their physical environment, of course, and with predators and parasites; but also, closest to home, they must compete with their fellows.

Darwin observed many examples of close cooperation in nature and indeed wrote a book about the interdependence of orchids and moths. For the most part this was easily explained in terms of mutual self-interest leading to co-evolution, but the occasional cases of altruism he came across—one creature apparently taking risks on another's behalf—puzzled him mightily. The individuals that survived the overall to-the-death struggle were the ones that were best adapted to the conditions—and hence, as the generations passed, the whole lineage became better adapted. In the 1860s, the philosopher Herbert Spencer summarized Darwin's argument as "the survival of the fittest," an expression that Darwin later adopted. This is how people tend to remember natural selection: "survival of the fittest." *Fit* in the Victorian sense commonly meant "appropriate," as in "fit for purpose." But it also, of course, had and has connotations of rude health and muscularity. Often it is interpreted to mean "survival of the strongest," from which it is easy to extrapolate: "might is right."

At the same time, Westerners in general have tended to believe that there is a clear gap between human beings and all other creatures. We see ourselves to be superior to the rest, and to have rights over the rest. The first book of the Bible, Genesis, reinforces this conceit. It speaks of "Man" being made "in the image of God," implying that all other creatures are not. The first man, Adam, was given "dominion" over the beasts of the field, which can imply stewardship, but also seems to imply the right simply to take charge and to use other creatures as we will. René Descartes, in the seventeenth century, reinforced this notion of human superiority when he confidently asserted that only human beings could really think, or feel any emotion of any subtlety. The rest were mere automata, who may look as if they think and feel, but in truth do not. Darwin offended many people, including clerics and scientists, when he suggested explicitly in *The Descent of Man*, in 1871, that human beings have descended from apes. He also suggested that there was no clear water between "us" and "them." For at some time in the historical past, he said, there must have been creatures that were half ape and half human—creatures that in the 1870s were still unknown. This hypothetical "missing link" achieved mythical status. By the mid-twentieth century, most biologists accepted that human beings have evolved from primitive ancestors in the same way that other creatures have—but the old conceit, that we are quite different from the rest, hung on. As late as the early 1980s, when DNA studies were already beginning to suggest that human beings and chimpanzees parted company a mere 5 million years ago, some frontline paleoanthropologists were arguing that, in fact, the chimp and the human lineages must have gone their separate ways in the very deep past—perhaps as long as 30 million years ago. They seemed frantic to maximize the distance between us and them, even in the face of scientific evidence of the kind that they themselves held to be sacrosanct.

What has all this got to do with the conservation of birds?

"Everything" is the answer.

In 2007, George W. Bush was working his way through an unimaginable *$6 trillion* as he attempted to extricate the United States from Iraq and Afghanistan. According to Britain's energy expert Sir John Houghton, that is the kind of sum the world needs to spend right now to stop global warming's getting absolutely out of hand, but instead we are spending it on making things worse. One-thousandth of that sum—$6 billion—could sort out most of the most pressing conservation problems of the world. One-millionth of that sum—$6 million—could make a huge difference to the fate of all the world's birds. I have met many a conservation biologist who is obliged to spend half his or her working life (and mostly free time) scrabbling round for grants of a few thousands—ten orders of magnitude short of what is now being spent on one awful but utterly pointless war.

But, of course, the war is not perceived to be utterly pointless. Destructive though they so obviously are, wars historically have been seen to be very good for GDP—gross domestic product—which measures the total wealth produced by a country in a year. Wars boost the arms industry, and arms are among the world's most lucrative businesses. Besides, to dominate the Middle East is to control the world's major sources of oil—the biggest prize of all in today's economy. The whole sorry enterprise is based on two premises that have always been in the background and truly came to the fore in the nineteenth century: materialism is one—the idea that the universe consists of nothing except what can be seen, touched, and measured—and the innate, brutal, relentless competitiveness of nature is the other. Materialism seems to justify the notion that nothing *counts* except material stuff, and the competitiveness of nature is commonly taken to justify the idea that human beings not only are bound to compete but that competition is the great virtue. Moral philosophers at least since St. Paul have been pointing out that what is natural is not necessarily right—but still, this notion that natural-is-right underpins much of our everyday behavior and is used to justify the competitiveness of the modern economy. There is a circular argument here. The brutal economics of his day to some extent inspired Darwin, and now Darwin is invoked to justify even more brutal economics.

Wildlife conservation in particular misses out on two counts. First, in our modern economy, conservationists are increasingly asked to justify their activities on economic grounds—which nowadays means on cash grounds, for economics these days is defined narrowly, in terms of cash. They feel obliged to show that the preservation of some wetland or forest in some ways contributes to the national or the corporate coffers. Up to a point conservation can contribute. To some extent "ecotourism" can be good for the local economy, and sometimes (as in Kenya) it contributes a significant slice of GDP. Many an earnest report seeks to justify conservation efforts on economic grounds. Some conservationists are so keen to align themselves with the modern economy that they seem to argue that conservation is a positive no-no, almost an evil, *unless* it pays its way.

Of course, we should not argue that wild creatures are more important than human beings. Sometimes ancient human societies have been swept aside to make way for nature reserves (and hence for ecotourism), and this surely is nonsense—and often vile. Always we need to find some compromise.

But it is very dangerous for conservationists to rely on economic arguments (meaning arguments that fit in with current economic norms). Most conservation projects cannot show a net financial gain. Any project that relies on economic argument is a hostage to fortune. Mangroves may

bring in the tourists, but it can be a great deal more lucrative to dig la-
goons and raise prawns; or fill in the whole lot with rubble and build a
casino. If it's just cash against cash, then it's obvious who must win. Conservationists cannot win out, long term, unless they bring quite different
values to bear—unless they point out, as was pointed out many years ago,
that we do not live by bread alone, and that cash has no value at all if there
is nothing left that is worth buying.

Wildlife conservation suffers, too, from the great human conceit that
"we" are very different from "them." Our intelligence, it is construed, is
the only intelligence; ours is the only consciousness. All other creatures
are clockwork toys and can reasonably be treated as such: as objects to be
employed or disposed of as we find convenient. To argue otherwise is
merely to be "sentimental," which is perceived to be the enemy of rationality. Rationality clearly means many different things even to philosophers, but it is generally construed to have a no-nonsense quality to it and
so we are persuaded to be guided by it—or by whatever is done in its
name—come what may.

In the early twentieth century, positivism re-emerged as "logical positivism," even more hard-nosed than the original. Science has ruled, and
science on the whole has been construed as the ultimate exercise in applied materialism. The brute competitiveness of nineteenth-century economics is now taken to be the norm, consolidated within the World Trade
Organization. Britain's current Prime Minister, Gordon Brown, speaks in
every other paragraph of the need to "compete," which means that, to
make more money than anybody else, many people have to be put out of
work so as to cut the short-term costs. Wilderness and the creatures that
live in it are very low, indeed, on the list of priorities, although Brown, like
all modern leaders, pays lip service to the idea of the "environment."

But as the twentieth century has moved into the twenty-first, we can
see that the prime conceits that held sway in the nineteenth century, and
have continued to dominate our lives and attitudes, are wrong. To a significant extent science in various ways led the world into its present mindset—emphasizing materialism, competitiveness, and the exclusiveness of
human beings, and a particular form of rationality based in mathematical
analysis. But to a significant extent, too, science is now providing the ideas
that are leading us out of it—perhaps, with luck, in the nick of time.

For whatever justification the ultra-competitive economy may claim,
it cannot claim to be rooted in natural law. Nature can be competitive, of

course—we all have to fight our corner—but that is less than half the story. Above all, nature is cooperative. We see this on the grandest scale. James Lovelock's idea that all the world is an organism—Gaia—is far from fatuous, far from an exercise in wishful thinking. The more that scientists look at the way the world operates—at the science of ecology—the more they perceive the key principle of interdependence, not only between creature and creature but also between life as a whole and the fabric of the planet. Our individual bodies—trillions of cells, working in harmony, each answering to the calls of 30,000 genes that together form what Bill Hamilton called a "parliament"—are a master class in cooperation. More and more, too, biologists are finding cooperativeness *within* populations of animals and between animals of different species that share the same habitat. If the universe were not more cooperative than it is competitive, it would fall apart. If we truly want to create an economy that is "natural," if we see innate virtue in this, then we should root it in cooperation.

It is clear, too, that the gap between humanity and the rest is not as absolute as so many people have wanted to believe. Darwin himself helped to blur the distinctions in various ways. First, as we have seen, he argued that we are descended from what the Victorians were wont to call "brute" apes. Second, he suggested that all the human traits that we claim to be unique in fact have echoes, or pre-echoes, in other creatures, including our complete gamut of emotions. Then, he argued that all living creatures share a common ancestor. St. Francis of Assisi (like many an Eastern thinker, including Mahatma Gandhi, and like many "pagans" from all around the world) saw the animals around him, birds and deer and mice, as his brothers and sisters. Presumably he meant this metaphorically— that we are all children of God. But Darwin showed that the relationship is literal. If sparrows are not exactly our sisters, they are at least our not-so-distant cousins; and mushrooms and oak trees and seaweeds, too, are within the outer reaches of our family tree, and we within theirs.

Above all, though, the last few decades of the twentieth century have produced a new breed of naturalists and animal psychologists who have begun to see that animals cannot be understood in purely materialist terms. It becomes impossible to avoid the idea that they are conscious and aware, that they think and feel, and, as Jane Goodall described in her work on chimps, that they have personalities. At the same time, scientists have been looking afresh at consciousness. Many psychologists of the hard-nosed, logical positivist kind still prefer to argue that consciousness is merely an "emergent property"—the feeling we get inside our own heads when our neurons are firing. This is the minimalist view, which is therefore perceived to be the most "rational." But a new generation of psychologists and physicists are putting a quite new spin on things. Consciousness, they say, is built into the fabric of the universe. The basic "stuff" of the universe is not just "matter," as the traditional materialists have it; nor is it just "mind," as some philosophers (such as England's Bishop Berkeley) maintained; nor do mind and matter jostle side by side, as Descartes proposed. The basic stuff is, in fact, "mind matter." In the same way, we should not think of time and space as separate entities (except, of course, for day-to-day practical purposes). Time and space are merely aspects of the fundamental "space time."

On this view, we need not—should not—suppose that human brains create consciousness and mind, and are uniquely capable of doing so. In truth we partake of the mind that is all around us, and indeed built into us. Our brains focus the universal mind in much the same way that a radio picks up and focuses electromagnetic waves—although, of course, unlike a radio our brains also contribute to the universal mind. Not only humans, of course; *all* sentient creatures work in just this way. All partake of the universal consciousness, and contribute to it.

Birds in all this are as wonderful and salutary as in all things. No sentient creatures are more obviously cooperative than birds. Many cooperate to bring up their children. Many, from vultures to avocets, cooperate to feed. Often, different species operate in partnership to feed, like the mixed flocks that follow columns of army ants; or to breed, like the terns that nest near gulls, which in general are far tougher than they are and can send the predators packing. Indeed, many cooperate both within species and between species to fend off predators. Of course, birds must compete, as we see in the feisty European Robin. But even robins cooperate to breed, and make sacrifices for their young. Without such cooperativeness they die. The mid-twentieth century saw a spate of literature on territoriality and aggression, including Konrad Lorenz's *On Aggression* (1966)—in part perhaps reflecting the Second World War and the cold war, just as Darwin reflected his own times. Humanity seemed predisposed to believe the Jeremiahs and to write off anyone who says that life doesn't have to be like this as a mere dreamer. The biology of our own time—and to some extent the worldview—is still dominated by Richard Dawkins's *The Selfish*

Gene (1976), another exercise in minimalist pessimism. But the pessimists tell less than half the story.

The equally damaging notion that animals are mere robots was always repellent and now it is discredited. Again, birds are showing the lie of it. Many birds do seem remarkably stereotyped in what they do, but others very clearly are not. Crows and parrots and chickadees may never be intellectuals in the human sense, but they are clearly very aware, and above all have an acute social sense, which is what in the end counts most.

In short, birds are wonderful to behold. They can bring us pleasure wherever they are. But also, the more we look at them, the more they tell us about ourselves and the way the world really is. St. Matthew's advice is well taken: consider the birds.

FURTHER READING

I will not bore you with an exhaustive list of all the books that have ever come my way. Here, instead, is a short list of books that I have found particularly useful and/or take continued pleasure in.

FIRST, A COUPLE THAT I FEEL MIGHT PROPERLY BE INCLUDED IN ANY BIRD LIBRARY:

Gill, Frank B. *Ornithology*. New York: W. H. Freeman, 2000. A fine basic textbook.

Perrins, Christopher (ed.). *The New Encyclopedia of Birds*. London: Oxford University Press, 2003. Surely the best up-to-date, one-volume overview of all the world's birds that has so far been published, with beautiful photographs and paintings for good measure.

WITHIN THE PARTICULAR FIELDS OF EVOLUTION AND PHYLOGENY, I AM PARTICULARLY FOND OF:

Benton, Michael J. *Vertebrate Palaeontology*, Second edition. London: Chapman and Hall, 1997. An excellent overview of the routes that have led to all the modern vertebrates.

Chiappe, Luis M., and Lawrence M. Witmer, (eds.). *Above the Heads of Dinosaurs*. Berkeley: University of California Press, 2003.

Cracraft, Joel, and Michael J. Donoghue. *Assembling the Tree of Life*. London: Oxford University Press, 2004. Includes a fine overview of bird phylogeny as now understood.

Feduccia, Alan. *The Origin and Evolution of Birds*. New Haven: Yale University Press, 1996. Professor Feduccia's views are now widely considered to be unorthodox, but he is one of the world's outstanding authorities and must be taken seriously. Orthodoxy is not always correct.

Fuller, Errol. *Extinct Birds*. London: Oxford University Press, 2000. Delivers what it promises: thorough accounts, beautifully illustrated; with scores of extinctions in historical times.

Sibley, Charles G., and Jon E. Ahlquist. *Phylogeny and Classification of Birds: A Study of Molecular Evolution*. New Haven: Yale University Press, 1990. The classic in the field, now overtaken by more recent techniques and more cladistic rigor, but posing many questions that demand to be resolved.

OUTSTANDING BOOKS ON PARTICULAR GROUPS OF BIRDS INCLUDE:

Ali, Salim. *The Book of Indian Birds. Bombay Natural History Society.* London: Oxford University Press, 1941, reprinted 1992. A classic by one of India's finest naturalists.

Davis, Lloyd Spencer. *The Plight of the Penguin.* Dunedin, NZ: Longacre Press, 2001. Fine, modern biology based largely on firsthand observations and described with a pleasingly light touch.

Nelson, Bryan. *Living with Seabirds.* Edinburgh: Edinburgh University Press, 1986. Brilliant close-up studies of gannets and other seabirds from islands all around the world.

Ogilvie, Malcolm. *Grebes of the World.* Uxbridge, UK: Bruce Colman, 2003. With brilliant illustrations by Chris Rose. Comprehensive and beautifuil to behold.

CLASSIC STUDIES OF BEHAVIOR, SEX, AND SOCIAL LIFE INCLUDE:

Cronin, Helena. *The Ant and the Peacock.* Cambridge, UK: Cambridge University Press, 1991. Excellent biology, and beautifully written. Everyone should have a copy.

de Waal, Frans. *The Ape and the Sushi Master.* London: Allen Lane, 2001. De Waal is a primatologist rather than an ornithologist, but his insights are of universal significance. All his books are well worth reading. This is just one.

Lack, David. *The Life of the Robin.* London: Witherby, 1943. An exemplar of experimental natural history.

Lorenz, Konrad. *King Solomon's Ring.* London: Methuen, 1952, reprinted 1964. A book that changed the course of modern biology (like Jane Goodall's books on chimpanzees).

Roughgarden, Joan. *Evolution's Rainbow.* Berkeley: University of California Press, 2004. Some excellent case histories of the strange sex and family lives of jacanas, Cliff Swallows, and many more, with many arresting insights.

Tinbergen, Niko. *The Herring Gull's World.* New York: Basic Books, 1960. Tinbergen and Lorenz between them established the modern science of ethology.

OTHERS WELL WORTH A LOOK INCLUDE:

Caro, Tim. *Antipredator Defenses in Birds and Mammals.* Chicago: University of Chicago Press, 2005.

Elphick, Jonathan, gen. ed. *Atlas of Bird Migration.* Ontario, CN: Firefly Books, 2007. A nicely illustrated, easy-to-follow guide.

Harrison, Kit and George. *Birds Do It Too.* Minocqua, WI: Willow Creek Press, 1997. The subtitle—The Amazing Sex Life of Birds—says it all. Fine illustrations by Michael James Riddet.

Mabey, Richard. *Gilbert White—A Biography.* London: Profile Books, 2006. Gilbert White is England's greatest naturalist and Richard Mabey is one of Britain's finest natural-history writers.

Finally, I cannot resist mentioning:

Austin, Oliver L. *Birds of the World*. London: Paul Hamlyn, 1961. Oliver Austin, curator emeritus at the Florida Museum of Natural History, died in 2008, at the age of eighty-five, but his great book, published nearly half a century ago (and inevitably a little out of date), lives on. If you see a copy—especially in its original, lavish, large-format guise—then do buy it. A classic piece of publishing.

On the Web, it is well worth searching for essays by:

Neal Smith, David Mcfarland, and Alex Kacelnik.

NOTES

PREFACE

1. The King James translation gives us "Behold the fowls of the air"; the New Jerusalem Bible offers "Look at the birds in the sky"; and other versions I have looked at offer one or the other. None says "consider." However, according to the *Greek--English New Testament Lexicon* (as originally edited by W. Bauer), Matthew in the "original" Greek script (that is, the text from which modern versions are taken) uses the verb *embleppon* (which I regret I don't know how to render in Greek script). This verb recurs throughout the New Testament with many connotations, including: what can be seen; look at; fix one's gaze upon; look for a way (to do something); note; and—*consider*.

The Oxford-based theologian I have consulted agrees that "consider the birds" is a perfectly valid translation for Matthew 6:26 and indeed gives more of the intended flavor than "Behold," or "Look at." So I feel fully justified in offering this variant.

CHAPTER 1

1. See Stephen M. Gatesy, "Locomotor Evolution on the Line to Modern Birds," in *Mesozoic Birds*, eds. Luis M. Chiappe and Lawrence M. Witmer, (Berkeley: University of California Press, 2002), pp 432--47.

2. Human babies can, however, suck and breathe at the same time without choking. In babies, the larynx is high in the throat. It falls to its final position as infancy progresses.

CHAPTER 2

1. In 1831 Patrick Matthew also wrote an account of evolutionary change in which he very clearly spells out the principle of natural selection. But for strange reasons he published it as an appendix to a book on naval architecture and, unsurprisingly, it went unnoticed. Neither did he have any kind of reputation as a naturalist. So Patrick Matthew remains as a historical footnote, like this one. History is not necessarily just.

2. In fact, the isolated feather (or rather its impression) was the first to be found—in 1860; and it was reported, by von Meyer, in 1861. But of course that

feather could not be linked with *Archaeopteryx* until *Archaeopteryx* itself came to light.

3. The common root *-apsida* in Synapsida, Diapsida, and Anapsida refers to the number of apertures in the side of the skull, to which the jaw muscles attach. The synapsids have only one such aperture; the diapsids have two; and the anapsids have no such apertures—their skull is basically a box. These differences are not of huge functional significance, synapsids and diapsids can both move their jaws perfectly well. But they are helpful for diagnosis: they enable zoologists to see which lineage a particular fossil belongs to.

4. I say "traditionally" because dinosaur classification has been much revised of late, in detail and overall. But there is no broad consensus at present, so it seems safest and easiest to go with the system that is still to be found in most textbooks. It might not strictly reflect reality, but it works for our purposes.

5. I will assume from now on that *Archaeopteryx* was indeed a feathered maniraptor, despite the doubts.

CHAPTER 3

1. *Bulletin of the British Ornithologists' Club*, 126, Suppl., 2006.

CHAPTER 4

1. Manoj Srinivasan and Andy Ruina, "Computer Optimization of a Minimal Biped Model Discovers Walking and Running," *Nature*, 409, (2001), pp. 72–75.

2. Cecile Mourer-Chauvire, Roger Bour, and Sonia Ribes, "Recent Avian Extinctions on Reunion (Mascarene Islands) from Paleontological and Historical Sources," *Bulletin of the British Ornithologists' Club*, 126 (Suppl. 2006), p. 40.

3. *The Auk*, 99, July 1982, pp. 431–45. 188.

4. Strict Hennig-style cladists will protest that if the Fringillidae did indeed give rise to these other two families, then it should include those other two families, otherwise it is paraphyletic. But it is often convenient, as here, to stretch a point.

CHAPTER 5

1. Georg Wilhelm Steller, for whom the eagle was named, was an eighteenth-century biologist and explorer who accompanied Vitus Bering to explore the Bering Straits, between Siberia and Alaska. Although he died at age thirty-seven (in 1746), he also gave his name to an albatross, an eider duck, a jay, a sea lion, and a sea cow—biggest of the dugong-manatee group of mammals, now, shamefully, shot to extinction.

CHAPTER 6

1. See K. L. Bildstein and J. I. Zalles, "Hawks Aloft Worldwide: A Cooperative Strategy for Protecting the World's Migratory Raptors. Raptor Migration Watch-Site Manual," *Journal of Wildlife Management*, 61, no. 3 (July 1997), pp. 982–83.

C H A P T E R 7

1. *Northanger Abbey* was, of course, written far earlier (in 1798), but the sentiment that Mr. Tilney expressed persisted well into the twentieth century, as is evident in many a Hollywood movie from the 1930s. In general, women could not be trusted to take the initiative.

C H A P T E R 9

1. Robert W. Mitchell and Nicholas S. Thompson, eds., "Deception in the Natural Communication of Chimpanzees," in *Deception: Perspectives on Human and Nonhuman Deceit* (Albany: State University of New York Press, 2006), pp. 221--44.
2. *Ibid.*

ACKNOWLEDGMENTS

I was first launched on the road to birddom by my cousin Peter Selwood, who lived by the sea and is a few years older than I am. He made me aware that not all birds are the sparrows, pigeons, and park-pond ducks of my native South London. There are pipits and hobbies and terns and goodness knows what else out there (not least in South London, had I but realized). When I was halfway through writing my book on trees, I was prompted to start writing this book by my old university friend Barrie Lees. "Trees are all very well," he said, "But birds do more. I want to know about birds." And I thought, "Why not?"

In general I am indebted to everyone who ever taught me biology, although for the particular purposes of this book I am especially grateful to the following, who gave me advice and in some cases read the various chapters. Roughly in the order in which I talked to them, these people include Luis Chiappe and Joel Cracraft, at the American Museum of Natural History, New York, with whom I had long and absorbing conversations on bird evolution and phylogeny; and Carl Jones, of the Mauritius Wildlife Foundation, taught me a great deal about bird conservation—his own work in saving the Mauritius Kestrel is a conservation classic. Roger Wilkinson, director of science at Chester Zoo (now the leading zoo in mainland Britain), was curator of birds when I first did some work for Chester in the 1990s, and he very kindly read the longest chapter for me. When I was attached to the Darwin Centre at the London School of Economics, I learned a great deal about the evolution of animal minds from Helena Cronin, Nick Humphrey, Oliver Curry, and Jennifer Scott (whose work on gorillas is quite brilliant). At Oxford, I have received generous help from Andrew Lack, of Brookes University, and from Alex Kacelnik, from the Department of Zoology at Oxford University. David McFarland, previously at Cambridge, also greatly improved my appreciation of animal psychology and the philosophy thereof. From Warwick Fox, at the

University of Central Lancashire, I have learned much of what I know about theory of mind. Neal Smith, a wonderfully original thinker at the Smithsonian Institution in Panama, has alerted me to quite new lines of thought over the past few years and again read passages of this book. I am very grateful to Richard Holdaway, of the University of Canterbury, for discussions on bird migration (and look forward to the publication of his present research, which promises to put a new slant on the whole subject). Many thanks, too, to Professor Tom Kemp, of Oxford University, who read and advised on matters of evolution.

My neighbor Ian Lees, an accomplished birder and bird photographer, read bits of the text for me and made some cogent comments. The best birding I have ever done was in the company of John Butler in Spain, but when I sought to remind myself of his address on his Web site, I was shocked to learn that he died in September 2007—a very sad loss indeed, for all kinds of reasons. (Brian Davies and Yolanda Davies-Papen are to continue his work; their Web site, at www.donanabirdtours, should surely be checked out.)

I am indebted, too, to my editors: Helen Conford of Penguin Books, who got me to do a lot of restructuring, and Lucinda Bartley at Crown. For the finishing touch, the text was then enormously enhanced by the brilliant illustrations of Jane Milloy.

Finally, I am absolutely indebted to my wife, Ruth West, off whom I bounce all ideas and without whose encouragement and organizational skills I would surely have lapsed into total inertia years ago.

COLIN TUDGE, *Wolvercote, February 9, 2009*

INDEX

NOTE: Page numbers in *italics* refer to illustrations.

woodpeckers *(cont.)*
  Black, 327–28, 418, 419
  and eating, 209–10
  Eurasian Green, 418, 419
  Great Spotted, 418
  Green, 167, 209, 328
  Ivory-billed, 80, 328, 418, 419
  Lesser Spotted, 418
  Pileated, 328
woodshrikes, 192
Woodstar, Amethyst, 237
woodswallows, 191–92, 238
wrens, 174
  Bush, 175
  gnatwrens, 182
  House, 352
  Jenny, 182
  nests of, 303
  New Zealand, 175–76, 401
  Rock, 175

scrubwrens, 174, 175
Stephen Island, 176, 401–3
Winter, 182
Wrentit, 418
wrynecks, 166
  Eurasian, 418

Xu Xing, 53

Yellowhammer, 185
Yerkes, Robert, 373

Zahavi, Amotz, 315, 316, 340
zoogeography, 171
Zusi, Richard, 77
zygodactylous species, 161, 171, 209

About the Author

Always interested in living creatures, COLIN TUDGE studied zoology at Cambridge, then began writing about science, first as features editor at the *New Scientist* and then as a documentary maker for the BBC. Now a full-time writer, he appears regularly as a public speaker. He is a fellow of the Linnean Society of London, has been a visiting research fellow at the Centre of Philosophy at the London School of Economics, and is heavily involved in issues of food and agriculture. His books include *The Link, The Tree, Feeding People Is Easy, So Shall We Reap,* and *The Variety of Life.* He lives in Oxford, England.

ALSO BY COLIN TUDGE

"Tudge writes in the great tradition of naturalists such as Humboldt and John Muir.... Eloquent and deeply persuasive."
—*Los Angeles Times*

In *The Tree,* Colin Tudge travels around the world—from the United States to the Costa Rican rain forest to New Zealand—bringing to life stories and facts about the trees around us: how they grow, reproduce, eat, and age; how they talk to one another (and they do); and why they came to exist in the first place. A marvelous blend of history, science, philosophy, and environmentalism, *The Tree* is destined to become a classic.

THE TREE
A Natural History of What Trees Are, How They Live, and Why They Matter
$15.95 paper (Canada: $17.95)
978-0-307-39539-9

AVAILABLE FROM THREE RIVERS PRESS WHEREVER BOOKS ARE SOLD

Printed in the United States
by Baker & Taylor Publisher Services